Mathematical Cognition and Learning

Language and Culture in Mathematical Cognition

Mathematical Cognition and Learning

Series Editors

Daniel B. Berch
David C. Geary
Kathleen Mann Koepke

VOLUME 4

Language and Culture in Mathematical Cognition

Volume Editors

Daniel B. Berch
David C. Geary
Kathleen Mann Koepke

Mathematical Cognition and Learning

Language and Culture in Mathematical Cognition

Edited by

Daniel B. Berch
Curry School of Education
University of Virginia
Charlottesville, VA, United States

David C. Geary
Psychological Sciences
University of Missouri
Columbia, MO, United States

Kathleen Mann Koepke
Eunice Kennedy Shriver National Institute of Child
Health and Human Development (NICHD)
National Institutes of Health (NIH)
Bethesda, MD, United States

ELSEVIER

ACADEMIC PRESS
An imprint of Elsevier

Academic Press is an imprint of Elsevier
50 Hampshire Street, 5th Floor, Cambridge, MA 02139, United States
525 B Street, Suite 1800, San Diego, CA 92101-4495, United States
The Boulevard, Langford Lane, Kidlington, Oxford OX5 1GB, United Kingdom
125 London Wall, London, EC2Y 5AS, United Kingdom

Library of Congress Cataloging-in-Publication Data
A catalog record for this book is available from the Library of Congress

British Library Cataloguing-in-Publication Data
A catalogue record for this book is available from the British Library

ISBN: 978-0-12-812574-8

ISSN: 2214-2568

Publisher: Nikky Levy
Acquisition Editor: Emily Ekle
Editorial Project Manager: Barbara Makinster
Production Project Manager: Anusha Sambamoorthy
Cover Designer: Matthew Limbert

Typeset by SPi Global, India

Contents

Contributors

David Barner, University of California, San Diego, CA, United States

Sieghard Beller, University of Bergen, Bergen, Norway

Andrea Bender, University of Bergen, Bergen, Norway

Daniel B. Berch, University of Virginia, Charlottesville, VA, United States

Ozlem Cankaya, Alberta Centre for Child, Family, and Community Research, Edmonton, AB, Canada

Danielle S. Dickson, The University of Texas at San Antonio, San Antonio, TX, United States

Chris Donlan, University College London, London, United Kingdom

Martin H. Fischer, University of Potsdam, Potsdam, Germany

David C. Geary, University of Missouri, Columbia, MO, United States

Silke M. Göbel, University of York, York, United Kingdom

Meghan C. Goldman, University of California, Irvine, Irvine, CA, United States

Dan Kim, The Ohio State University, Columbus, OH, United States

Kathleen Mann Koepke, NICHD/CDBB, Rockville, MD, United States

Jo-Anne LeFevre, Carleton University, Ottawa, ON, Canada

Carolina Jiménez Lira, Carleton University, Ottawa, ON, Canada

Elisabeth Marchand, University of California, San Diego, CA, United States

Alejandro Martínez, Basque Center on Cognition, Brain and Language, San Sebastián, Spain

Amanda Martinez-Lincoln, The University of Texas at Austin, Austin, TX, United States

James Negen, Durham University, Durham, United Kingdom

Yukari Okamoto, University of California, Santa Barbara, CA, United States

John E. Opfer, The Ohio State University, Columbus, OH, United States

Jike Qin, The Ohio State University, Columbus, OH, United States

Elena Salillas, Basque Center on Cognition, Brain and Language, San Sebastián, Spain

Barbara W. Sarnecka, University of California, Irvine, Irvine, CA, United States

Geoffrey B. Saxe, University of California, Berkeley, Berkeley, CA, United States

Samuel Shaki, Ariel University, Ariel, Israel

Nicole Y. Wicha, The University of Texas at San Antonio, San Antonio, TX, United States

Chang Xu, Carleton University, Ottawa, ON, Canada

Foreword: Mathematical Cognition, Language, and Culture: Understanding the Links

John Towse
Lancaster University, Lancaster, United Kingdom

Psychologists have long had a fascination with the twin notions of "homogeneity" and "heterogeneity." *Homogeneity* has a technical and statistical meaning that describes the uniformity or similarity of a sample or population. For example, if all children arrived at exactly the same answer to a mathematical problem and did so through exactly the same computations, there would be homogeneity of process and output. If all the children received entirely the same teaching about such problems (no less, no more), that would also give rise to homogeneity, on the input side. The flip side to this is *heterogeneity*, where there is nonuniformity and variation. For example, if different children use different strategies or the same children use different strategies for solving different problems, even for something as apparently simple as addition (Siegler, 1987), we have heterogeneity of process. If some children get private tuition outside school (Huang, 2013), we have heterogeneity on the input side.

Psychologists have navigated a troubled relationship with these twin terms. The purported methodological advantages of each can quickly crumble and present themselves as limiters or hurdles in addressing important questions of validity—the extent to which research findings accurately describe the state of the world, its processes, and relationships. What we have needed is the right perspective on how both homogeneity and heterogeneity, in the present case for mathematics across cultural and linguistic contexts, can best be harnessed and exploited. The current volume takes a confident, authoritative, and highly skilled approach to addressing the consequences of homogeneity and heterogeneity in order to better understand mathematical cognition.

Sampling uniformity or homogeneity can be crucial in order to improve the signal-to-noise ratio in the detection of an important psychological effect, both for within-participant and between-participant designs. Yet a long-standing concern has been the potential distortion that this uniformity produces on real effects in the world. For example, a sample comprising university students at a Western campus might be asked to carry out single-digit multiplication problems. Phenomena such as the problem-size effect (questions with larger numbers take longer to solve and are more error-prone) and 5-operand advantage (multiplication combinations with operands of 5 (e.g., 5×4) are solved faster than comparable combinations that don't include a 5 (e.g., 7×3)) appear robust and consistent and have been documented repeatedly. Yet, it turns out that findings might be biased due to participant homogeneity—when performances of Canadian and Chinese students are examined separately, some important differences emerge (LeFevre & Liu, 1997). However, only when we understand the relevant variables—here, the educational experience of the participants—can such differences be teased apart. The same limitation may be true for a sample of children who, for example, attend a school that is conveniently close and accessible to a university campus of a key researcher on a project; there may be biases introduced by unwittingly drawing on, for example, children from highly educated and prosperous families.

The implications of such issues are well recognized, at least in principle, even though we cannot always know in advance which variables are going to be important. However, more subtle effects may be occurring that are harder to recognize due to the broader forces and shifts that can occur over time. In terms of the number of school leavers who go on to study university degrees, the university student cohort in the United Kingdom is now more diverse than it was, say, 40 years ago. The representativeness of a student sample in the United Kingdom today (relative to the population of mostly young adults) is different from the representativeness of the sample in a previous generation. The higher educational sector is also much more international than it was 40 years ago, for a host of economic and financial reasons. This again can affect the demographic student profile that is sampled.

Demographic changes on a university campus, for example, can be seen, audited, and tracked, especially where they may involve explicit institutional plans. But less obvious changes may take place in schools also. School children may become more or less homogeneous following changes over time in local migration, family size, and economic status. So even when we understand the principle of how biases can arise, recognizing the potential relevance of changes in a sample can be extremely difficult—all the more so if we become anchored to out-of-date representations of the sample profiles.

Sufficient heterogeneity is of course a prerequisite for studies of individual differences (Cronbach, 1957). Without a spread of scores on the variables of interest, there is inevitable attenuation in the potential for tracking mediating processes and associations that might actually be relevant in a population. One of the problems though is that, *a priori*, it might not be feasible to identify all the dimensions that are pertinent. Without the empirical and conceptual grounding to understand the associations that matter, heterogeneity can lead to true effects being masked.

In summary, research design can be a tricky business. It takes critical thinking and careful interpretation to put together the right mixture of homogeneity and heterogeneity so as to yield coherent and useful findings. And herein lies the essential beauty of this volume—the demonstration of how science can benefit from the ties that bind and the dissociations that distinguish mathematical cognition, language, and culture.

Mathematics is an abstract code, and mathematical concepts and relationships are lawful, predictable, and often cumulative. In other words, the mathematical domain contains strong regularities. Yet even so, we know that the configurations through which children experience mathematical information can vary substantially. The ways in which children are taught can vary across generations (such that parents, used to representing multidigit subtractions in a vertical format, view with bemusement how their children have been taught to tackle the same problems by displaying them in a horizontal format). This is exemplified in a fascinating review by *Opfer*, *Kim*, & *Qin* of the link between mathematics performance, cognitive abilities, and curriculum specifications. They point out that the oft-cited "learning gap" between U.S. educated and Asian educated children may be a recent historical phenomenon. *Okamoto* also makes a profound point about the potential impact of cultural and linguistic differences on a smaller timescale. The potential for language-coding characteristics to shape learning and understanding of number names and early number processing is not necessarily the same as the potential impact of these exact same differences on other, later mathematical processes. Of course, there might be cascading effects of competence and confidence, as relative mastery with one mathematical process transforms into challenges to be faced in acquiring the next, different process. Nonetheless, the point at which these exogenous influences and sources of heterogeneity impact children's learning process is crucial. So for *Okamoto*, the path of core numerical knowledge need not run in parallel with the path of specific skills.

Part of the key to research progress is thus the successful harnessing of heterogeneity. Where learners experience mathematical ideas in different ways, whether through the lens of different languages and behavioral evidence (*Marchand* & *Barner*) or multiple languages—through the study of brain activity in bilingual children (*Wicha*, *Dickson*, & *Martinez-Lincoln*) and adults (*Salillas* & *Martinez*), we can gain valuable clues as to the processes of mathematical development and the constraints that operate on mathematical processing. As *Sarnecka*, *Negen*, & *Goldman* remind us, these are complex issues, as they present evidence that language and cultural differences can be accompanied by other relevant variables, such as socioeconomic status.

How do culture and language interact to shape number processing and how do we properly tease apart these influences? It is an issue beautifully captured in the chapter by *Saxe*, whose work focuses on a remote tribe in Papua New Guinea. Separately, *Bender and Beller* discuss an exciting program of work that blends anthropological insights with state-of-the-art methodological and analytic innovations supporting a distributed cognition model of the cultural evolution of number systems.

Making use of the full range of heterogeneity (or, at least, a *fuller* range) may also be important. A number of previous empirical papers have exploited the contrast between, on the one hand, transparent coding of number values in East Asian languages such as Japanese and Chinese and, on the other hand, the nontransparent or idiosyncratic coding of number names in English (e.g., Saxton & Towse, 1998). One of the successful features of the contribution by *Lefevre, Cankaya, Xu, and Lira* is that they exploit the true multidimensional range of language-coding mappings. While there is a binary contrast to be made between East Asian-based and English languages, there is more we can learn from studying other languages such as French and Turkish, which have different points of confluence and divergence as compared with the traditional binary opposites. As such, *LeFevre et al.* increase our awareness of the richness or heterogeneity of the environment. In the context of the contribution of *Bender and Beller* already noted above, making the case for the variety and distinctiveness of number name systems across cultures, we are now at a point where science can take advantage of these natural variables.

In a number of ways, analysis of mathematical skills has been less well supported than analysis of reading skills as a topic of psychological study (see Towse, 2015, to see how these two compare as research topics). So it is good to remember that while these two domains differ in important ways and likely diverge at various developmental phases, there are some noteworthy connections and intersections. In an elegant program of work, *Göbel* shows how and when reading affects and molds number development and when it does not. This complements the more specific focus considered by *Shaki and Fischer* who reveal subtleties in how reading habits (for many, left to right but, for some, right to left) can shape symbolic number associations. There are not only long-term impacts but also short-term or transient effects that combine into a compelling account of how empirical research can make sense of heterogeneous findings. Donlan also complements the understanding of links between language and number by studying the trajectory of what has come to be known as the approximate number system in children with specific language impairments, further emphasizing the connections between skill domains and the potential for atypical development, to reveal important insights for cognition.

Put simply, this is a simply tremendous book. It is a tour de force of how innovative, rigorous, and thoughtful research can really make a difference in shining a light on what we understand about mathematical cognition. The authors bring enormous energy and passion to the work, and despite the difference in topics across authors, there is genuine coherence in the project. If you want concerted and convincing evidence that language and culture can offer the keys to making progress in the field, this is it.

REFERENCES

Cronbach, L. J. (1957). The two disciplines of scientific psychology. *American Psychologist, 12*, 671–684. https://doi.org/10.1037/h0076829.

Huang, M.-H. (2013). After-school tutoring and the distribution of student performance. *Comparative Education Review*, *57*(4), 689–710. https://doi.org/10.1086/671346.

LeFevre, J., & Liu, J. (1997). The role of experience in numerical skill: Multiplication performance in adults from Canada and China. *Mathematical Cognition*, *3*(1), 31–62. https://doi.org/10.1080/135467997387470.

Saxton, M., & Towse, J. N. (1998). Linguistic relativity: The case of place value in multi-digit numbers. *Journal of Experimental Child Psychology*, *69*, 66–79. https://doi.org/10.1006/jecp.1998.2437.

Siegler, R. S. (1987). The perils of averaging data over strategies: An example from children's addition. *Journal of Experimental Psychology: General*, *116*(3), 250–264. https://doi.org/10.1037/0096-3445.116.3.250.

Towse, J. (2015). Finding a voice for numerical cognition. *Journal of Numerical Cognition*, *1*(1), 1–6. https://doi.org/10.5964/jnc.v1i1.16.

Preface

The intersections between language, culture, and mathematics are many and varied. For example, the "mathematics of language" provides the mathematical foundations of linguistics by describing language (and languages) with the use of mathematical models and methods. In contrast, the "language of mathematics" denotes the system of technical nomenclatures and grammatical conventions unique to mathematical discourse used by mathematicians for communicating mathematical ideas. Rather than being directly concerned with either of these topics, the present volume focuses in large part on how language in general and different languages in particular influence the acquisition of basic numerical skills, arithmetic processing, and, more broadly, mathematics achievement. This being said, how children acquire the foundational vocabulary and grammatical conventions of mathematics is indeed within the purview of this volume.

But what can we make of the relation between language and culture? For the most part, scholars working in this area, such as applied linguists, consider that language and culture influence each other in a continuously interactive manner (Jiang, 2000) and that language expresses, embodies, and symbolizes cultural reality (Kramsch, 2009). Although this kind of interaction is fascinating and worthy of study in its own right, it is by no means the direct focus of the present volume. And in apparent contradiction to the notion that language and culture are inseparable, authors of several of our chapters (see especially Chapter 3) delineate both empirical and statistical methods that can be used by researchers for disentangling the effects of these factors on mathematical cognition.

Finally, what kinds of associations exist between mathematics and culture? On the one hand, according to the influential mathematician, David Hilbert, "Mathematics knows no races or geographic boundaries; for mathematics, the cultural world is one country" (quoted in Eves, 1971). In stark contrast to this assertion, the somewhat controversial province of mathematics education known as *ethnomathematics* is by definition concerned with the study of relations between mathematics and culture, especially the mathematics that is practiced by particular cultural groups. Although such interactions between mathematics and culture are not nominally covered in the present volume, the chapter by Saxe does provide an example of a distinctive collective practice involving children selling candy on the streets of an urban center in Brazil and an analysis of transactions between Oksapmin trade store clerks and customers in a remote region of Papua New Guinea (Chapter 12). Furthermore, Bender and Beller (Chapter 13) discuss how several different numeration systems have served as cultural tools for numerical cognition and examine the implications of cross-cultural variability in the different structural properties of these systems for number representation and processing.

Whereas the first three volumes of this series dealt with the evolutionary origins and early development of number processing, the neural underpinnings of and genetic influences on basic numerical cognition, and the learning of more complex arithmetic and higher-order math skills, in this fourth volume, we focus on the substantial contributions that language and culture can make to mathematical cognition, development, learning, and achievement. Although it is widely known that the systematic study of such phenomena began over 30 years ago and continues to this day, reports of this work are primarily published in a diverse range of scholarly journals and occasionally in solitary chapters within all-encompassing handbooks of mathematical cognition and other kinds of compendiums. As a consequence, reviews of contemporary research along with summaries of the relevant prior literature have not been collected and organized in a manner that provides both senior researchers and nonspecialized readers with a comprehensive picture capturing the rich history and current vitality of these areas within one and the same volume. That was the goal of the current effort.

We expect this volume will be of interest not only to researchers and scholars, as well as graduate and undergraduate students in the general field of cognitive development, but also to those who work in the areas of psycholinguistics, cross-cultural psychology, anthropology, educational psychology, and special education. Concomitantly, as the current volume covers a wide range of theoretical concepts, methodological issues, empirical phenomena, and pedagogical implications associated with linguistic and cultural influences on mathematical cognition, this volume could certainly be used as a textbook for a variety of courses taught in many different kinds of university courses.

Regarding the organization of the chapters appearing here, we expressly chose not to separate them into different sections, as the majority of the contributions weave together facets of both linguistic and cultural influences. In addition to the topics already mentioned, other chapters cover how the construction of analogical mappings between linguistic and nonverbal representations may provide the foundation for the learning of number words and their use in estimating large sets of objects (Chapter 2); the effects of mathematical language on children's mathematics achievement and central conceptual knowledge of numbers (Chapter 4); how symbolic numerical magnitude may serve as the mechanism underlying cross-national differences in mathematics achievement (Chapter 5); the kinds of mathematical challenges encountered by children with specific language impairment (Chapter 6); how employing event-related brain potentials reveals that both the language of learning arithmetic facts and the extent of use of such facts in a language can influence the manner and efficiency with which bilinguals access these facts in each of their languages (Chapter 7); how electrophysiological studies of linguistic traces in core number processing deficits at the neural level have been used to test whether the input language of numerical learning for bilingual children with developmental dyscalculia will matter even more than it does for typically developing bilingual children (Chapter 8); the influence of Head Start programs on the learning of symbolic numerical skills by preschool, dual-language learners (Chapter 9); the shaping of numerical processing by number word construction as contrasted with reading direction (Chapter 10); and the role of culture and language in creating associations between number and space (Chapter 11).

Finally, we thank the Child Development and Behavior Branch of the *Eunice Kennedy Shriver* National Institute of Child Health and Human Development, NIH, for their continuing support of the annual conferences that have served as the foundation for this book series. And we are especially grateful to Dr. Lindy Crawford and the ANSERS Institute at Texas Christian University and Dr. Jacqueline Huscroft-D'Angelo, University of Nebraska-Lincoln, the hosts of the conference on which this volume is based. Their guidance and support from the onset was critical to making this conference such a success. In addition, we extend a special thanks to Dr. Mary Martin Patton, dean of TCU's College of Education, who generously provided both supplementary funding for the event and a first-class venue for holding the meeting there.

Daniel B. Berch
David C. Geary
Kathleen Mann Koepke

REFERENCES

Eves, H. W. (1971). *Mathematical circles revisited: A second collection of mathematical stories and anecdotes*. Boston, MA: Prindle, Weber, & Schmidt.

Jiang, W. (2000). The relationship between culture and language. *ELT Journal*, *54*, 328–334. https://doi.org/10.1093/elt/54.4.328.

Kramsch, C. (2009). Discourse, the symbolic dimension of intercultural competence. In A. Hu & M. Byram (Eds.), *Interkulturelle kompetenz und fremdsprachliches lernen. Modelle, Empirie, Evaluation* (pp. 107–121). Tübingen: Gunter Narr.

Chapter 1

Introduction: Language and Culture in Mathematical Cognitive Development

Daniel B. Berch*, David C. Geary[†], and Kathleen Mann Koepke[‡]
*University of Virginia, Charlottesville, VA, United States
[†]University of Missouri, Columbia, MO, United States
[‡]NICHD/CDBB, Rockville, MD, United States

INTRODUCTION

The contributions of linguistic and cultural factors to the development and learning of numerical, arithmetic, and other mathematical concepts and skills are broad and varied, ranging from different types of linguistic influences to contrasting cultural beliefs about the relative importance of ability and effort in learning mathematics, to a traditional society whose language possesses only a limited numerical vocabulary. Indeed, the expansive collection of studies reviewed by the authors in this edited volume is a testament to the diversity of themes, critical issues, theoretical perspectives, research methods, and even the participants (e.g., dual-language learners, children with specific language impairments, and members of a remote indigenous population in New Guinea) that exemplify the fascinating and important spheres of linguistic and cultural influences on mathematical cognition.

In this introductory chapter, we begin by presenting a brief historical review of research on the roles of language and culture in mathematical cognition. Next, we provide readers with a useful framework for categorizing specific linguistic levels of influence on numerical processing (e.g., lexical, semantic, and syntactic), followed by a brief discussion of one of the major questions that has guided research on the role of language in numerical development. We then describe other considerations in studying linguistic influences on mathematical cognition, including the neural substrates of verbal, language-based representations of numerical information and the impact of bilingualism on mathematical thinking and learning.

Mathematical Cognition and Learning, Vol. 4. https://doi.org/10.1016/B978-0-12-812574-8.00001-8
2018 Published by Elsevier Inc.

With respect to the role of cultural factors that influence mathematical cognition, we begin by briefly reviewing many of the classic studies on this topic. Next, we present some purported myths of cultural psychology as a framework for discussing specific kinds of cultural factors that could potentially influence mathematical learning and achievement. We then discuss the complexities associated with disentangling linguistic and cultural influences on numerical processing and development and conclude by acknowledging how studying the intersection between language, culture, and mathematics has flourished since the early groundbreaking research on these topics, exemplified by the work described in this volume that has certainly produced a deeper understanding of this intersection.

A BRIEF HISTORICAL REVIEW OF RESEARCH ON THE ROLES OF LANGUAGE AND CULTURE IN MATHEMATICAL COGNITION

The systematic, large-scale study of cross-national differences in students' mathematical development began in 1964 with the International Project for the Evaluation of Educational Achievement (IEA; Husén, 1967). The goal was to assess a broad range of mathematical skills of 13- and 17-year olds from developed nations and to examine potential influences on these skills, such as the amount of homework and family background, as we elaborate in "The Role of Culture in Mathematics Achievement". The results revealed substantive cross-national differences in mathematics achievement at both ages. The second IEA study was conducted about two decades later and confirmed substantive cross-national variation in mathematical competencies (Crosswhite, Dossey, Swafford, McKnight, & Cooney, 1985). These cross-national studies have expanded over the years and are now conducted on a regular basis through the Program for International Student Assessment (PISA; https://www.oecd.org/pisa/aboutpisa/) and Trends in International Mathematics and Science Study (TIMSS; http://timssandpirls.bc.edu/). The story remains the same; there is substantive cross-national variation in mathematical competencies.

The systematic study of language influences on these cross-national differences began several decades after the first IEA and focused largely on the transparency of the base-10 structure of Arabic numerals in the corresponding number words (e.g., Miller & Stigler, 1987; Miura, 1987; Miura, Okamoto, Kim, Steere, & Fayol, 1993). For example, number words in most Asian languages are a direct reflection of the base-10 system, whereas number words in most European-derived languages are more arbitrary, at least until the 100s. For instance, the Chinese number word for 27 is translated as two ten seven, which makes it obvious to children that 27 is composed of 2 tens and 7 ones. This structure is less obvious in English or in most other European-based languages, especially for number words in the teens. These differences in turn can influence students' learning of some aspects of number and arithmetic (Fuson & Kwon, 1992). Studies of this type continue today, as illustrated in Chapter 4 by Okamoto and Chapter 10 by Göbel.

More recently, the study of number words has shifted to those that emerge in traditional populations (e.g., Butterworth, Reeve, & Reynolds, 2011; Gordon, 2004). These studies have revealed that people in many of these groups have a limited set of number words (e.g., the equivalent of "one, two, and three"), which allows researchers to separate the influence of number words from children's and adults' inherent understanding of relative quantity, as noted in Section "How Do Children Learn the Meaning of Number Words?" Cultures with more developed trading systems generally have an expanded set of number words that are often tailored to aid in specific functional tasks (Beller & Bender, 2008), as contrasted with the systematic mathematical system that children now learn with formal education. Within economically developed nations with multilingual populations, there is now a focus on how learning number words and other aspects of mathematics in two different languages influence children's mathematical development, as illustrated in Chapter 7 by Wicha et al., Chapter 8 by Salillas and Martínez, and Chapter 9 by Sarnecka et al.

SPECIFIC LINGUISTIC LEVELS OF INFLUENCE ON NUMERICAL PROCESSING

In a recent editorial for a special issue of *Frontiers in Psychology* concerning linguistic influences on mathematical cognition, Dowker and Nuerk (2016) contend that the preponderance of research in this area has not gone beyond simplistic demonstrations that language influences number, thus failing to explore the specific level at which particular language effects operate. As a consequence, these authors devote their editorial to differentiating seven linguistic levels. We adopt these here as a framework both for introducing the various ways in which language can influence mathematical processing and for previewing the topics to be covered in the present volume. The specific levels identified by Dowker and Nuerk include lexical, semantic, syntactic, phonological, visuospatial orthographic, conceptual, and other language related skills.

Lexical: The Composition of Number Words

The study of lexical factors in mathematical cognition has focused primarily on the effects of number words. As Dowker and Nuerk (2016) point out, the extent to which a language's number word structure is *transparent* (i.e., compatible with its written/digital number system) can influence numerical performance, even when number words are not involved in a mathematical problem.

One such lexical property is what Dowker and Nuerk refer to as *power* transparency—the degree to which the construction of number words provides a clear and consistent representation of the base-10 system in the language. For example, in East Asian languages, the power of any number beyond 10 can be derived directly from the number word (e.g., "ten-one" for 11). And since such correspondences are more direct (i.e., more transparent) than in languages such as English where many number words are irregular (e.g., "eleven"), some researchers have maintained that the comparatively superior counting and arithmetic skills of Asians may be attributable, at least in part, to the greater transparency of their number words (see Chapter 13 by Bender and Beller, for a linguistic typological analysis of regularity in different numeration systems). However, as Dowker and Nuerk contend, findings based upon such comparisons are often confounded by numerous other cultural and educational differences between countries. (See Section "The Role of Culture in Mathematical Cognition" below for a delineation of these potentially confounding factors, along with the section on "Disentangling Linguistic and Cultural Influences".)

Dowker and Nuerk (2016) also discuss another type of lexical influence known as the inversion property. For example, as Chapter 10 by Göbel describes it, in some languages, the number word starts with the unit followed by the decade (e.g., drei-und-zwanzig, three and twenty), which conflicts with the order of the digits in the Arabic form (e.g., 23). However, in languages such as English that contain noninverted number words, the number word sequence for two-digit numbers is decade first followed by unit (e.g., twenty-three), consistent with the Arabic notation (23). As Göbel describes it, the evidence suggests that children whose language contains decade-unit number word inversions make frequent inversion errors in numerical transcoding—writing numbers in response to spoken number words (e.g., writing 32 for three and twenty)—which seldom occurs for children whose language does not possess number word inversion (see also Chapter 4 by Okamoto, for a discussion of the role of inversion in number-line estimation).

Despite the strong evidence demonstrating that number word inversions can lead to numerical transcoding errors in children, as Dowker and Nuerk (2016) summarize, other research has shown that the inversion property does not influence all numerical and arithmetic skills. That being said, Göbel (Chapter 10 by Göbel) found that for 7- to 9-year olds, number word inversion can impact symbolic arithmetic when addition problems require carry operations (e.g., 27+6), in contrast to those that do not (e.g., 31+6). Moreover, Lonnemann and Yan (2015) recently demonstrated that inverted number words can complicate arithmetic processing even for adults in countries where arithmetic processing is an accomplished skill.

Before leaving the topic of inversion to review other lexical influences on numerical and arithmetic processing treated in the present volume, it is worth noting that Dowker and Nuerk (2016) discuss the results of an interesting study in their special issue. Here, Prior, Katz, Mahajna, and Rubinsten (2015) examined inversion in Arabic-Hebrew bilinguals, since Arabic words possess the inversion property while Hebrew words do not. Administering arithmetic problems orally, the authors found that the participants were better at solving these problems when the language structure paralleled that of the arithmetic operation, regardless of whether the problem was presented in their native language or the one they acquired later on. As a consequence, the authors concluded that the language in which math facts are originally learned does not entirely govern arithmetic processing.

As a final example of the role of lexical factors in mathematical cognition, Chapter 4 by Okamoto discusses how differences between the English and Chinese terms for geometric shapes may influence the ease with which young children learn the shape names, depending on how directly they refer to the visuospatial properties of such shapes. More specifically, with respect to the etymology of English terms for geometric shapes, Okamoto points out that these shape names, like "rectangle" and "pentagon," are composed of Greek and Latin root words and are comparatively lengthy and difficult for young children to learn. She then contrasts such words with Chinese geometric terms (written in the form of logograms—characters that stand for words or phrases), describing how these not only are much less complicated but also have embedded in them the meanings of the shapes to which they refer (e.g., "three-angle shape" for triangle and "five-angle shape" for pentagon). And although Okamoto speculates that children who learn the simple terms may find it easier to associate such words with visual shapes and their properties as compared with those who have to memorize the longer and more complex words, she acknowledges that no such cross cultural investigations have been carried out to date.

Semantic: The Meaning of Words

Semantics is the branch of linguistics concerned with the study of meaning in language. As Brysbaert (2005) has pointed out, there are several reasons for distinguishing between the lexical system, the word-form level, and the semantic system, the word meaning level. Included among these are some cases in which one-to-one mappings between words and their meanings are lacking, where, for example, the meaning of a given word may vary depending in part on the context in which it is presented—as in the meaning of "big" within the phrase the "big mouse" as compared with its meaning in the phrase the "big building." Furthermore, as many words possess multiple meanings (i.e., homonyms), such as point, right, change, and watch, it is challenging to explain how the ambiguities in the mappings from word-form to word-meaning can be resolved within one and the same level of representations (Brysbaert, 2005). In discussing how the semantic level or word meanings may influence mathematical processing, Dowker and Nuerk (2016) introduce a study by

Daroczy, Wolska, Meurers, and Nuerk (2015) who reviewed evidence showing that semantic and numerical properties frequently interact in word problems. More specifically, when the requisite mathematical operation is lexically consistent with the words in the text (e.g., having to carry out addition for a problem in which the words "more" or "buy" appear or having to subtract is associated with the words "less" or "sell"), the word problem is easier to solve than when the words and the required operations are inconsistent.

The words "more" and "less," along with others such as "few," "most," "many," and "some," are examples of essentially quantitative terms. Purpura and Reid (2016) have argued that although there has been a good deal of research on the role of language in the early development of numerical knowledge, most of this work has concentrated on the use of domain-general language measures—such as vocabulary or phonological awareness—rather than on domain-specific measures of quantitative or mathematical language (Harmon, Hedrick, & Wood, 2005). Furthermore, they point out that even though general language skills are consistently related to numeracy performance, enhancing these skills has not been shown to improve children's numeracy knowledge (Jordan, Glutting, Dyson, Hassinger-Das, & Irwin, 2012). Consequently, Purpura and Reid suggested that as mathematical language is a more proximal (i.e., to math content) type of linguistic skill, individual differences in young children's mathematical language may be uniquely predictive of their numeracy performance (i.e., independent of their differences in general language skills).

To test this hypothesis, Purpura and Reid (2016) administered both a general language measure (a definitional vocabulary test) to preschoolers and kindergartners along with an author-developed, content-specific test of mathematical language that included items consisting of quantitative terms, such as "take away" and "more" and spatial terms, such as "nearest" and "middle." Fifteen of the 16 items were receptive and were tested with pictures, and one of the quantitative items was expressive and involved the use of manipulatives. As predicted, mathematical language accounted for more unique variance in numeracy skills than did general language. As a consequence, these authors reasoned that these more content-specific language skills are probably more modifiable than domain-general factors and thus may constitute a more viable target for instructional interventions.

Following up on this supposition, Purpura, Napoli, Wehrspann, and Gold (2017) randomly assigned 3- to 5-year-old children in a Head Start program to either a mathematical language intervention group or a comparison (business as usual) condition. A so-called dialogic storybook reading intervention was administered throughout an eight-week period, in which the children became the storytellers while an adult interventionist prompted and questioned them by placing emphasis exclusively on the book's mathematical language—the quantitative and spatial terms and pictures (e.g., more and near)—by repeating, expanding on, etc. the children's descriptions. Students in the intervention group subsequently performed significantly better on a mathematical language test than students in the comparison group, and more importantly, were also superior on a test of mathematical knowledge.

Although several authors in the present volume discuss some aspect of semantic properties in numerical cognition, the reader is directed to Chapter 6 by Donlan for an interesting discussion regarding the kinds of quantitative modifiers discussed above and differences between their logical and pragmatic interpretations in children and adults, where the latter is concerned with how context influences meaning.

Syntactic: The Grammatical Structure of Languages Beyond Word-Level Influences

The terms for cardinal values used to count quantities, such as "one" and "two," are known as lexical numbers. However, as Dowker and Nuerk (2016) point out, numerical information can also be communicated through the use of so-called grammatical numbers—morphological influences on numerical processing that are principally characterized by singular/plural language markings (e.g., the "s" for plural in English) that differentiate sets of one from sets of two or more (Sarnecka, 2014; see Chapter 6 by Donlan, for other kinds morphological influences on number processing). Based upon a review by Sarnecka (2014) in their special issue, Dowker and Nuerk conclude that the development of numerical understanding occurs later for children whose language does not distinguish between singular and plural (see also Chapter 9 by Sarnecka et al.).

Almoammer et al. (2013) provided a particularly compelling outcome in support of the importance of grammatical number. These researchers investigated the acquisition of number words in young children (2-, 3-, and 4-year olds) who were learning Slovenian and Saudi Arabic—two unrelated languages that possess not only singular/plural marking but also dual marking (i.e., for sets of two). These researchers found that learning a dual marking was associated with more rapid acquisition of the cardinal meaning of the number word "two" than in English-speaking children (or in any other previously studied language), even when having had less experience with counting. As the authors assert, these findings are particularly striking given that children learn the Slovenian and Saudi Arabic languages in completely different cultural contexts (see Chapter 6 by Donlan, for a more detailed description of this study).

Finally, drawing on the intriguing study by Overmann (2015), Chapter 13 by Bender and Beller affirms that although lexical numbers are almost universal in the world's languages, grammatical number is less prevalent and is neither necessary nor sufficient for the (evolutionary) emergence of linguistic numbers—despite its role in facilitating children's acquisition of cardinal numbers when it exists as a feature of their language.

Phonological: The Structure and Sequencing of Speech Sounds

Phonological awareness refers to the conscious awareness of the sound structure of words, including phonemes, the smallest unit of sounds. These skills are tested in a variety of ways, including the ability to rhyme, to match words with beginning sounds, and to blend sounds into words. Based on the articles in their special issue, Dowker and Nuerk (2016) claim that although phonological skills are not directly related to general mathematical functioning, they are specifically associated with verbal representations of and operations on numbers. For example, Jordan, Wylie, and Mulhern (2015) found that the impairments in phonological processing experienced by children with reading disabilities and co-occurring reading and mathematical difficulties played a negligible role in mediating performance on a battery of seven mathematics tasks. However, Pixner, Leyrer, and Moeller (2014) demonstrated that deaf and profoundly hearing-impaired children with cochlear implants who had experienced atypical language development exhibited specific difficulties with place-value understanding, which appears to be directly mediated by language. Chapter 6 by Donlan describes how phonological processing difficulties in children with specific language impairment can adversely impact numerical transcoding—the formation of associations between spoken number words and Arabic numerals. Finally, Chapter 13 by Bender and Beller discusses how phonological codes can mediate the effect of one of our numeration systems on the acquisition of number words. Namely, they explain that learning to count on one's fingers actually helps the learner distinguish between number words by associating their different phonological patterns with unique finger patterns.

Visuo-Spatial-Orthographic: Conventions for Writing and Reading Directions of Languages

As Dowker and Nuerk (2016) characterize it, visual-spatial-orthographic influences on numerical processing largely entail the predominant reading and/or writing direction of a given culture's script or its complexity. They point out that these kinds of factors are most frequently associated with space-number relations, such as the spatial-numerical association of response codes (SNARC) effect, where participants tend to associate relatively small numbers with the left side of space and relatively larger numbers with the right side. Dowker and Nuerk also make reference to Rodic et al.'s (2015) study in their special issue. These researchers reported that 6- to 9-year-old Russian children who were learning a second language that makes use of a spatially complex, character-based writing system—Chinese or Japanese—over one school year did not exhibit a greater improvement in arithmetic than Russian children who spent the same amount of time learning English and/or Spanish. Based upon these findings, Dowker and Nuerk conclude that while visuospatial orthographic skills appear to influence the direction of space-number relations, they do not have an effect on the learning of arithmetic skills per se.

More detailed reviews of work concerning spatial-numerical associations in general and the SNARC effect in particular can be found in Chapter 10 by Göbel and Chapter 11 by Shaki and Fischer. For the most part, the work reviewed here not only corroborates in part the conclusions arrived at by Dowker and Nuerk (2016) but also expands upon and in some cases qualifies them. For example, Shaki and Fischer review evidence suggesting that the acknowledged influence of culturally-based reading direction on the SNARC effect is not as persistent and long-lasting as once believed. Furthermore, these authors contend that the effects that reading direction has on spatial-numerical associations often occur in combination with the influences of other factors and are also dependent on the situational context.

Likewise, Chapter 10 by Göbel reports that despite strong evidence that one's culturally dominant reading direction indeed shapes number processing, this outcome does not necessarily mean that reading direction influences individual differences in mathematical performance and development. Furthermore, she reports data demonstrating that neither children's nor adults' overall performance on standardized mathematics tests appears to be related to the degree to which they have been influenced by their culture's dominant reading direction.

Conceptual: Conceptual Properties of Language

Dowker and Nuerk (2016) also call attention to another albeit comparatively unusual category of linguistic influences on mathematical cognition, specifically, conceptual properties of language—attributes that go beyond basic linguistic structures such as phonemes and words, by operating at what they see as the highest linguistic level of abstract verbal concepts (see Nuerk, Iversen, & Willmes, 2004, for a more detailed account of why they consider the conceptual level to be the highest linguistic level, based in large part on the production model of Levelt, Roelofs, & Meyer, 1999). A primary example is known as linguistic markedness: the condition of being conspicuously unusual as compared with a regular or more common form. With respect to an unmarked-marked oppositional pair of adjective terms, the unmarked one is considered the more comprehensive, dominant, and default form; in contrast, the marked one is irregular and is sometimes marked or derived by a negative prefix—as in the pair "happy-unhappy." More generally, the first or unmarked term of such pairs is the one that is posed in questions, such as "how successful" or "how clean" are you? Alternatively, the use of marked terms in such queries—"how unsuccessful" or "how dirty" are you?—would presume that the person being questioned is in fact unsuccessful or dirty. In other words, unmarked-marked adjectives pairs often possess an evaluative nonequivalence.

In the area of mathematical cognition, linguistic markedness has primarily been studied with respect to the SNARC effect, focusing on parity tasks in which the relevant adjective pairs are "even-odd" and "right-left," where even and right are considered unmarked and odd and left marked. As has been demonstrated, when the linguistic markedness of the numerical stimuli and the response side codes are both congruent, that is, the correct response to an even digit is to the right side and to an odd digit the left side, performance is facilitated, whereas incongruent codes (even-left and odd-right) produce interference (Nuerk et al., 2004). The evidence also suggests that this so-called markedness of response codes (MARC) effect can attenuate the SNARC effect (Berch, Foley, Hill, & Ryan, 1999).

Although linguistic markedness is not discussed in the present volume, Shaki and Fischer's (Chapter 11 by Shaki and Fischer) "harmony" model of the SNARC effect can explain a recent finding pertaining to the role of linguistic markedness in the parity task for native speakers of Hebrew. First, according to Shaki and Fischer, for the SNARC effect to emerge in a given task, participants' directional habits for reading text (from left to right or right to left) must be consistent or in harmony with their directional habits for reading numbers (i.e., from left to right or right to left, respectively). Until recently, there was no evidence of a SNARC effect for parity (odd-even) judgments in native speakers of Hebrew—who read and write words from right to left, but numbers from left to right—presumably due in large part to the conflict that arises between their alphabetic and numeric-writing systems. As Zohar-Shai, Tzelgov, Karni, and Rubinsten (2017) recently reasoned, because the Hebrew word for odd, *ecpzugi*, means "not even" (*zugi* is even), the morphological prefix negates the unmarked word root, making this linguistic markedness factor more salient than for their corresponding English-term equivalents (even vs odd), thereby enhancing the MARC effect with Hebrew participants. Based upon this analysis, Zohar-Shai et al. (2017) designed a series of studies that revealed the first evidence of a reliable SNARC effect in Hebrew participants after managing to reduce the MARC effect.

To understand how these researchers accomplished this feat, it is important to point out first that under standard parity task conditions, a within-participants design is employed, wherein the two types of parity-to-response-side mapping blocks—congruent (i.e., even-right/odd-left) and incongruent (even-left/odd- right) with respect to markedness—are run within the same single session. In contrast, Zohar-Shai et al. (2017) ran these blocks on two different days separated by a one week interval, reasoning that this approach might reduce the coactivation of the two mappings, thereby attenuating the MARC effect, which in turn could counteract its typical overriding of the SNARC effect, permitting it to emerge. This is indeed what happened, presumably because the time interval allowed the activation of the first mapping to decay, thereby effectively eliminating the standard conflict (i.e., lack of harmony) that arises between the two parity-to-response-side mappings in the standard, single-session judgment task.

Other Language-Related Skills: Verbal Working Memory and Other Cognitive Skills Related to Language Representations

Working memory is generally considered to be a limited capacity system responsible for temporarily storing, maintaining, and mentally manipulating information over brief time periods to serve other ongoing cognitive activities and operations. As Dowker and Nuerk (2016) acknowledge, it has been known for at least 35 years that verbal working memory is associated with mental arithmetic (Ashcraft & Stazyk, 1981). Nevertheless, Soltanlou, Pixner, and Nuerk (2015) demonstrated that although verbal working memory is related to multiplication skills in Grade 3, visual-spatial working memory plays a more important role in this kind of mathematical processing in Grade 4. Readers should also note that even beyond this developmental change, the association between working memory and mathematical processing is not always straightforward. For example, as Raghubar, Barnes, and Hecht (2010) have noted, language of instruction is one of the numerous factors that may influence and therefore complicate the interpretation of findings pertaining to the relations between working memory and math performance.

Some of the chapters in the current volume corroborate the importance of verbal working memory and other cognitive skills associated with language representations. For example, Chapter 6 by Donlan concludes that for children with specific language impairment, the increased cognitive load resulting from the combination of verbal memory and phonological processing needed to learn the number word sequence can be overwhelming, limiting how effectively associations can be formed between spoken words and Arabic numerals. Chapter 7 by Wicha et al. discusses how, in the context of bilingualism, theories differ with respect to whether arithmetic facts are encoded in a verbal memory store limited to the language in which they were learned or whether these facts exist in verbal memory stores that are distinct for each language even though the facts were predominantly learned solely in one of the languages. Finally, Chapter 13 by Bender and Beller discusses how numeration systems that differ by modality (verbal, notational, body-based, and material) may differentially impact working memory load, which in turn can influence their use as cognitive tools for enabling the representation and development of numerical and arithmetic processing.

Finally, we think that another example of a language-related cognitive skill that can influence numerical cognition is that of analogical mapping—the construction of correspondences between two representations that share similar structures—as discussed in Chapter 2 by Marchand and Barner. First, however, as Ferry, Hespos, and Gentner (2015) have acknowledged, even though possessing relational language is not necessary for analogical ability, language (e.g., the use of common labels) has been shown to enhance children's relational capacity by strengthening attention to specific relations and by inviting comparison processes (Namy & Gentner, 2002). In their chapter, Marchand and Barner review evidence supporting their contention that learning the meaning of number words is essentially reliant on the construction of analogical mappings between language and nonverbal representations. They also provide compelling evidence that learning to rapidly assign labels to large arrays of objects when estimating is likewise dependent on a process of analogical mapping.

A KEY QUESTION DRIVING RESEARCH ON NUMERICAL DEVELOPMENT

How Do Children Learn the Meaning of Number Words?

Number words are different from other types of words. They refer to the abstract property of *sets* of individuals, such as five dogs, rather than attributes of the individuals themselves, such as large dogs. Number words are also children's first foray into symbolic mathematics and as such set the foundation for their long-term mathematics achievement. Given their unique linguistic properties and their importance for symbolic mathematics, it is not surprising that children's learning of the meaning of number words—their cardinal value (e.g., that "five" represents a set of five things)—has been the subject of theoretical debate and empirical study for many decades (Carey, 2009; Gelman & Gallistel, 1978; Le Corre & Carey, 2007; Lipton & Spelke, 2005; Wynn, 1992), and continues to this day (e.g., Spelke, 2017; vanMarle et al., in press). At the center of the debate are the processes that support children's learning of the cardinal value of number words and reasons for the protracted development of this knowledge. This insight is thought to be based, at least in part, on the acquisition of the successor principle, that each successive number word in the count list is exactly one more than the word before it.

It is now well established that many 2-year olds can recite at least part of the verbal count list (e.g., "one, two, three, four...") but may take another 2 years before they understand the cardinal value of these number words (Gelman, 1972; Wynn, 1990, 1992). The development of this knowledge is revealed in the give-N task, whereby children are asked to give an experimenter a particular number of objects. Children who understand the cardinal value of one—"one knowers"—will give the experimenter one object when asked to do so and a random number of objects for all other number words. Over the next 3–6 months, these children will become "two knowers" and, several months later, "three knowers." Once children understand the meaning of "four," they transition to "cardinal principle knowers" (CP-knowers); that is, they provide to the experimenter the exact number of requested objects up to the limit of their count list (Carey, 2004; Le Corre & Carey, 2007; Wynn, 1992). Children's transition to CP-knowers is quite remarkable, because even our closest relatives do not achieve this conceptual insight, even with extensive training (Beran, Parish, & Evans, 2015).

One key and still debated question is the extent to which this insight is dependent on the evolutionarily ancient approximate number system (ANS). The ANS supports the representation and comparison of the relative quantity of collections of discrete objects and for performing simple arithmetic operations on these representations (Feigenson, Dehaene, & Spelke, 2004; Gallistel & Gelman, 2005; for recent reviews, see Geary, Berch, & Mann Koepke, 2015). In other words, the ANS is thought to provide an intuitive understanding of aspects of number, counting, and arithmetic (Gelman & Gallistel, 1978), but this implicit knowledge must still be linked to the child's count list and number words. One possibility is that young children first form associative mappings between numerals and ANS representations as they learn the meaning of number words and the ordinal nature of the counting system (Lipton & Spelke, 2005). And, children and adults do appear to have a mapping between number words and corresponding ANS representations of quantity, but these mappings may only be for the first five or six number words, at least in children (Sullivan & Barner, 2014). Moreover, ANS representations of quantity are approximate, whereas the cardinal values of number words are exact. vanMarle and colleagues (in press) proposed that this mismatch can be bridged if children's learning of the cardinal value of the first few number words is supported by both the ANS and the object tracking system (OTS). The latter supports the individuation and tracking of up to three or four objects as they move through space. The OTS is not a quantitative system per se but does provide an exact representation that, in combination with the ANS and number words, may contribute to children's understanding of the cardinal value of the first three to four number words (see also Spelke, 2017).

It has also been proposed that perhaps children can deduce the cardinal value of number words through the same language mechanisms that support the learning of other words, without reliance on the ANS or OTS (Bloom & Wynn, 1997); specifically, the syntax of a sentence and the semantics of other words in the sentence allow children to narrow the range of possible meanings of an unknown word used in that sentence. Bloom and Wynn argued there are some linguistic features that are consistently associated with use of number words and that these features, when averaged across many spoken examples, allow children to deduce the cardinal value of specific number words. Number words are generally used with count (e.g., three dogs) not mass (e.g., half a glass) nouns; they do not appear with modifiers (e.g., very two), suggesting absolute properties; they precede but never follow adjectives (two big dogs vs big dogs two), and they can occur in partitive contexts (X of the Ys). Syrett, Musolino, and Gelman (2012) evaluated a corpus of parent-child utterances to determine if partitive phrases occurred frequently enough and in contexts that unambiguously allowed children to infer the exact quantities of number words. They found that most of the partitive phrases that conveyed quantitative information referred to general amounts (get some of the cookies), and less than 14% included exact amounts (get three of the cookies).

Two follow-up experiments also showed that children use this syntactic feature of language to make inferences about specific amounts, only under restrictive conditions. They did not assess the other aspects of language proposed by Bloom and Wynn, but their results and other factors suggest that aspects of natural language may aide children's learning the cardinal value of number words but is probably not sufficient in and of itself (Barner, 2012; Butterworth, 2012; Spelke, 2017).

At this point, the developmental pattern of children's learning of the cardinal value of number words is well established (Wynn, 1992), and there is recent evidence suggesting that the ease with which children learn number words is critically important to their early mathematics achievement (Geary & vanMarle, 2016), but exactly how this learning occurs is not fully understood. Most likely, some combination of the ANS, OTS, and the system for natural language learning are key to this learning, but the relative timing and mix of these processes to learning the meaning of number words remains to be determined.

OTHER CONSIDERATIONS IN STUDYING LINGUISTIC INFLUENCES ON NUMERICAL PROCESSING

Language and Number in the Brain

Much is now known about the representation of numerical information in the brain, including involvement of the right and left intraparietal sulcus and areas of the prefrontal cortex (for reviews, see Berch, Geary, & Mann Koepke, 2016; Nieder & Dehaene, 2009). The former are not classic language areas and, as noted earlier, people in traditional cultures without extensive number words have an intuitive sense of approximate number (e.g., Gordon, 2004). Clearly, the basic aspects of people's inherent number sense are independent of language, but language is critical for aspects of children's learning about symbolic number and arithmetic. On top of the earlier described relation between language and children's learning of the cardinal value of number words, children rely on counting when first learning to solve arithmetic problems (Siegler & Shrager, 1984). Use of fingers to represent cardinal value and use of verbal counting using fingers may provide an early bridge between spatial representations of quantity, language, and numerical representations in the brain (Bertelleti & Booth, 2016). Indeed, early studies of brain injury involving areas of the parietal cortex around the angular gyrus revealed a link between dyscalculia and finger agnosia (Gerstmann, 1940); these studies were with educated adults and involved symbolic arithmetic, not an intuitive sense of number. Over the course of development, the repeated use of finger representations and verbal counting in quantitative contexts appears to result in an integration of brain regions involved in hand movements, speech, and those involved in number representation, including the intraparietal sulcus.

Dehaene and Cohen (1995) argued that some aspects of arithmetic, especially addition and multiplication facts, eventually become stored in long-term memory as a verbal, language-based code, which will be dependent in part on the left angular gyrus and left temporal cortex associated with storage of word meanings. The storage as a verbal code follows from the use of verbal counting and other language-related strategies to solve these problems (Siegler & Shrager, 1984); the formation of these types of long-term memories does not occur as easily for subtraction and division because they are not commutative (9−5 does not equal 5−9). There are however some differences between addition and multiplication, with long-term memory representations of multiplication facts showing more involvement of the temporal cortex associated with memory for words than addition facts (Bertelleti & Booth, 2016). In any case, these studies indicate that children's use of verbal counting, with and without fingers, eventually results in an integration of the intraparietal sulcus with other areas of the brain, including some language areas, at least for a few aspects of symbolic arithmetic.

Domain-General and Domain-Specific Influences of Bilingualism on Cognition

A Brief History of the Study of Bilingualism and Its Effects on Cognitive Development

The impacts of bilingualism on cognitive development and capacities have been of interest at least since the late 1800s and been studied for a corresponding length of time by psychologists and educators. Investigators initially compared the IQs or other standardized general cognitive measures of monolinguals with bi- or multilingual individuals and typically found an advantage for monolinguals (Jones & Stewart, 1951; Laurie, 1890; Saer, 1923), generally hypothesizing that multilingual individuals or "polyglots" are constantly battling the interference of competing languages (Weinreich, 1953) and greater isolation from the majority language and experiences (Lambert, Hodgson, Gardner, & Fillenbaum, 1960). Researchers and educators alike believed that the bilingual's "confusion" and presumed need to inhibit competing words and potentially competing representations would inhibit cognitive development (Saer, 1923).

But as neutral or positive data on the benefits of bilingualism emerged along with awareness of appropriate control variables (Peal & Lambert, 1962) such as socioeconomic status (McCarthy, 1954) and degree of bilingualism (Saer, 1931), and the advent of multiple factor analysis, new research suggested the possibility of distinct cognitive factors (Guilford, 1956) and a possible role for cognitive organization and/or cognitive inhibition (Bialystok & Ryan, 1985; Green, 1998; Peal & Lambert, 1962) in the polyglot intellect. In particular, evidence of changes in interference and cross language facilitation during simple language tasks in some polyglots has been attributed to a variety of cognitive capacities including *levels of processing* (Garrett, 1975), *working memory* (Morales, Calvo, & Bialystok, 2012; Soliman, 2014), *attention* (Astheimer, Berkes, & Bialystok, 2016; Chung-Fat-Yim, Sorge, & Bialystok, 2017), and *executive function* (Bialystok, 2017; Wiseheart, Viswanathan, & Bialystok, 2016).

In contrast to early conjectures, bilingualism has more recently been demonstrated to produce positive benefits in each domain identified above. Despite this mounting evidence for a "bilingual advantage," many still debate the unique contributions of bilingualism, the consistency of these effects, and decry the publication bias in favor of finding such an effect (Paap & Greenberg, 2013; Valian, 2015). Even one of the earliest and most vocal proponents of a bi-/multilingual advantage admits that a diverse array of experiences have the power to positively impact behavior and brain structure and function (Bialystok, 2016; Bialystok & Werker, 2017). It is nevertheless noteworthy that the above list of likely impacted cognitive domains is largely domain-general; it would therefore be prudent for students and investigators of any cognitive capacity to consider the potential role that bi-/polylingualism might have on their both verbal and nonverbal, symbolic and nonsymbolic cognitive domains of interest.

Further, the prevalence of multilingual environments and the complexities associated with trying to accurately characterize these and the participants sampled should give additional pause to all cognitive and behavioral scientists; it is estimated that more than half the world is at least bilingual (Grosjean, 2010 as quoted by Ellen Bialystok & Werker, 2017) to some degree. Even countries that, for political or social reasons, have declared a single "official language" and resisted learning other languages typically have high rates of bilingualism. This list includes the United States, where although English is the official language, the US Census Bureau reported that 382 individual languages and language groups were most commonly spoken in U.S. homes in 2013, and more than 20% of the population (over the age of 5 years) spoke a language other than English at home (Ryan, 2013), with some urban areas as high as 70% (see www.lep.gov).

The complexity of defining "bilingualism" was recognized early on, as not all bilinguals are equally proficient in both languages. Hywela Saer (1931) attempted to create a standardized measure of the degree of bilingualism by creating a ratio of proficiency in the two spoken languages where equivalent proficiency (or a ratio = 1) was considered true bilingualism. But bilingualism is seldom perfectly balanced; it often changes over the course of a lifetime, and it is sensitive to environmental factors such as the density of other speakers. And rarely are any but so-called simultaneous bilinguals (i.e., who learn two languages from birth) "true bilinguals." Thus our capacity to study, as well as to accurately account for and statistically adjust our analyses for, the potential impact of bilingualism on cognitive factors is most likely more challenging than is typically recognized. Finally, it should be noted that any domain-general benefits of bilingualism may be dependent on factors such as cross lingual lexical similarity (Duyck & Brysbaert, 2008), articulatory speeds (Elliott, 1991), or other domain-specific factors. Here again, the complexity of defining "bilingualism" and understanding its effects demands additional study both in breadth and depth.

Bilingualism and Mathematical Cognition

As Chapter 8 by Salillas and Martínez discusses, the study of bilingualism can serve as a window to various forms of language influences on numerical and arithmetic processing in several ways, for example, for testing the distinction between approximate and exact calculation; by investigating the effect of having two accessible, bilingual linguistic contexts on calculation; and to determine the relative dominance of two languages for mathematics learning (see their chapter for a review of such studies). Along these lines, by recording event-related potentials (a time-sensitive measure of neural electrical activity) with Spanish-English bilingual children and adults, Chapter 7 by Wicha et al. demonstrates that both the language in which math facts were initially learned and the frequency with which these facts are used in the second language can affect the manner in which bilinguals access them in each of their languages—and the efficiency with which they do so. Also employing the event-related potentials technique, Chapter 8 by Salillas and Martínez reports new data suggesting that brain differences in the management of numerical judgments between bilingual dyscalculic children and matched bilingual controls without dyscalculia are regulated by the input language of the number words that is, either the language in which they originally learned math or their other language.

THE ROLE OF CULTURE IN MATHEMATICAL COGNITION

The earlier noted cross-national differences in mathematics achievement are important findings, because the mathematical competencies of children and adolescents will have a long-term impact on their later educational and employment opportunities (Ritchie & Bates, 2013; Rivera-Batiz, 1992) and influence their ability to deal with now routine quantitative tasks (Reyna, Nelson, Han, & Dieckmann, 2009). The sources of national differences in success in fostering children's mathematical development are not fully understood but are related at least in part to the opportunity to learn. The latter is related to national differences in the cultural valuation of mathematics that in turn contributes to differences in the quantity and quality of children's exposure to mathematics at home and in school (Geary, 1994). These differences are nicely illustrated by Harold Stevenson's and colleagues' studies of the mathematics achievement of American (USA), Japanese, and Chinese (Taiwan) students (Stevenson, Chen, & Lee, 1993, 1986, Stevenson, Lee, & Stigler, 1986, Stevenson et al., 1990; Stevenson & Stigler, 1992). The gist of these studies is that Chinese and Japanese students have an advantage over their American peers by first grade, and the gap widens across grades and perhaps even across generations (Geary, Salthouse, Chen, & Fan, 1996, Geary et al., 1997).

Among other things, these differences and the differences noted in the large-scale international studies are related to curriculum. Based on the first IEA, Husén concluded "thus, national differences [in mathematical achievement] can in part be explained by differences in emphasis in curriculum" (Husén, 1967, Vol. 2, p. 300). Generally, achievement in different mathematical areas (e.g., algebra) varied directly with the relative degree of emphasis on that area in the mathematics curriculum. Stevenson and colleagues provided a fine-grained analysis of these differences, with extensive classroom observations. In one study of classroom behavior, American first graders were engaged in academic activities about 70% of the time, compared with 85% and 79% of the time for their Chinese and Japanese peers, respectively (Stevenson et al., 1986). The gap widened by the fifth grade. In the first grade, American teachers spent 1–2 hours less per week on mathematics than did teachers in Taiwan and Japan (Stevenson et al., 1990). Again, the gap widened by fifth grade. In addition to time, there are also substantive differences in teaching approaches across nations (Stigler, Gonzales, Kwanaka, Knoll, & Serrano, 1999) and substantial differences in the amount of time spent on homework per week. American first graders spent very little time on homework, whereas their peers in Japan and China spent, on average, 4–8 hours per week on homework. The gap narrowed a bit by fifth grade but was still significant.

The differences in mathematics classrooms and homework are compounded by national differences in the rigor of the mathematics curriculum. Stevenson and Bartsch (1992), for instance, found that Japanese and American elementary and secondary school mathematics textbooks generally presented on the same topics, but the material was conceptually more difficult and presented in an earlier grade in Japanese, in relation to American, textbooks. American textbooks were also cumbersome and included too much unnecessary, nonmathematical material, a concern echoed 16 years later by the National Mathematics Advisory Panel (2008). In a related study, Fuson, Stigler, and Bartsch (1988) found that arithmetic topics and problems presented in American fifth–sixth grade textbooks were presented two–three grades earlier in East Asian textbooks and in textbooks from the former Soviet Union. The common core initiative in the United States was, in part, designed to help to address these gaps (http://www.corestandards.org/), but its success at doing so is uncertain due to resistance to the standards.

There are also wider cultural beliefs that may influence children's investment in learning mathematics. For instance, American students and parents generally believe that success in mathematics is more strongly influenced by some type of inherent ability or propensity to learn mathematics, whereas parents from East Asian nations emphasize the importance of hard work over ability (Stevenson et al., 1990). Interventions that shift students' emphasis from ability to hard work are associated with gains in mathematics achievement (Blackwell, Trzesniewski, & Dweck, 2007), suggesting these attributions about what it takes to become skilled in mathematics matter. These differences in beliefs and a high valuation of mathematics in Asian culture (Hatano, 1990) translate into differences in parental expectations for their children's mathematical development and associated differences in emphasis on informal learning of mathematics in the home (Crystal & Stevenson, 1991).

In all, there are clearly myriad factors that influence students' opportunity to learning mathematics in one nation or another, and these influences translate into large and very persistent cross-national differences in students' mathematics achievement. These influences range from national or local policy for the mathematics curriculum to the rigor of the mathematics presented in textbooks to students' and teachers' engagement with mathematics in the classroom to students' and parents' attitudes about what it takes to excel in mathematics. These differences in turn may reflect deeper cultural differences in the valuation of mathematics, the relative focus on individual agency (e.g., resulting in more or less emphasis on student-directed learning), and perceived opportunities for second chances, that is, the extent to which the early learning trajectory influences later educational opportunities (Geary, 1996; Hatano, 1990).

Five Myths of Cultural Psychology

In an effort to extol the virtues of studying cultural influences in psychological research, Wang (2017) recently acknowledged that there are a number of myths regarding cultural psychology that frequently diminish the likelihood that psychologists will incorporate such factors in their research. More specifically, Wang offers what she considers to be five false assumptions that she then critiques, largely by providing counterexamples from her own seminal research on the role of culture in the development of autobiographical memory, autonomous and relational self-goals, knowledge of emotions, and future thinking. Here, we briefly describe each of these myths and then point readers toward the chapters in this volume that further demonstrate how, in direct opposition to these erroneous assertions, investigating the impact of cultural factors on mathematical cognitive development can not only increase the generalizability of findings obtained from conventionally studied participants in Western societies but also provide unique methodological and theoretical contributions.

Assumption 1. Cultural Psychological Science Focuses Only on Finding Group Differences

Wang (2017) argues that although studying the kinds of differences that exist between cultures is important, cultural psychology can go beyond such basic outcomes by investigating *why* and *how* those differences occur. She points out that one way this can be achieved is by manipulating variables in experimental settings that can make individuals from one culture behave more like those from another culture. Readers can find such examples pertaining to mathematical performance in the chapter by Göbel. Other authors in this volume also discuss evidence from different cultures that goes beyond foundational findings of group differences, thereby making progress toward uncovering the mechanisms responsible for such differences and the factors responsible for giving rise to such outcomes. These include the chapters by LeFevre, Cankaya, Xu, and Lira; Okamoto; Opfer, Kim, and Qin; and Shaki and Fischer.

Assumption 2. Cultural Psychology Disregards Group Similarities

With respect to the second false assumption, Wang (2017) contends that far from being a failure, a cross-cultural psychological study that yields no cultural differences can reveal that a psychological construct or process may be closely coupled with shared sociocultural experiences, neurobiological constraints, or an interaction between these factors. We would add that developmental changes can also be informative, as children from different cultures may, for example, exhibit similarities at an earlier age but differences at a later one. Taken together, several of the chapters in the present volume reveal how both group similarities and differences can inform our understanding of mathematical cognitive development, including the contributions by Göbel, LeFevre et al., Okamoto, Opfer et al., and Shaki & Fischer. Basically, as Wang states, "cultural similarities may suggest universality in the underlying biological cognitive mechanisms on the one hand, and shared human conditions and life circumstances on the other" (p. 22).

Assumption 3. Cultural Psychology Concerns Only Group-Level Analysis

According to Wang (2017), although it is true that cultural psychological research frequently entails comparisons of cultural groups, this should not be interpreted as diminishing the importance of individual differences. Indeed, as she cogently points out, investigating individual differences is often essential for revealing the factors that can explain the group-level cultural differences that have been observed (see Chapter 5 by Opfer et al. and Chapter 10 by Göbel for examples derived from the study of mathematical cognitive development).

Assumption 4. Cultural Psychology Is Irrelevant to Basic Psychological Processes

Wang (2017) summarizes evidence demonstrating that contrary to this generally accepted assumption, various basic psychological processes, including neuronal functioning, sensation, visual illusions, face processing, color perception, and event segmentation, are indeed sensitive to cultural influences. Several chapters in this volume demonstrate further how basic cognitive processes underlying numerical representations and processing may likewise be responsive to cultural influences (see Chapter 3 by LeFevre et al., Chapter 4 by Okamoto, Chapter 5 by Opfer et al., Chapter 10 by Göbel, Chapter 11 by Shaki and Fischer, Chapter 12 by Saxe, and Chapter 13 by Bender and Beller).

Assumption 5. Cultural Psychological Research Only Confirms the Generalizability of Theories

According to Wang (2017), one of the principal contributions that cultural psychology can make to psychological science is to enable researchers to test their theories and hypotheses beyond their customary pool of participants—specifically, those sampled from western, educated, industrialized, rich, and democratic societies (Henrich, Heine, & Norenzayan, 2010). However, she also argues that in to addition to confirming the generalizability of theories, cross-cultural research can play an important role in modifying and augmenting extant theories, and possibly even permit the discovery of mechanisms that may only be found in non-Western samples. An illustration of the former with respect to numerical processing can be found in Chapter 11 by Shaki and Fischer, while a prime example of the latter is described in Chapter 12 by Saxe concerning his work on the 27-body part counting system used by the Oksapmin of Papua New Guinea. We would add that the study of cultural influences on the acquisition of numerical and arithmetic skills can also yield unique findings that can inform the design of instructional interventions (e.g., see Chapter 5 by Opfer et al. and Chapter 10 by Göbel).

Before moving on to the next section, it is important to point out that both the cross-cultural and within-culture research methods discussed by Wang (2017) fall under the rubric of what Chapter 12 by Saxe refers to as *dichotomous* approaches to the study of culture and cognition. As he cogently points out, these constitute both an epistemological stance and a set of empirical techniques wherein researchers treat cognitive processes as occurring within the heads of individuals and as dependent variables that are influenced by a distinct and separate set of surrounding cultural elements. These cultural influences are regarded as independent variables, consisting of factors such as early numeracy experiences, numerical language characteristics, and parental attitudes toward mathematics. In contrast, Saxe adopts a methodological framework that he refers to as an *intrinsic relations* approach where the emphasis is on studying the dynamic and complex interleaving of cultural factors (especially "collective practices" that are cultural microcosms) and the development of numerical cognition in the daily activities of individuals as "processes in motion" (see the Saxe chapter for more details).

DISENTANGLING LINGUISTIC AND CULTURAL INFLUENCES ON NUMERICAL PROCESSING AND MATHEMATICS ACHIEVEMENT

Dowker and Nuerk (2016) point out that attributing the superiority of East Asians' numerical and arithmetic skills over that of students who speak European languages to linguistic (power) transparency may be moderated by the confounding of many other differences between countries with respect to their cultural and educational practices. They also contend that one approach for gathering more evidence of language effects per se is to compare children who are schooled in different languages, albeit in the same country and educational system. The example they provide is Wales, where both the English (irregular) and the Welsh (regular/transparent) counting systems are in use. By comparing children receiving these different counting systems within the same country, cultural environment, educational system, and curriculum, Dowker, Bala, and Lloyd (2008) found that children in Welsh-medium primary schools exhibited specific advantages in reading and comparing two-digit numbers, even though they did not perform better in arithmetic overall. Dowker and Nuerk discuss an extension of this study using the same kind of design and another study employing a related design using additional controls with children in Hong Kong who can learn mathematics using both the Chinese (regular/transparent) and English (irregular) languages (Mark & Dowker, 2015). Taken together, these results show that in spite of some selective numerical processing benefits associated with a transparent counting system, this linguistic property does not have an all-encompassing impact on cross-national differences in arithmetic (Dowker & Nuerk, 2016).

Chapter 3 by LeFevre et al. discusses in further detail the studies just described, acknowledging that cultural, experiential, and other demographic factors have frequently been confounded with linguistic differences in studies that have compared the influences of language on disparities in the acquisition of early numeracy skills. Such factors have included, among others, the amount of practice children have had in their respective languages, the amount of direct instruction in math at home, and the level of encouragement to engage in math-related activities. Furthermore, LeFevre et al. provide a comprehensive review and analysis of these kinds of factors with respect to her (and her colleagues') own studies of young children's counting across five different languages, taking into account the role of language, country, parents' education, and home numeracy factors. Likewise, Chapter 5 by Opfer et al. examines numerous contrasts in attitudes toward and educational practices pertaining to the learning of math that may contribute to the cross-national gap between East Asian and American students' mathematics proficiency.

As we described earlier, these factors include beliefs concerning whether ability or effort is more important in learning mathematics, both the quantity and quality of mathematics instruction in school, and the quality of teacher knowledge, among others. In addition, Chapter 4 by Okamoto discusses the contribution of qualitatively different problem-solving experiences between children in different cultures to cross-national gaps in mathematics achievement.

As Chapter 3 by LeFevre et al. concludes, researchers who investigate differences in the acquisition of early numeracy skills among languages should also measure other prospective causal factors and then control for as many of them as is feasible, either by developing other procedures for testing language-related differences or by means of statistical methods. Interestingly, Chapter 5 by Opfer et al. discusses how they have applied both of these recommended approaches to the exploration of cross-national differences in mathematics achievement.

CONCLUSIONS

The concepts and procedures that define modern mathematics as an academic discipline may be universal, but the language, educational, and broader cultural supports (e.g., beliefs about the factors that facilitate mathematical learning, such as hard work or native ability) that influence people's access to this discipline and the ease with which they learn it are not. Many of these language and cultural supports are invisible in the sense that the ways in which their structure facilitates (or does not) children's early learning of symbolic number and arithmetic are not obvious to the casual observer. Few parents and for that matter not many teachers understand that something as simple as the names of number words, such as eleven versus ten-one, can provide cues to children about the underlying structure of the number system, the base-10 structure in this case (Miura, 1987). And, teachers can use these words to teach this structure, at least in languages in which it is transparently reflected in number words (e.g., as in ten-one for 11 or two-ten-one for 21). Research on the mapping between number words and children's learning of the mathematical number system was among the first studies to highlight how language can influence children's access to and understanding of symbolic mathematics. Since this early pioneering research, the study of the intersection between language, culture, and mathematics has blossomed and, as illustrated by the chapters in this volume, has yielded a richer and more nuanced understanding of this intersection.

Included among these advances and building on Saxe's (1981) seminal work with the Oksapmin are ever-increasing studies of number words and concepts among people living in traditional cultures who are not exposed to formal education and analyses of the cultural evolution of number systems other than the Hindu-Arabic system that is the basis for academic mathematics as we know it. This burgeoning cross-cultural research is providing insights into similarities and differences in how human's construct mathematical systems, depending on needs (e.g., trade) within the culture, and how this knowledge can be transmitted across cultures. These types of studies add to the more familiar comparisons of the relative academic achievement of students receiving formal education in economically developed societies (e.g., Stevenson et al., 1986). Not surprisingly, the latter differences are related in part not only to differences in the content and rigor of mathematics instruction but also to broader social beliefs about mathematics learning (e.g., that it takes hard work and practice) and the valuation (or not) of mathematics achievement (Stevenson & Stigler, 1992). A fuller understanding of all of these factors—language, education, and cultural—will provide insights into how the human mind creates and learns mathematics, and this in turn will provide guidance on how to better structure (e.g., sequence grade-level content) and teach mathematics.

REFERENCES

Almoammer, A., Sullivan, J., Donlan, C., Marusic, F., Zaucer, R., O'Donnell, T., et al. (2013). Grammatical morphology as a source of early number word meanings. *Proceedings of the National Academy of Sciences of the United States of America*, *110*, 18448–18453.

Ashcraft, M. H., & Stazyk, E. H. (1981). Mental addition: A test of three verification models. *Memory & Cognition*, *9*, 185–196.

Astheimer, L. B., Berkes, M., & Bialystok, E. (2016). Differential allocation of attention during speech perception in monolingual and bilingual listeners. *Language, Cognition and Neuroscience*, *31*(2), 196–205. https://doi.org/10.1080/23273798.2015.1083114.

Barner, D. (2012). Bootstrapping numeral meanings and the origin of exactness. *Language Learning and Development*, *8*(2), 177–185.

Beller, S., & Bender, A. (2008). The limits of counting: Numerical cognition between evolution and culture. *Science*, *319*, 213–215.

Beran, M. J., Parrish, A. E., & Evans, T. A. (2015). Numerical cognition and quantitative abilities in nonhuman primates. In D. C. Geary, D. B Berch, & K. Mann Koepke (Eds.), *Evolutionary origins and early development of number processing* (Vol. 1, *Mathematical Cognition and Learning*, pp. 91–119). San Diego, CA: Elsevier/Academic Press.

Berch, D. B., Foley, E. J., Hill, R. J., & Ryan, P. M. (1999). Extracting parity and magnitude from Arabic numerals: Developmental changes in number processing and mental representation. *Journal of Experimental Child Psychology*, *74*, 286–308.

Berch, D. B., Geary, D. C., & Mann Koepke, K. (Eds.), (2016). *Development of mathematical cognition: Neural substrates and genetic influences*. San Diego, CA: Elsevier/Academic Press.

Bertelleti, I., & Booth, J. R. (2016). Finger representations and finger-based strategies in the acquisition of number meaning and arithmetic. In D. B. Berch, D. C. Geary, & K. Mann Koepke (Eds.), *Development of mathematical cognition: Neural substrates and genetic influences* (pp. 109–139). San Diego, CA: Elsevier/Academic Press.

Bialystok, E. (2016). The signal and the noise: Finding the pattern in human behavior. *Linguistic Approaches to Bilingualism*, *6*, 517–534.

Bialystok, E. (2017). The bilingual adaptation: How minds accommodate experience. *Psychological Bulletin*, *143*(3), 233–262. https://doi.org/10.1037/bul0000099.

Bialystok, E., & Ryan, E. B. (1985). A metacognitive framework for the development of first and second language skills. In G. E. MacKinnon, D. L. Forrest- Pressley, & T. G. Waller (Eds.), *Meta-cognition, cognition, and human performance* (pp. 207–252). New York, NY: Academic Press.

Bialystok, E., & Werker, J. F. (2017). Editorial: The systematic effects of bilingualism on children's development. *Developmental Science. 20*(1). https://doi.org/10.1111/desc.12535.

Blackwell, L. S., Trzesniewski, K. H., & Dweck, C. S. (2007). Implicit theories of intelligence predict achievement across an adolescent transition: A longitudinal study and an intervention. *Child Development, 78*, 246–263.

Bloom, P., & Wynn, K. (1997). Linguistic cues in the acquisition of number words. *Journal of Child Language, 24*, 511–533.

Brysbaert, M. (2005). Number recognition in different formats. In J. I. D. Campbell (Ed.), *Handbook of mathematical cognition* (pp. 23–42). New York, NY: Taylor & Francis.

Butterworth, B. (2012). Commentary on "how syntax can support number word acquisition?" by Kristen Syrett, Julien Musolino and Rochel Gelman. *Learning and Development, 8*, 186–189.

Butterworth, B., Reeve, R., & Reynolds, F. (2011). Using mental representations of space when words are unavailable: Studies of enumeration and arithmetic in indigenous Australia. *Journal of Cross-Cultural Psychology, 42*, 630–638.

Carey, S. (2004). Bootstrapping and the origins of concepts. *Daedalus, 133*, 59–68.

Carey, S. (2009). *The origin of concepts*. Oxford: Oxford University Press.

Chung-Fat-Yim, A., Sorge, G. B., & Bialystok, E. (2017). The relationship between bilingualism and selective attention in young adults: Evidence from an ambiguous figures task. *Quarterly Journal of Experimental Psychology, 70*(3), 366–372. https://doi.org/10.1080/17470218.2016.1221435.

Crosswhite, F. J., Dossey, J. A., Swafford, J. O., McKnight, C. C., & Cooney, T. J. (1985). *Second International Mathematics Study summary report for the United States*. Champaign, IL: Stipes.

Crystal, D. S., & Stevenson, H. W. (1991). Mothers' perceptions of children's problems with mathematics: A cross-national comparison. *Journal of Educational Psychology, 83*, 372–376.

Daroczy, G., Wolska, M., Meurers, W. D., & Nuerk, H.-C. (2015). Word problems: A review of linguistic and numerical factors contributing to their difficulty. *Frontiers in Psychology, 6*, 348. https://doi.org/10.3389/fpsyg.2015.00348.

Dehaene, S., & Cohen, L. (1995). Towards an anatomical and functional model of number processing. *Mathematical Cognition, 1*, 83–120.

Dowker, A., Bala, S., & Lloyd, D. (2008). Linguistic influences on mathematical development: How important is the transparency of the counting system? *Philosophical Psychology, 21*, 523–538.

Dowker, A., & Nuerk, H.-C. (2016). Editorial: Linguistic influences on mathematics. *Frontiers in Psychology, 7*, 1035. https://doi.org/10.3389/fpsyg.2016.01035.

Duyck, W., & Brysbaert, M. (2008). Semantic access in number word translation: The role of crosslingual lexical similarity. *Experimental Psychology, 55*(2), 102–112. https://doi.org/10.1027/1618-3169.55.2.102.

Elliott, M. (1991). Is language important in mental arithmetic? *Singapore Journal of Education, 11*(2), 35–44. https://doi.org/10.1080/02188799108547670.

Feigenson, L., Dehaene, S., & Spelke, E. S. (2004). Core systems of number. *Trends in Cognitive Sciences, 8*, 307–314.

Ferry, A. L., Hespos, S. J., & Gentner, D. (2015). Prelinguistic relational concepts: Investigating analogical processing in infants. *Child Development*, *86*, 1386–1405.

Fuson, K. C., & Kwon, Y. (1992). Korean children's understanding of multidigit addition and subtraction. *Child Development*, *63*, 491–506.

Fuson, K. C., Stigler, J. W., & Bartsch, K. (1988). Grade placement of addition and subtraction topics in Japan, Mainland China, the Soviet Union, Taiwan, and the United States. *Journal for Research in Mathematics Education*, *19*, 449–456.

Gallistel, C. R., & Gelman, R. (2005). Mathematical cognition. In K. Holyoak & R. Morrison (Eds.), *The Cambridge handbook of thinking and reasoning* (pp. 559–588). New York, NY: Cambridge University Press.

Garrett, M. F. (1975). The analysis of sentence production. In G. H. Bower (Ed.), *The psychology of learning and motivation*: *Vol. 9* (pp. 133–177). New York, NY: Academic Press.

Geary, D. C. (1994). *Children's mathematical development: Research and practical applications*. Washington, DC: American Psychological Association.

Geary, D. C. (1996). International differences in mathematical achievement: Their nature, causes, and consequences. *Current Directions in Psychological Science*, *5*, 133–137.

Geary, D. C., Berch, D. B., & Mann Koepke, K. (Eds.), (2015). *Evolutionary origins and early development of number processing*. San Diego, CA: Elsevier/Academic Press.

Geary, D. C., Hamson, C. O., Chen, G. P., Liu, F., Hoard, M. K., & Salthouse, T. A. (1997). Computational and reasoning abilities in arithmetic: Cross-generational change in China and the United States. *Psychonomic Bulletin & Review*, *4*, 425–430.

Geary, D. C., Salthouse, T. A., Chen, G. P., & Fan, L. (1996). Are East Asian versus American differences in arithmetical ability a recent phenomenon? *Developmental Psychology*, *32*, 254–262.

Geary, D. C., & vanMarle, K. (2016). Young children's core symbolic and non-symbolic quantitative knowledge in the prediction of later mathematics achievement. *Developmental Psychology*, *52*, 2130–2144.

Gelman, R. (1972). The nature and development of early number concepts. In H. W. Reese (Ed.), *Advances in child development and behavior*: *Vol. 7* (pp. 115–167). New York, NY: Academic Press.

Gelman, R., & Gallistel, C. R. (1978). *The child's understanding of number*. Cambridge, MA: Harvard University Press.

Gerstmann, J. (1940). Syndrome of finger agnosia, disorientation for right and left, agraphia and acalculia: Local diagnostic value. *Archives of Neurology and Psychiatry*, *44*, 398–408.

Gordon, P. (2004). Numerical cognition without words: Evidence from Amazonia. *Science*, *306*, 496–499.

Green, D. W. (1998). Mental control of the bilingual lexico-semantic system. *Bilingualism: Language and Cognition*, *1*(2), 67–81. https://doi.org/10.1017/S1366728998000133.

Grosjean, F. (2010). *Bilingual: Life and reality*. Cambridge, MA: Harvard University Press.

Guilford, J. P. (1956). The structure of intellect. *Psychological Bulletin*, *53*(4), 267–293. https://doi.org/10.1037/h0040755.

Harmon, J. M., Hedrick, W. B., & Wood, K. D. (2005). Research on vocabulary instruction in the content areas: Implications for struggling readers. *Reading & Writing Quarterly*, *21*, 261–280.

Hatano, G. (1990). Toward the cultural psychology of mathematical cognition. Commentary. In H. W. Stevenson, S. Y. Lee, C. Chen, J. W. Stigler, C. C. Hsu, & S. Kitamura (Eds.), *Monographs of the Society for Research in Child Development: 55(1–2). Contexts of achievement: A study of American, Chinese, and Japanese children* (Serial No. 221).

Henrich, J., Heine, S. J., & Norenzayan, A. (2010). The weirdest people in the world? *Behavioral and Brain Sciences*, *33*, 61–83.

Husén, T. (1967). *International study of achievement in mathematics: A comparison of twelve countries* (Vols. 1 & 2). New York, NY: Wiley.

Jones, W. R., & Stewart, W. A. C. (1951). Bilingualism and verbal intelligence. *British Journal of Statistical Psychology*, *4*(1), 3–8. https://doi.org/10.1111/j.2044-8317.1951.tb00300.x.

Jordan, N. J., Glutting, J., Dyson, N., Hassinger-Das, B., & Irwin, C. (2012). Building kindergartners' number sense: A randomized controlled study. *Journal of Educational Psychology*, *104*, 647–660.

Jordan, J. A., Wylie, J., & Mulhern, G. (2015). Mathematics and reading difficulty subtypes: Minor phonological influences on mathematics for 5–7-years-old. *Frontiers in Psychology*, *6*, 221. https://doi.org/10.3389/fpsyg.2015.00221.

Lambert, W. E., Hodgson, R. C., Gardner, R. C., & Fillenbaum, S. (1960). Evaluational reactions to spoken languages. *The Journal of Abnormal and Social Psychology*, *60*(1), 44–51. https://doi.org/10.1037/h0044430.

Laurie, S. S. (1890). Lectures on language and linguistic method in the school (Delivered in the University of Cambridge Easter Term 1889). *Lectures on language and linguistic method in the school (Delivered in the University of Cambridge Easter Term 1889)*. New York, NY: Macmillan & Co.

Le Corre, M., & Carey, S. (2007). One, two, three, four, nothing more: An investigation of the conceptual sources of the verbal counting principles. *Cognition*, *105*, 395–438.

Levelt, W. J. M., Roelofs, A., & Meyer, A. S. (1999). A theory of lexical access in speech production. *Behavioral and Brain Sciences*, *22*, 1–75.

Lipton, J. S., & Spelke, E. S. (2005). Preschool children's mapping of number words to nonsymbolic numerosities. *Child Development*, *76*, 978–988.

Lonnemann, J., & Yan, S. (2015). Does number word inversion affect arithmetic processes in adults? *Trends in Neuroscience and Education*, *4*, 1–5.

Mark, W., & Dowker, A. (2015). Linguistic influence on mathematical development is specific rather than pervasive: Revisiting the Chinese number advantage in Chinese and English children. *Frontiers in Psychology*, *6*, 203. https://doi.org/10.3389/fpsyg.2015.00203.

McCarthy, D. (1954). Language development in children. In L. Carmichael (Ed.), *Manual of child psychology*. New York, NY: Wiley.

Miller, K. F., & Stigler, J. W. (1987). Counting in Chinese: Cultural variation in a basic cognitive skill. *Cognitive Development*, *2*, 279–305.

Miura, I. T. (1987). Mathematics achievement as a function of language. *Journal of Educational Psychology*, *79*, 79–82.

Miura, I. T., Okamoto, Y., Kim, C. C., Steere, M., & Fayol, M. (1993). First graders' cognitive representation of number and understanding of place value: Cross-national comparisons-France, Japan, Korea, Sweden, and the United States. *Journal of Educational Psychology*, *85*, 24–30.

Morales, J., Calvo, A., & Bialystok, E. (2012). Working memory development in monolingual and bilingual children. *Journal of Experimental Child Psychology*, *114*, 187–202. https://doi.org/10.1016/j.jecp.2012.09.002.

Namy, L. L., & Gentner, D. (2002). Making a silk purse out of two sow's ears: Young children's use of comparison in category learning. *Journal of Experimental Psychology: General*, *131*, 5–15.

National Mathematics Advisory Panel (2008). *Foundations for success: Final report of the National Mathematics Advisory Panel*. Washington, DC: United States Department of Education.

Nieder, A., & Dehaene, S. (2009). Representation of number in the brain. *Annual Review of Neuroscience*, *32*, 185–208.

Nuerk, H.-C., Iversen, W., & Willmes, K. (2004). Notational modulation of the SNARC and the MARC (linguistic markedness of response codes) effect. *Quarterly Journal Experimental Psychology, Section A*, *57*, 835–863.

Overmann, K. A. (2015). Numerosity structures the expression of quantity in lexical numbers and grammatical number. *Current Anthropology*, *56*, 638–653.

Paap, K. R., & Greenberg, Z. I. (2013). There is no coherent evidence for a bilingual advantage in executive processing. *Cognitive Psychology*, *66*(2), 232–258. https://doi.org/10.1016/j.cogpsych.2012.12.002.

Peal, E., & Lambert, W. E. (1962). The relation of bilingualism to intelligence. *Psychological Monographs: General and Applied*, *76*(27), 1–23. https://doi.org/10.1037/h0093840.

Pixner, S., Leyrer, M., & Moeller, K. (2014). Number processing and arithmetic skills in children with cochlear implants. *Frontiers in Psychology*, *5*, 1479. https://doi.org/10.3389/fpsyg.2014.01479.

Prior, A., Katz, M., Mahajna, I., & Rubinsten, O. (2015). Number word structure in first and second language influences arithmetic skills. *Frontiers in Psychology*, *6*, 266. https://doi.org/10.3389/fpsyg.2015.00266.

Purpura, D. J., Napoli, A. R., Wehrspann, E. A., & Gold, Z. S. (2017). Causal connections between mathematical language and mathematical knowledge: A dialogic reading intervention. *Journal of Research on Educational Effectiveness*, *10*, 116–137.

Purpura, D. J., & Reid, E. E. (2016). Mathematics and language: Individual and group differences in mathematical language skills in young children. *Early Childhood Research Quarterly*, *36*, 259–268.

Raghubar, K. P., Barnes, M. A., & Hecht, S. A. (2010). Working memory and mathematics: A review of developmental, individual difference, and cognitive approaches. *Learning and Individual Differences*, *20*, 110–122.

Reyna, V. F., Nelson, W., Han, P., & Dieckmann, N. F. (2009). How numeracy influences risk comprehension and medical decision making. *Psychological Bulletin*, *135*, 943–973.

Ritchie, S. J., & Bates, T. C. (2013). Enduring links from childhood mathematics and reading achievement to adult socioeconomic status. *Psychological Science*, *24*, 1301–1308.

Rivera-Batiz, F. (1992). Quantitative literacy and the likelihood of employment among young adults in the United States. *Journal of Human Resources*, *27*, 313–328.

Rodic, M., Tikhomirova, T., Kolienko, T., Malykh, S., Bogdanova, O., Zueva, D. Y., et al. (2015). Spatial complexity of character-based writing systems and arithmetic in primary school: A longitudinal study. *Frontiers in Psychology*, *6*, 333. https://doi.org/10.3389/fpsyg.2015.00333.

Ryan, C. (2013). Language use in the United States: 2011. In U. S. C. Bureau (Ed.), *ACS-22 ed.* (pp. 1–16): U.S. Census Bureau.

Saer, D. J. (1923). The effects of bilingualism on intelligence. *British Journal of Psychology*, *14*(1), 25–38.

Saer, H. (1931). Experimental inquiry into the education of bilingual peoples. In W. T. R. Rawson (Ed.), *Education in a changing commonwealth (Report of a British Commonwealth Education Conference Held in London in July, 1931)* (pp. 116–121). London: New Education Fellowship.

Sarnecka, B. W. (2014). On the relation between grammatical number and cardinal numbers in development. *Frontiers in Psychology*, *5*, 1–4.

Siegler, R. S., & Shrager, J. (1984). Strategy choice in addition and subtraction: How do children know what to do? In C. Sophian (Ed.), *Origins of cognitive skills* (pp. 229–293). Hillsdale, NJ: Erlbaum.

Soliman, A. M. (2014). Bilingual advantages of working memory revisited: A latent variable examination. *Learning and Individual Differences*, *32*, 168–177. https://doi.org/10.1016/j.lindif.2014.02.005.

Soltanlou, M., Pixner, S., & Nuerk, H.-C. (2015). Contribution of working memory in multiplication fact network in children may shift from verbal to visuo-spatial: A longitudinal investigation. *Frontiers in Psychology*, *6*, e1062.

Spelke, E. S. (2017). Core knowledge, language, and number. *Language Learning and Development*, *13*, 147–170.

Stevenson, H. W., & Bartsch, K. (1992). An analysis of Japanese and American textbooks in mathematics. In R. Leetsma & H. Walberg (Eds.), *Japanese educational productivity* (pp. 103–133). Ann Arbor, MI: Center for Japanese Studies.

Stevenson, H. W., Chen, C., & Lee, S. Y. (1993). Mathematics achievement of Chinese, Japanese, and American children: Ten years later. *Science*, *259*, 53–58.

Stevenson, H. W., Lee, S. Y., Chen, C., Stigler, J. W., Hsu, C. C., & Kitamura, S. (1990). Contexts of achievement: A study of American, Chinese, and Japanese children. *Monographs of the Society for Research in Child Development*, *55(1-2)*. (Serial No. 221).

Stevenson, H. W., Lee, S. Y., & Stigler, J. W. (1986). Mathematics achievement of Chinese, Japanese, and American children. *Science*, *231*, 693–699.

Stevenson, H. W., & Stigler, J. W. (1992). *The learning gap: Why our schools are failing and what we can learn from Japanese and Chinese education*. New York, NY: Summit Books.

Stigler, J. W., Gonzales, P., Kwanaka, T., Knoll, S., & Serrano, A. (1999). *The TIMSS videotape classroom study: Methods and findings from an exploratory research project on eighth-grade mathematics instruction in Germany, Japan, and the United States. A Research and Development Report*. Washington, DC: National Center for Educational Statistics.

Sullivan, J., & Barner, D. (2014). Inference and association in children's early numerical estimation. *Child Development*, *85*, 1740–1755.

Syrett, K., Musolino, J., & Gelman, R. (2012). How can syntax support number word acquisition? *Language Learning and Development*, *8*, 146–176.

Valian, V. (2015). Bilingualism and cognition. *Bilingualism: Language and Cognition*, *18*(1), 3–24. https://doi.org/10.1017/S1366728914000522.

vanMarle, K., Chu, F. W., Mou, Y., Seok, J. H., Rouder, J. N., & Geary, D. C. (in press). Attaching meaning to the number words: Contributions of the object tracking and approximate number systems. *Developmental Science*.

Wang, Q. (2017). Five myths about the role of culture in psychological research. *APS Observer*, *30*, 21–24.

Weinreich, U. (1953). *Languages in contact—Findings and Problems*. New York, NY: Linguistic Circle of New York.

Wiseheart, M., Viswanathan, M., & Bialystok, E. (2016). Flexibility in task switching by monolinguals and bilinguals. *Bilingualism*, *19*(1), 141–146. https://doi.org/10.1017/S1366728914000273.

Wynn, K. (1990). Children's understanding of counting. *Cognition*, *36*, 155–193.

Wynn, K. (1992). Children's acquisition of the number words and the counting system. *Cognitive Psychology*, *24*, 220–251.

Zohar-Shai, B., Tzelgov, J., Karni, A., & Rubinsten, O. (2017). It does exist! A left-to-right spatial–numerical association of response codes (SNARC) effect among native Hebrew speakers. *Journal of Experimental Psychology: Human Perception and Performance*, *2017*(43), 719–728.

Chapter 2

Analogical Mapping in Numerical Development

Elisabeth Marchand and David Barner
University of California, San Diego, CA, United States

INTRODUCTION

Mathematical thinking relies on a wide range of abstract representations that are unique to human cognition. The ability to create abstract numerical representations such as variables, matrices, and infinite magnitudes is related to learning a symbolic number system, beginning with a count list. Humans have created a host of different counting systems, including nonlinguistic systems such as the abacus, tallies, and body-count systems and linguistic systems, like verbal count lists (Ifrah, 2000; Menninger, 1969). Over the past several decades, there has been increasing evidence that language provides an especially powerful tool for representing number, beginning early in a child's life. Studies have shown that the acquisition of a verbal counting system allows children and adults to precisely quantify large sets of objects and that verbal working memory is essential to carry out arithmetic operations (Ashcraft & Stazyk, 1981; Barner, Chow, & Yang, 2009; Dehaene, Spelke, Pinel, Stanescu, & Tsivkin, 1999; Frank, Everett, Fedorenko, & Gibson, 2008; Gordon, 2004; Pica, Lemer, Izard, & Dehaene, 2004). However, the precise role of language and the mechanisms by which it is related to nonverbal numerical representations remains controversial.

In this chapter, we argue that a key component to the human ability to represent number with natural language is what Gentner and colleagues (Forbus, Gentner, & Law, 1995; Gentner, 1983; Gentner & Markman, 1997) have called "analogical mapping," defined as the construction of correspondences between two representations that share similar structures. Specifically, we review evidence that both the acquisition of number words and their use when estimating large sets depend fundamentally on the construction of analogical mappings between language and nonverbal representations. To support the claim that number-word learning relies on analogical mapping, we analyze Carey's (2004, 2009) account of number learning, address the problems with

Mathematical Cognition and Learning, Vol. 4. https://doi.org/10.1016/B978-0-12-812574-8.00002-X

this account, and offer alternatives for how children acquire the meaning of counting based on analogical mapping. Second, we argue that very similar processes of analogical mapping support numerical estimation—for example, the labeling of large sets of rapidly presented objects. To make this claim, we review evidence from calibration effects during estimation and from the role of ordinality in the representation of symbolic numbers. We conclude that analogy is key to understanding how perception and language become related in development and how this allows humans to acquire fundamental mathematical knowledge.

ANALOGICAL MAPPING IN INTEGER LEARNING

The central aim of this chapter is to understand how language becomes related to nonverbal numerical knowledge and whether mapping numerical content to language fundamentally changes the types of numerical thoughts humans can entertain. Our proposal is that learning the meanings of number words and learning how to rapidly assign labels to large arrays of objects when estimating both rely on the so-called analogical mapping. In this first section, we explore the role of language and analogical mapping in the acquisition of counting logic and the concept of natural numbers.

ACQUISITION OF NATURAL NUMBERS

A central challenge when children learn how to count is the problem of discovering how counting represents cardinality. Studies of number-word learning have found that it is indeed a difficult process that takes place over the course of many years. In the United States, English-speaking children typically begin by learning the count list at around the age of 2 as though it were a single expression, much like the alphabet (one two three four...), and at this point attach no meaning to the individual words (Carey, 2001; Carey & Sarnecka, 2006; Le Corre & Carey, 2007; Wynn, 1990, 1992). For example, if a child is asked, "how many objects are there?", they may be able to point at the objects while reciting the count list but nevertheless fail to recognize that the last number word in the count sequence represents the total amount of objects in the set (Fuson, 1988). Also, if asked to give a particular number of objects—even a small number like one or two—children at this stage often give random amounts. When children do begin to associate number words with meanings, they do so in sequence, learning the meanings of *one*, *two*, *three*, and *four* in order over many months (Le Corre, Van de Walle, Brannon, & Carey, 2006; Sarnecka & Carey, 2008; Sarnecka & Lee, 2009; Wynn, 1990, 1992).

This sequence has been demonstrated using the give-a-number task, in which children are asked to provide *N* number of objects (e.g., "can you give me four marbles?"; Fuson, 1988; Wynn, 1992). Sometime, after beginning to recite the count list, children who are asked to give one object are able to give one, and do not give one for larger numbers. However, they are unable to reliably give appropriate amounts for larger numbers like *two*, *three*, and so on. These children are often called "one-knowers." Around 6–9 months later, children learn an exact meaning for *two* and can give one and two objects when asked for *one* and *two* but give incorrect responses for larger numbers. These children are generally called "two-knowers." Next, children become "three-knowers" and sometimes "four-knowers." Finally, after learning *three* (and sometimes *four*), children appear to make a remarkable breakthrough: they seem to realize that the last number word used when counting objects refers to the cardinality of the set and can use their counting routine to generate and give sets larger than three or four. At this stage, children are referred to as "cardinal-principle knowers" (CP-knowers).

By some accounts, becoming a CP-knower involves learning the logic of counting and in particular the successor function. The Peano-Dedekind axioms, which describe the core logical foundation for natural number and arithmetic, define the successor function as including a combination of (1)–(3):

1. One is a natural number.
2. For every natural number n, the successor of n is a natural number.
3. Every natural number has a successor.

The successor function sits at the core of natural number because it defines all possible numbers. For children, acquiring it requires learning that, for any number, no matter how large, a larger number can be created by adding "one" and, following this logic, that numbers are infinite (for discussion, see Cheung, Rubenson, & Barner, 2017).

As we discuss in the following section, a number of researchers, including Carey (2009), have proposed that children become CP-knowers when they notice that the successor function describes the relations between the numbers *one*, *two*, and *three*. On Carey's account, this conceptual change is the product of a type of analogical mapping between relations among number words and relations among cardinalities.

ANALOGICAL MAPPING IN ACQUIRING NATURAL NUMBERS

In the present section, we begin by defining what we mean by analogical mapping, describe Carey's account of how analogy might guide the acquisition of the successor function (e.g., as per Carey, 2009), and argue that, while Carey is right to invoke analogy in this case, she is likely incorrect about when children learn the successor function and thus which information they use to do so.

According to Gentner's (1983) "structure-mapping" theory, learning abstract concepts like number words often involves a process of establishing a structural alignment between two representations—a base and a target—and then projecting inferences from the base to the target (and vice versa).

To illustrate how structural alignment is deployed and how children can learn from analogical mapping, we present here an example similar to the stimuli used by Goswami and Brown (1990): nest:bird::hive:bee or in common language "a nest is to a bird what a hive is to bees." In this example, the analogical mapping aligns two distinct instances of a biological relation—that between an animal and its corresponding habitat. Since the child is more likely to be familiar with birds and nests, they are also more likely to discover the relation between the two elements: that nests are the habitats of birds where they lay eggs. The relation nest:bird can therefore constitute the base of the structural alignment. In contrast, the relation between bees and hives, which the child is less familiar with, constitutes the target.

The structural alignment takes place as the child notices that the two biological relations share an isomorphic structure: the two elements of the base and the target—animal and habitat—can be put into correspondence such that the two animals go with their two habitats (Fig. 1). Spontaneously, the child can compare the two biological relations and find the similarities and differences between them. For instance, whereas hives contain the bee's food source (honey), nests do not contain the chick's food source. This comparison process allows the child to make inferences about the target (hive) from what he knows about the base (nest), for example, that the hive houses bee eggs. From this analogical mapping, the child can also discover abstract, less obvious commonalities between many animal species (e.g., that most animals build habitats that are adapted to their needs in order to raise their offspring). This example illustrates how analogical mapping can promote learning: it supports abstraction, highlights differences and common underlying relational structures, and facilitates inferential thinking and the transfer of information between two representations.

According to Wynn (1992) and Carey (2004, 2009), analogy plays a similar role in children's discovery of the successor function. On this view, children begin by learning small number words via item-based associative mappings between the words *one*, *two*, *three*, and *four* and representations of small sets. For Carey, these small sets are represented by what she calls "parallel individuation" (PI) system, which allows human to keep track of three to four individual objects in parallel and to compare sets via

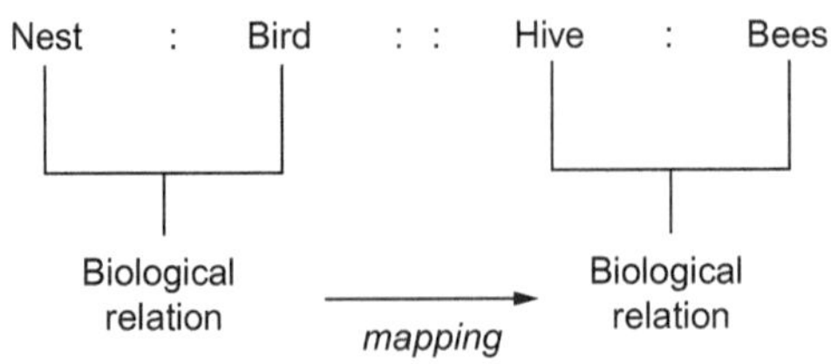

FIG. 1 Analogical mapping applied to the case of nest, bird and hive, bees. The mapping takes place between the two biological relations nest:bird (the base) and hive:bees (the target).

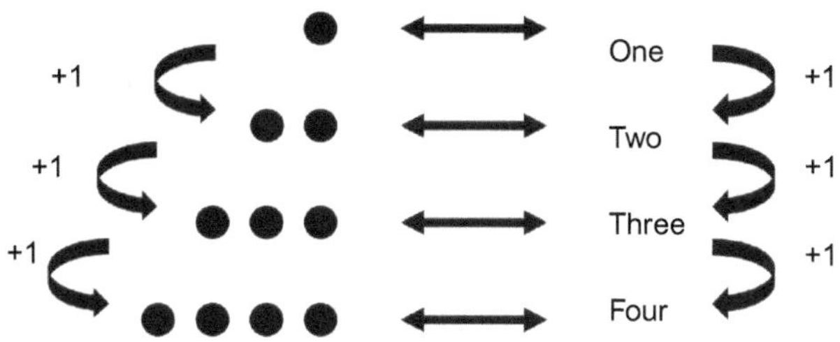

FIG. 2 Relation between set's magnitudes (left) and number words (right). Analogical mapping allows children to notice the logic of the counting list: further by one in the count list is mapped onto increasing the +1 relation in numerical magnitude.

one-to-one correspondence.[1] Once children store associations between the number words *one*, *two*, and *three* and the sets that they represent in long-term memory, they are in a position to "make a wild analogy—that between the order of a particular quantity within an ordered list and that between this quantity's order in a series of sets related by additional individuals" (Carey, 2004, p. 67)—or what Wynn (1992) described as "the analogy between the magnitudinal relationships of their own representations of numerosities and the positional relationships of the number words" (p. 250).

In other words, children notice that the labels for small numbers (1, 2, and 3) comprise an ordered sequence and that this sequence is isomorphic to the ordered sequence of quantities that the words denote; specifically, each step forward in the count list corresponds to the addition of exactly one to the cardinality of the set (Fig. 2). According to Carey, "if the child recognizes this analogy, she is in the position to make the crucial induction: for any word on the list whose quantificational meaning is known, the next word on the list refers to a set with another individual added" (Carey, 2001, p. 67). Based on this, Carey argues that children next infer that this parallel increase of one holds for all numbers in the count list, an inductive generalization that is confirmed with experience counting larger numbers. Hence, according to Wynn (1992) and Carey (2001, 2004), analogical mapping allows children to acquire the logic that governs number words (successor function), which not only determines the cardinalities of all number words but also provides the logical foundations on which to build other mathematical knowledge.

PROBLEMS WITH THE CURRENT ACCOUNT OF ANALOGICAL MAPPING IN INTEGER LEARNING

According to Carey's (2004, 2009) account, children form an analogy between the count list and representations of small sets, which allows them

1. For a review of Carey's account and alternatives to it, see Barner (2017), Margolis and Laurence (2008), and Rips, Asmuth, and Bloomfield (2008).

to discover the successor function. On this account, children who have made this analogy understand the cardinal principle—that is, the last word in a count labels the cardinality of the counted set—and are thus referred to as CP-knowers. One problem with this theory is that, according to recent studies, many children classified as CP-knowers do not appear to have knowledge of the successor function (Cheung et al., 2017; Davidson, Eng, & Barner, 2012). Below, we review the original evidence that supported Carey's account and describe in more detail the problems with this account.

As evidence for the hypothesis that becoming a CP-knower is driven by acquiring the successor function, Sarnecka and Carey (2008) tested 2- to 4-year-old children with what they called the "unit task." In this task, an experimenter places a set of objects into a box and labels it with a number word—for example, "OK, I'm putting FOUR frogs in here." Next, they add either one or two more objects to the box and ask children how many there are in a two-alternative forced choice task—for example, "now, is it FIVE or SIX?". Overall, Sarnecka and Carey (2008) found that, whereas subset knowers (one-, two-, three-, and four-knowers) performed at chance and chose randomly between the two options, CP-knowers selected the correct number 67% of the time, on average. Based on these results, they concluded that CP-knowers, but not subset knowers, have knowledge of the successor function, and therefore, learning the successor function is what causes children to become CP-knowers.

However, there are several problems with this conclusion. First, most of the CP-knowers in Sarnecka and Carey's (2008) study (16 out of 29) did not perform better than chance on the unit task. If mastery of the successor function were really what caused children to become CP-knowers, then we should expect them to be able to identify successors relatively systematically either before or at the moment they become CP-knowers. The fact that most children failed suggests that children may not learn the successor function until *after* they become CP-knowers. Second, whereas Sarnecka and Carey (2008) claimed that children had made an inductive inference—that is, an inference that applies to *all* numbers—they only tested very small, familiar numbers, making it possible that the small group of CP-knowers who did succeed at the task only did so because they had learned item-specific knowledge regarding the difference between, for example, *four* and *five*. To demonstrate that children have made an inductive inference over all numbers, much stronger evidence is required and at the least evidence that children can perform the task for more than one or two small numbers.

To explore these issues, Davidson et al. (2012) presented CP-knowers with different tasks measuring their knowledge of the order of numbers in the count list, the successor function (via the unit task used by Sarnecka & Carey, 2008) and the count list. Overall, the authors found that most CP-knowers didn't know which of two numbers in their count list denotes the greater quantity and most importantly failed to understand the +1 rule for even very small numbers such as 5 or 6 (well within their counting range). Interestingly, performance on the unit task (which probed the successor function) was related to children's knowledge of the count list—a proxy for experience counting: children who could count up to 19 or less performed at chance for small numbers like 4 and 5, while children who could count up to 30 showed a better understanding of the relation between small numbers but were nevertheless at chance for numbers in the teens. Only the highest counters, who could count to 30 or more, performed well for small numbers, though these children continued to struggle with numbers greater than 10, even though these numbers were well within their familiar counting range. Together, these findings suggest that CP-knowers have difficulty generalizing the successor principle to all numbers within their productive counting range, as an inductive inference account would predict. In other words, knowledge of the successor function does not arise when children become CP-knowers. Instead, according to Davidson et al. (2012), becoming a CP-knower appears to involve learning simple procedures for counting and constructing sets but with very little insight into how or why this works.

While Davidson et al. (2012) showed that young CP-knowers have not yet learned the successor function, their study did not describe when children learn this knowledge nor how. In fact, as noted by Cheung et al. (2017), the unit task is not an appropriate measure of this knowledge when used in isolation, because the task does not measure the successor function as it is typically defined. In past studies, the successor function is generally defined as a relation between numbers—for example, that the successor of a particular number n is defined as $n+1$. Critically, however, the successor function as defined in the Peano axioms states that for any number, n, this number has a unique successor whose value is $n+1$, such that there is no highest number. This means that in order to have a complete understanding of the integer concept, children must learn that *all* numbers have a successor and thus that numbers never end. Thus, Sarnecka and Carey's (2008) study was limited not only by the relatively small range of numbers that they tested but also by the fact that the unit task is limited to probing a finite set of numbers—that is, whichever numbers children happen to be familiar with.

In an effort to address this problem, Cheung et al. (2017) presented 4- to 6-year-old children with the unit task of Sarnecka and Carey (2008) and an infinity task—a set of open-ended questions that tested whether children thought all possible numbers have a successor (adapted from Evans, 1983; Hartnett & Gelman, 1998). In this task, children were asked, for example, "What is the biggest number you can think about?", "If I keep counting, will I ever get to the end of numbers, or do numbers go on forever?", and "If we thought of a really big number, could we always add to it and make it even bigger, or is there a number so big, we couldn't add any more?". Cheung et al. (2017) found that CP-knowers' ability to identify successors was highly variable and depended upon their age and their counting skills, such that experienced counters, who are also older, are more likely to find a successor for a presented number and know that every number n has a successor $n+1$. In the infinity task, only 24 out of 100 CP-knowers judged that it is possible to always add 1 to a number and that there is no largest number. In light of these results, the experimenters concluded that it is only after becoming CP-knowers and mastering numbers up to around 80 that children develop a complete understanding of the successor function, which occurs sometime between 5½ and 6 years of age. These findings also suggest that children only have item-based knowledge of the successor function when they become CP-knowers and that experience counting to large numbers may contribute to learning the successor function.

How can these findings be reconciled with the account proposed by Carey (2004, 2009)? As explained by Cheung et al. (2017), their data pose a problem to Carey's analogical mapping account only if we assume, as Carey does, that the successor function is acquired from an induction over mappings between small sets and their labels. However, it is also possible that the same mechanism operates over different input available to children after they become CP-knowers. As described in Cheung et al., after becoming CP-knowers, children continue to be exposed to large numbers and are placed in various situations where they have to count large quantities. At this point, CP-knowers have access to a counting procedure (i.e., the ability to generate sets of any sizes), which provides them with the opportunity to place numbers in an isomorphic relation with many different cardinalities, thus allowing them to notice that larger numbers, like small ones, are related by a difference of 1. This broader range of experience may allow children to infer that the successor function holds across their productive count list. Furthermore, as children acquire new sequences of the count list, they may become more likely to notice the recursive structure of the count list (e.g., that the same units from one to nine are repeated within every decade). Given such an observation, children would then be in the position of knowing (1) that the successor function applies to all words in their known count list and (2) that the count list has a recursive structure and thus can be extended indefinitely. Together, these observations may allow them to infer that an infinite number of numbers are related by the successor function.

An alternative possibility is that children do not discover the successor function themselves but instead acquire it chiefly through explicit instruction, when they begin to learn arithmetic in school. At around 5½ to 6 years old—around the time that Cheung et al. (2017) argued that children learn the successor function—children begin to receive exposure to explicit arithmetic facts, like $4+1=5$ and $5+1=6$. These facts, which are directly relevant to acquiring the successor function, might alone serve as the inputs to acquiring the broader generalization that the successor function applies to all numbers, without direct appeal to the representations of perceptual magnitudes. For example, having learned that $1+1=2$, $2+1=3$, $3+1=4$, etc., children might form the generalization that for all n, $n+1=S(n)$—i.e., that adding 1 to a set labeled by the word *four* produces a set labeled by the successor of *four* in the count list. Whereas the first alternative is based on an analogical mapping, this second alternative for how children acquire the successor function does not.

A third possibility is that a complete understanding of the successor function might be supported by a combination of different types of information. It is possible, for example, that counting experience enables children to learn, via analogical mapping, that the successor function applies to all possible sets that they can count (since a count of 9 and a count of 10 differ by exactly 1, just as all counts that differ by one numeral). This might allow the child to extend the successor function to the largest number or set that they can think of, which, assuming the child cannot imagine infinite magnitudes, might be limited to a hypothesis space of finite sets and finite number words. To transcend this, children might need to learn a set of rules that consider numbers as distinct abstract entities, disconnected from perception or the count list, which might support the intuition that numbers can be added endlessly—for example, as is the case with arithmetic. Counting experience and memorizing arithmetic facts are two situations that have similar structure: They rely on learning procedures and constructing isomorphisms between these procedures and cardinalities in the world—for example, that the equation $2+2=4$ is isomorphic to a situation in which two sets of 2 are combined to create a set of 4. By noticing this isomorphism and thereby making an analogical mapping between symbols and sets, children might learn the successor function directly from arithmetic. On this hypothesis, the successor function might be learned from a combination of analogical mapping and experience with formal mathematics.

In summary, while several alternative accounts remain viable regarding how children learn the logic of counting, most have one aspect in common: they rely, at least in part, on analogical mapping. Future studies should test the predictions made by these three accounts to determine the specific experience and knowledge needed for this analogical mapping: counting with large numbers, recursive count list, and/or memorizing arithmetic facts. In the next section, we argue that analogical mapping plays an additional crucial role in numerical cognition: allowing children to build relations between verbal and nonverbal systems of approximate numerical representation.

ANALOGICAL MAPPING IN MAGNITUDE ESTIMATION

As children learn to count, they also begin to associate number words with large sets in parallel, allowing them to make rapid estimates without counting. For example, without counting, 5-year-old children can give nonrandom verbal estimates for large sets, though these estimates follow a logarithmic pattern early on, and become more linear at ages 6 and 7 (Barth & Paladino, 2011; Lipton & Spelke, 2005; Mundy & Gilmore, 2009). How do children construct this mapping between the verbal and nonverbal number systems? One possibility is that these systems become related to each other via associative learning (i.e., item-based association between number words and their magnitude). However, in this section, we propose that the same analogical mapping mechanism that is involved in acquiring the logic of counting also supports children's developing ability to make estimates of large sets. Specifically, in this section, we provide two pieces of evidence in support of the idea that the count list and nonverbal representations are related via an analogical mapping during estimation. First, we present findings revealing the considerable flexibility and interdependence of participants' responses when estimating. This evidence suggests that number words are not rigidly associated to numerical magnitudes in an associative, item-based way but instead that children construct a holistic, analogical mapping between the list as a whole and the ordered set of nonverbal magnitudes. Second, again consistent with the idea that there are no item-based associations between the linguistic and nonlinguistic number systems, we discuss evidence that ordinality is essential to the representation of symbolic numbers. Finally, we relate facts regarding two different types of estimation tasks—number-line estimation in addition to dot-array estimation—which provide converging evidence of the role of analogical mapping in estimation processes.

ANALOGICAL MAPPING IN ESTIMATION: EVIDENCE FROM THE FLEXIBILITY AND INTERDEPENDENCE OF ESTIMATES

The approximate number system allows us to represent numbers in a nonsymbolic format (see Feigenson, Dehaene, & Spelke, 2004, for a review), and major accounts agree that the approximate number system is involved in the process of estimation (Booth & Siegler, 2006; Dehaene, Piazza, Pinel, & Cohen, 2003; Le Corre & Carey, 2007), though the exact role of the ANS in the computation often remains disputed. Here, we propose that verbal estimation is based on an analogical mapping between the count list and the ordered set of nonverbal numerical magnitudes encoded by the approximate number system.

The linguistic and nonlinguistic number systems share an isomorphic structure, which is favorable to the elaboration of a holistic, broad, analogical mapping. For instance, both the count list and the nonverbal number system represent increasing ordinal magnitude; the number word 50 comes after 25 just as a set of 50 objects is more numerous than a set containing 25 items. Moreover, both representations encode relative distance and proportions between symbols and approximate magnitudes. For example, the number word 50 is twice the distance from 25 in the count list, just as a set of 50 objects is twice as large as a set of 25.

According to the analogical mapping hypothesis, mappings between the constituents of the two systems are holistic. In the case of estimation, this means that there are no individual associations between specific words but instead one single mapping between the entire count system as a whole and the magnitude system as a whole. Critically, this idea makes strong predictions about estimation behavior: if estimation is driven by a global analogical mapping where one system is associated to another, then learning about one number word would be expected to influence all other number words. In contrast, if number words were individually associated with magnitudes, we would expect feedback about one number word to be relatively independent of feedback for another number—that is, we would not expect the structure of either the count list or the magnitude system to play a role in calibrating these associations. To tease apart these two accounts, Sullivan and Barner (2013, 2014) used a method developed by Izard and Dehaene (2008), whereby adults were presented with rapidly flashed arrays of dots and were asked to produce a verbal estimate of the total number of dots. In the absence of feedback, participants' estimates were largely inaccurate, underestimating the actual numerosity, but their estimates nevertheless increased with numerosity. Interestingly, when feedback (whether correct or misleading) was provided to the participants for a single dot array—for example, telling them that an array of 25 dots contained 50 items—estimates shifted considerably not only for the trained dot array but also for a wide range of numbers. The fact that all estimates were shifted as a group supports the idea of a broad mapping between the entire count list and perceived magnitudes and speaks to the major role that analogical thinking plays in the representation of large quantities.

Several recent studies suggest that, across development, both adults and children rely on an analogical mapping between the count list and the nonverbal representations of number while also building associative item-based mappings for a handful of small numbers. Sullivan and Barner (2013) asked adult participants to estimate flashed arrays of dots but told them before the experiment that there would be an upper limit on the number of dots they would see. Critically, although this limit was actually always 350 dots, different groups were told that they would see up to 75, 375, or 750 dots. Participants were also presented with a number-matching task in which they heard a number word and then saw two arrays of dots presented sequentially and had to judge which array best matched the word. Sullivan and Barner (2013) showed that for small magnitudes (up to about 12), participants matched number words to arrays very accurately and were highly resistant to calibration, whereas for larger sets they were highly influenced by misleading calibration and struggled to correctly match number words to arrays despite the fact that the quantities always exhibited a highly discriminable 2:1 ratio. From these findings, the authors argued that estimation depends on a combination of associative mappings—individual words mapped to their specific magnitudes, which support small numbers—and analogical mapping (or "structure mapping"), which support estimation for larger numbers. In a similar study with children, Sullivan and Barner (2014) found a similar pattern of results:

children exhibited strong item-based associative mappings for small numbers (up to around six) but relied primarily on structure mapping when making estimates for larger numbers.

Besides, making predictions about malleability, the structure-mapping hypothesis also makes a second prediction: because of the interdependency of the number-word mappings, the magnitudes and verbal estimates of previous trials should influence the responses for successive trials. Two different studies tested this prediction (Cicchini, Anobile, & Burr, 2014; Vul, Barner, & Sullivan, 2013). In the first study, Vul et al. (2013) presented adults with a calibrated dot-array estimation task and a noncalibrated task and showed that even if participants are not given explicitly misleading feedback, their estimates are mutually constraining, such that estimation error carries forward from trial to trial. From these findings, the authors claimed that the variability found in estimation arises from the variability in the mapping function that connects the verbal and nonverbal number systems. Using a similar logic, Cicchini et al. (2014) measured intertrial dependency and demonstrated that any particular current response was influenced by up to 15% by the immediately preceding response. Again, this suggests that participants' estimates are based on a dynamic process that is more consistent with an analogical account than with associative mappings.

Studies using the number-line paradigm also provide evidence for the analogical mapping hypothesis. In a typical number-line task, participants are presented with a black horizontal line, most frequently marked at the beginning and the end by two numerals (e.g., 1 and 100), and are asked to mark the proper position of a given number. Early studies using this technique found that young children (e.g., 5 years of age) have nonlinear representations, spacing out estimates for small numbers on the line in a relatively linear way but making logarithmic estimates for larger numbers. In contrast, older children (e.g., 7-year olds) make more linear estimates (Booth & Siegler, 2006; Opfer & Siegler, 2007; Siegler & Booth, 2004; Siegler & Opfer, 2003). Recently, additional studies have argued that in these tasks, children estimate using a form of proportional reasoning that resembles analogical mapping, rather than reliance on linear versus logarithmic number representations as previously believed (Barth & Paladino, 2011; Barth et al., 2016; Cohen & Blanc-Goldhammer, 2011). Using the number-line paradigm with children aged 6–8, Barth and Paladino (2011) showed that patterns of estimation were better explained by a proportion-judgment model than a linear model. In a more recent study, Rouder and Geary (2014) developed a set of hierarchical statistical models to contrast the pattern of variability in number-line placements predicted by different theories of number-line representation. Most first graders' placements suggested the use of end points, compatible with the proportional reasoning view. Also, as children got older (fifth grade), their pattern of estimates reflected the incorporation of a midpoint anchor and thus a modified proportional reasoning strategy.

Like analogical mapping, proportional reasoning is based on an understanding of the order and the distance between the number words of the count list: for example, if a child knows that 50 is between 1 and 100 and that 100 is twice as far as 50, he can infer quite easily the position of 50 on the number line.

ANALOGICAL MAPPING IN ESTIMATION: THE ROLE OF ORDINALITY

Thus far, we have argued that analogical mapping links the verbal and nonverbal number systems during estimation. This account makes one additional prediction that we explore here. Specifically, it predicts that familiarity with the count list and in particular familiarity with how the list is structured should affect children's estimation performance, since analogical mapping is fundamentally a process of aligning structures from different domains of knowledge.

Previous studies suggest that estimation abilities are strongly related to children's knowledge of the symbolic system and how it is structured (Ebersbach, 2016; Ebersbach, Luwel, Frick, Onghena, & Verschaffel, 2008; Le Corre & Carey, 2007; Lipton & Spelke, 2005). For example, Lipton and Spelke (2005) presented 5-year-old children with three tasks that tested their knowledge of the count list. In a free estimation task, children who were classified as skilled counters produced estimates that were linearly related to numerosity, unlike children classified as unskilled counters. In the number-word comprehension task, where children had to choose which set corresponded to a given number word, only skilled counters were able to perform the task above chance level for small and large numbers. Finally, in the number ordering task, children had to estimate an array of dots based on the cardinal value of one array presented simultaneously. As expected, skilled counters were able to produce ordinal responses—response in the right direction—unlike unskilled counters.

Consistent with the analogical mapping hypothesis, these results suggest that knowledge of the count list is fundamental to estimation. However, this is of course compatible with any alternative accounts, including associative mapping. A unique prediction made by an analogical mapping account is that *structural knowledge* of the count list should be especially important. Consistent with this, previous studies have found that ordinality is key to understanding how numbers are represented symbolically (Lyons & Ansari, 2015; Lyons, Ansari, & Beilock, 2011; Lyons & Beilock, 2009, 2011, 2013; Roggeman, Verguts, & Fias, 2007; Turconi, Campbell, & Seron, 2006). For example, Lyons et al. (2011) asked participants to compare magnitudes expressed in different formats—that is, arrays of dots, Arabic digits, and verbal numbers. The authors showed that mixing representational formats (such as arrays of dots and Arabic digits) increased reaction times, suggesting that symbolic representations do not automatically trigger representations of quantity. These results are compatible with the analogical mapping view in which the mappings between the linguistic and nonlinguistic number systems are dynamic and rooted in an understanding of the structure of the count list, rather than being driven by automatic, item-based associations between number words and approximate magnitudes. As noted by Lyons et al. (2011), "a crucial facet of numerical symbols is how they relate to other symbols; indeed it may even be the case that with repeated exposure to numerical symbols, symbol-symbol relations in literate adults come to usurp symbol-quantity relations" (p. 1515).

Relatedly, as part of their investigation of analogical mapping in estimation, Sullivan and Barner (2014) explored the relation between ordinal knowledge and estimation abilities in 5- and 7-year-old children. The authors presented children with an ordinality task (e.g., which is more, 20 or 40?) and an estimation task (dot arrays). In the estimation task, children's responses tended to be ordinal (i.e., in the correct direction relative to previous estimates) regardless of estimation accuracy, suggesting that knowledge of ordinality is a precursor to accurate estimation. Moreover, accuracy at the ordinality improved substantially with age. Interestingly, the authors did not find a relation between accuracy in the verbal ordering task and the likelihood that a participant would provide an ordinal response on the estimation task. The authors suggested that "although possessing ordinal number-word representations is a prerequisite for forming structure mappings (or analogical mappings), the presence of relatively good verbal ordering does not ensure that estimates will be ordinal" (Sullivan & Barner, 2014, p. 1751). However, this study used only one task of ordinal knowledge, and further work is needed to shed light on the relation between ordinal knowledge and estimation accuracy.

OTHER EVIDENCE OF ANALOGY AS A SOURCE OF REPRESENTATIONAL CHANGES

In addition to linking number words to approximate magnitudes, analogy may also play additional roles in the creation of adult-like estimation abilities. For example, Thompson and Opfer (2010) propose an interesting case in which children might analogically extend knowledge of small numbers to inform estimates for much larger numbers. They first evaluated second graders' number-line estimation with lines of identical physical length but that represented the scales of either 0–100, 0–1000, or 0–10,000. On this first task, children's estimates exhibited a linear pattern of estimates on the 1–100 scale but a logarithmic pattern on both of the larger scales. Next, they asked whether explaining the similarities between the more familiar, smaller scale (0–100) and the less familiar scales (0–1000 and 0–10,000) might invite children to form an analogy between the two types of scale and thereby improve their estimates for the larger scales. Consistent with this, they found that after training, children's pattern of estimates on the larger number lines became linear. This suggests that, besides creating an initial link between number words and magnitudes, analogy may also play an important role in gradually extending knowledge of small numbers to larger ones, as children come to understand how the count list is structured.

GENERAL CONCLUSION

In the current chapter, we explored the role of analogical mapping, as defined by Gentner (1983, 2010), in numerical development. We argued, following past accounts (Carey, 2009; Wynn, 1990, 1992), that analogical mapping contributes considerably to the acquisition of the logic of counting but that this learning takes place much later than generally assumed and using different inputs than proposed in previous studies. Also, based on arguments provided by Sullivan and Barner (2013, 2014), we argued that the same mechanism of analogical mapping is also involved in mapping number words to nonverbal approximate number representations in the service of estimation. Based on this, we conclude that analogical mapping provides a key mechanism to acquiring both the logic of counting and its association to perceptual representations, like those provided by the approximate number system.

REFERENCES

Ashcraft, M. H., & Stazyk, E. H. (1981). Mental addition: A test of three verification models. *Memory & Cognition*, *9*(2), 185–196.

Barner, D. (2017). Language, procedures, and the non-perceptual origin of number word meanings. *Journal of Child Language*, *44*(3), 553–590.

Barner, D., Chow, K., & Yang, S. J. (2009). Finding one's meaning: A test of the relation between quantifiers and integers in language development. *Cognitive Psychology*, *58*(2), 195–219.

Barth, H. C., & Paladino, A. M. (2011). The development of numerical estimation: Evidence against a representational shift. *Developmental Science*, *14*(1), 125–135.

Barth, H., Slusser, E., Kanjlia, S., Garcia, J., Taggart, J., & Chase, E. (2016). How feedback improves children's numerical estimation. *Psychonomic Bulletin & Review*, *23*(4), 1198–1205.

Booth, J. L., & Siegler, R. S. (2006). Developmental and individual differences in pure numerical estimation. *Developmental Psychology*, *42*(1), 189.

Carey, S. (2001). Cognitive foundations of arithmetic: Evolution and ontogenesis. *Mind & Language*, *16*(1), 37–55.

Carey, S. (2004). Bootstrapping & the origin of concepts. *Daedalus*, *133*(1), 59–68.

Carey, S. (2009). Where our number concepts come from. *The Journal of Philosophy*, *106*(4), 220.

Carey, S., & Sarnecka, B. W. (2006). The development of human conceptual representations: A case study. In Y. Munakata & M. Johnson (Eds.), *Processes of change in brain and cognitive development: Attention and performance XXI* (pp. 473–496). Oxford, UK: Oxford University Press.

Cheung, P., Rubenson, M., & Barner, D. (2017). To infinity and beyond: Children generalize the successor function to all possible numbers years after learning to count. *Cognitive Psychology*, *92*, 22–36.

Cicchini, G. M., Anobile, G., & Burr, D. C. (2014). Compressive mapping of number to space reflects dynamic encoding mechanisms, not static logarithmic transform. *Proceedings of the National Academy of Sciences of the United States of America*, *111*(21), 7867–7872.

Cohen, D. J., & Blanc-Goldhammer, D. (2011). Numerical bias in bounded and unbounded number line tasks. *Psychonomic Bulletin & Review*, *18*(2), 331–338.

Davidson, K., Eng, K., & Barner, D. (2012). Does learning to count involve a semantic induction? *Cognition*, *123*(1), 162–173.

Dehaene, S., Piazza, M., Pinel, P., & Cohen, L. (2003). Three parietal circuits for number processing. *Cognitive Neuropsychology*, *20*(3-6), 487–506.

Dehaene, S., Spelke, E., Pinel, P., Stanescu, R., & Tsivkin, S. (1999). Sources of mathematical thinking: Behavioral and brain-imaging evidence. *Science*, *284*(5416), 970–974.

Ebersbach, M. (2016). Development of children's estimation skills: The ambiguous role of their familiarity with numerals. *Child Development Perspectives*, *10*(2), 116–121.

Ebersbach, M., Luwel, K., Frick, A., Onghena, P., & Verschaffel, L. (2008). The relationship between the shape of the mental number line and familiarity with numbers in 5 to 9-year-old children: Evidence for a segmented linear model. *Journal of Experimental Child Psychology*, *99*(1), 1–17.

Evans, D. W. (1983). *Understanding zero and infinity in the early school years.* Unpublished doctoral dissertation, Philadelphia, University of Pennsylvania. Dissertations available from ProQuest. AAI8326285.

Feigenson, L., Dehaene, S., & Spelke, E. (2004). Core systems of number. *Trends in Cognitive Sciences*, *8*(7), 307–314.

Forbus, K. D., Gentner, D., & Law, K. (1995). MAC/FAC: A model of similarity-based retrieval. *Cognitive Science*, *19*(2), 141–205.

Frank, M. C., Everett, D. L., Fedorenko, E., & Gibson, E. (2008). Number as a cognitive technology: Evidence from Pirahã language and cognition. *Cognition*, *108*(3), 819–824.

Fuson, K. (1988). *Children's counting and concepts of number.* New York, NY: Springer-Verlag.

Gentner, D. (1983). Structure-mapping: A theoretical framework for analogy. *Cognitive Science*, *7*(2), 155–170.

Gentner, D. (2010). Bootstrapping the mind: Analogical processes and symbol systems. *Cognitive Science*, *34*(5), 752–775.

Gentner, D., & Markman, A. B. (1997). Structure mapping in analogy and similarity. *American Psychologist*, *52*(1), 45.

Gordon, P. (2004). Numerical cognition without words: Evidence from Amazonia. *Science*, *306*(5695), 496–499.

Goswami, U., & Brown, A. L. (1990). Melting chocolate and melting snowmen: Analogical reasoning and causal relations. *Cognition*, *35*(1), 69–95.

Hartnett, P., & Gelman, R. (1998). Early understandings of numbers: Paths or barriers to the construction of new understandings? *Learning and Instruction*, *8*, 341–374.

Ifrah, G. (2000). *The universal history of numbers: From prehistory to the invention of the computer.* New York: John Wiley & Sons, Inc.

Izard, V., & Dehaene, S. (2008). Calibrating the mental number line. *Cognition*, *106*(3), 1221–1247.

Le Corre, M., & Carey, S. (2007). One, two, three, four, nothing more: An investigation of the conceptual sources of the verbal counting principles. *Cognition*, *105*(2), 395–438.

Le Corre, M., Van de Walle, G., Brannon, E. M., & Carey, S. (2006). Re-visiting the competence/performance debate in the acquisition of the counting principles. *Cognitive Psychology*, *52*(2), 130–169.

Lipton, J. S., & Spelke, E. S. (2005). Preschool children's mapping of number words to nonsymbolic numerosities. *Child Development*, *76*(5), 978–988.

Lyons, I. M., & Ansari, D. (2015). Numerical order processing in children: From reversing the distance-effect to predicting arithmetic. *Mind, Brain, and Education*, *9*(4), 207–221.

Lyons, I. M., Ansari, D., & Beilock, S. L. (2011). Evidence against a strong association between numerical symbols and the quantities they represent. In *CogSci*.

Lyons, I. M., & Beilock, S. L. (2009). Beyond quantity: Individual differences in working memory and the ordinal understanding of numerical symbols. *Cognition*, *113*(2), 189–204.

Lyons, I. M., & Beilock, S. L. (2011). Numerical ordering ability mediates the relation between number-sense and arithmetic competence. *Cognition*, *121*(2), 256–261.

Lyons, I. M., & Beilock, S. L. (2013). Ordinality and the nature of symbolic numbers. *The Journal of Neuroscience*, *33*(43), 17052–17061.

Margolis, E., & Laurence, S. (2008). How to learn the natural numbers: Inductive inference and the acquisition of number concepts. *Cognition*, *106*(2), 924–939.

Menninger, K. W. (1969). *Number words and number symbols: A cultural history of numbers*, translated by P. Broneer. Boston: MIT Press.

Mundy, E., & Gilmore, C. K. (2009). Children's mapping between symbolic and nonsymbolic representations of number. *Journal of Experimental Child Psychology*, *103*(4), 490–502.

Opfer, J. E., & Siegler, R. S. (2007). Representational change and children's numerical estimation. *Cognitive Psychology*, *55*(3), 169–195.

Pica, P., Lemer, C., Izard, V., & Dehaene, S. (2004). Exact and approximate arithmetic in an Amazonian indigene group. *Science*, *306*(5695), 499–503.

Rips, L. J., Asmuth, J., & Bloomfield, A. (2008). Do children learn the integers by induction? *Cognition*, *106*(2), 940–951.

Roggeman, C., Verguts, T., & Fias, W. (2007). Priming reveals differential coding of symbolic and non-symbolic quantities. *Cognition*, *105*(2), 380–394.

Rouder, J. N., & Geary, D. C. (2014). Children's cognitive representation of the mathematical number line. *Developmental Science*, *17*(4), 525–536.

Sarnecka, B. W., & Carey, S. (2008). How counting represents number: What children must learn and when they learn it. *Cognition*, *108*(3), 662–674.

Sarnecka, B. W., & Lee, M. D. (2009). Levels of number knowledge during early childhood. *Journal of Experimental Child Psychology*, *103*(3), 325–337.

Siegler, R. S., & Booth, J. L. (2004). Development of numerical estimation in young children. *Child Development*, *75*(2), 428–444.

Siegler, R. S., & Opfer, J. E. (2003). The development of numerical estimation evidence for multiple representations of numerical quantity. *Psychological Science*, *14*(3), 237–250.

Sullivan, J., & Barner, D. (2013). How are number words mapped to approximate magnitudes? *The Quarterly Journal of Experimental Psychology*, *66*(2), 389–402.

Sullivan, J., & Barner, D. (2014). Inference and association in children's early numerical estimation. *Child Development*, *85*(4), 1740–1755.

Thompson, C. A., & Opfer, J. E. (2010). How 15 hundred is like 15 cherries: Effect of progressive alignment on representational changes in numerical cognition. *Child Development*, *81*(6), 1768–1786.

Turconi, E., Campbell, J. I., & Seron, X. (2006). Numerical order and quantity processing in number comparison. *Cognition*, *98*(3), 273–285.

Vul, E., Barner, D., & Sullivan, J. (2013). Slow drift of individuals' magnitude-to-number mapping. In *Proceedings of the 35th annual conference of the Cognitive Science Society*. Retrieved from http://mindmodeling.org.

Wynn, K. (1990). Children's understanding of counting. *Cognition*, *36*(2), 155–193.

Wynn, K. (1992). Children's acquisition of the number words and the counting system. *Cognitive Psychology*, *24*(2), 220–251.

Chapter 3

Linguistic and Experiential Factors as Predictors of Young Children's Early Numeracy Skills

Jo-Anne LeFevre*, Ozlem Cankaya†, Chang Xu*, and Carolina Jiménez Lira*
*Carleton University, Ottawa, ON, Canada
†Alberta Centre for Child, Family, and Community Research, Edmonton, AB, Canada

Young children vary substantially in the early numeracy skills that they bring to learning mathematics (Duncan et al., 2007; Geary, 2011). Two potential sources of this variability are their number experiences at home or in child-care settings (Skwarchuk, Vandermaas-Peeler, & LeFevre, 2016) and the language(s) that they speak (Cankaya & LeFevre, 2016). Numeracy-related experiences and language have often been studied separately as potential predictors of differences in later mathematical achievement. However, effects of language and experience are often closely connected (see Chapter 4 by Okamoto). In this chapter, we use data from eight studies conducted in our laboratory over the last 15 years to explore the relative contributions of language and other numeracy-related factors as predictors of 3- to 7-year-old children's early numeracy skills. We focus on verbal counting because the structure of the counting words varies across languages and because counting performance is highly correlated with other early numeracy measures (e.g., Aunio & Niemivirta, 2010; Aunola, Leskinen, Lerkkanen, & Nurmi, 2004; Purpura, Baroody, & Lonigan, 2013). Wiese (2003) characterized the counting sequence as a fundamental numerical tool that underlies children's development of the number concept.

Cross-cultural comparisons of the mathematics performance of American and Asian children (reviewed by Wang & Lin, 2009) were an early source of information about the role of number languages in counting (e.g., Miller, Smith, Zhu, & Zhang, 1995). In many studies, Chinese children perform better than English-speaking children on early numeracy measures, including

Mathematical Cognition and Learning, Vol. 4. https://doi.org/10.1016/B978-0-12-812574-8.00003-1

verbal counting (Miller et al., 1995), place-value understanding (Muira, Okamoto, Kim, Steere, & Fayol, 1993), and locating numbers on a number line (Siegler & Mu, 2008). Although these differences have been attributed to aspects of the Chinese language that are assumed to enhance performance (i.e., shorter number words and simple number word structure; Chan, 2014), the Chinese children's experiences at home, in preschool, and as part of a culture that values mathematics are also distinct from those of American children and may mediate the Chinese advantage (Huntsinger, Jose, Liaw, & Ching, 1997; Luo, Jose, Huntsinger, & Pigott, 2007). Nevertheless, language is a plausible source of some of the differences between Chinese and American children's performance because the Chinese number word system is more regular than that of English and corresponds directly to the base-10 structure used in written digits; these structural differences may allow Chinese-speaking children to acquire the counting words more easily than English-speaking children (Chan, 2014; Mark & Dowker, 2015; Miller et al., 1995; Chapter 4 by Okamoto).

Table 1 compares the various languages spoken by children in the studies described in this chapter. We focus on differences in the composition of number words and specifically how closely they correspond to the written Arabic notation.[1] Among the languages shown in Table 1, the Chinese counting system is the simplest. Up to 100, Chinese number words are composed of only 10 unique words plus a simple rule that directly reflects the 10-unit (i.e., base-10) structure used in the written Arabic digits. Turkish uses an equivalent 10-unit structure to 19. In contrast, English has unique words to 12 and then uses an implicit unit-ten structure (i.e., with the word "teen" reflecting ten) from 13 to 19 (e.g., thirteen or three-ten). Spanish and French have unique words to 16 and 17, respectively, subsequently switching to a 10-unit structure that is equivalent to Chinese and Turkish (e.g., dix-sept or ten-seven). From 21 on, Turkish, English, French, and Spanish use a decade-unit structure (e.g., thirty-one, quarante-deux); however, French switches from a decade to a base-20 structure from 70 to 99 (e.g., soixante-dix-huit or sixty-ten-eight for 78). Thus, based on the relative complexity of the number words from 1 to 100, the Chinese system should be easy to learn, Turkish may be easier than English or Spanish, and French may be the most difficult. Other research provides some support for these predictions (e.g., Miller et al., 1995; Seron & Fayol, 1994). Despite such research, however, relatively few direct comparisons have been done on a variety of languages. Most studies compare only two languages, and few have simultaneously considered other aspects of children's environments that could influence their learning. In this chapter, we

1. Note that languages also vary in the order in which decade and unit names are spoken, for example, 34 is thirty-four in English but vier-en-dertig or four and thirty in Dutch. Chapter 10 by Göbel thoroughly addresses the effects of number word inversion on children's transcoding errors.

TABLE 1 Comparison of Verbal Number Words Across Five Languages

	English	French	Spanish	Turkish	Chinese
1	One	Un (une)	Uno	Bir	Yi
2	Two	Deux	Dos	Iki	Er
3...	Three	Trois	Tres	Üç	San
10	Ten	Dix	Diez	On	Shi
11	Eleven	Onze	Once	**On** bir	**Shi** yi
12	Twelve	Douze	Doce	**On** iki	**Shi** er
13	Thir**teen**	Treize	Trece	**On** üç	**Shi** san
14	Four**teen**	Quatorze	Catorce	**On** dört	**Shi** si
15	Fif**teen**	Quinze	Quince	**On** beş	**Shi** wu
16	Six**teen**	Seize	**Dieci**seis	**On** altı	**Shi** liu
17	Seven**teen**	**Dix**-sept	**Dieci**siete	**On** yedi	**Shi** qi
18	Eigh**teen**	**Dix**-huit	**Dieci**ocho	**On** sekiz	**Shi** ba
19	Nine**teen**	**Dix**-neuf	**Dieci**nueve	**On** dokuz	**Shi** jiu
20	Twen**ty**	Vingt	Veinte	Yirmi	Er **shi**
21	Twen**ty**-one	Vingt et un	Veintiuno	Yirmi bir	Er **shi** yi
22...	Twen**ty**-two	Vingt-deux	Veintidos	Yirmi iki	Er **shi** er
60	Six**ty**	Soix**ante**	Ses**enta**	Altmış	Liu **shi**
70	Seven**ty**	Soix**ante-dix**	Set**enta**	Yetmiş	Qi **shi**
80	Eigh**ty**	Quatre-vingts	Och**enta**	Seksen	Ba **shi**
90	Nine**ty**	Quatre-vingt-**dix**	Nov**enta**	Doksan	Jiu **shi**
96	Nine**ty**-six	Quatre-vingt-seize	Nov**enta** y seis	Doksan altı	Jiu **shi** liu
100	One hundred	Cent	Cien	Yüz	Yi bai
101	One hundred and one	Cent un	Ciento uno	Yüz bir	Yi bai ling yi

The presence of the word, affix, or suffix representing 10 in each language is highlighted to illustrate the (ir)regularity of each system.

illustrate the complexities of contrasting language versus experience as predictors of early numeracy performance differences between diverse groups of children and define a role for language complexity as one factor out of many that influence children's early numeracy learning.

NUMBER LANGUAGE AND EARLY NUMERACY PERFORMANCE

In a classic study, Miller et al. (1995) asked 3-, 4-, and 5-year-old Chinese and American children to count as high as possible. There were no differences across language groups for 3-year olds, but 4- and 5-year-old Chinese children counted higher than American children. Furthermore, there were no differences in counting performance to 10, but fewer American children could count to 20 than Chinese children. From 20 onward, both groups showed a similar pattern, where performance was stable within a decade but declined when the decade changed. This scalloped pattern of counting performance across decades reflects children's gradual acquisition of the decade words (Siegler & Robinson, 1982). Overall, the two groups of children did not differ on other early numeracy skills, specifically, their understanding of cardinality for small sets of objects (i.e., the knowledge that the final count of a set represents the quantity of that set) or their procedural ability to count objects within their counting range. Hence, the American children's performance in verbal counting between 10 and 20 seemed to be evidence for a language-specific skill difference. How specific is this pattern to the comparison of Chinese and English? To answer this question, in a series of studies, we compared the counting knowledge of English-speaking Canadian children with that of children who spoke other languages.

In the literature on differences in numeracy skills for children speaking different languages, several other factors have been identified as potential explanations for language effects. For example, Rasmussen, Ho, Nicoladis, Leung, and Bisanz (2006) found that the amount of practice children had in each language predicted bilingual children's performance in English and Chinese. Cultural values may also influence children's exposure to mathematical activities: Huntsinger and colleagues found dramatic differences in preschool children's home experiences, with children of Chinese immigrant parents experiencing more direct instruction in mathematics and more encouragement to participate in mathematical activities and expressing more positive views of mathematics compared with children whose parents had been born in the United States (Huntsinger et al., 1997). Thus, the attitudes and beliefs of parents, when they result in differences in children's early numeracy experiences, can influence children's early numeracy development (see also Huntsinger, Jose, & Luo, 2016; Kleemans, Peeters, Segers, & Verhoeven, 2012; LeFevre et al., 2009; LeFevre, Polyzoi, Skwarchuk, Fast, & Sowinski, 2010; Segers, Kleemans, & Verhoeven, 2015; Skwarchuk, 2009; Skwarchuk, Sowinski, & LeFevre, 2014; Skwarchuk et al., 2016; Chapter 5 by Opfer et al.). When these factors are confounded with language, it can be very difficult to determine an accurate explanation for the differences observed across groups.

Few studies have assessed cultural and experiential factors in groups differing in language (cf. Huntsinger et al., 1997; LeFevre, Clarke, & Stringer, 2002). Even when information about cultural and experiential factors is collected, it is not always easy to separate the effects of these factors from those of language differences, as illustrated in this chapter. We first describe comparisons between English-speaking children and groups of children speaking Turkish, Chinese, Spanish, and French, asking whether performance can be linked directly to language differences. In all of these studies, we also collected information about parents' education, their expectations for children's early numerical achievement, and their reports of home numeracy experiences. Second, we present the results of three studies that used other methods to address the question of whether number language had specific effects on children's counting performance.

COMPARISONS OF COUNTING ACROSS LANGUAGES

English- Vs Chinese-Speaking Canadian Children

Can the differences in verbal counting between English- and Chinese-speaking children be replicated in a somewhat different sample? Counting performance of Chinese-speaking children living in Canada ($n=44$) was compared with other Canadian children who spoke English ($n=45$; Xu & LeFevre, 2015). Chinese-speaking children were recruited through a variety of sources, including personal contacts and an online message board; Canadian children were recruited from daycares. Children in both groups ranged from 3 to almost 6 years of age, with means of 4.2 and 4.3 years for the Chinese- and English-speaking groups, respectively. Consistent with other research, the Chinese-speaking children had better early numeracy skills in that they counted higher, correctly identified more Arabic digits, labeled small quantities more quickly (i.e., subitized), and performed better on a standardized numeracy test (the KeyMath numeration subtest; Connolly, 2000). As shown in Fig. 1, the patterns of verbal counting across groups were similar to those reported by Miller et al. (1995) in that fewer English- than Chinese-speaking children counted successfully to 20. Also, similar to Miller et al., after 20, the pattern of performance showed typical decrements at decade names. Thus, these results provide support for the view that language differences can translate into a specific effect on verbal counting performance (see also Chapter 4 by Okamoto).

More generally, despite the plausibility of the language-based explanation for the verbal counting differences, many other factors might differentiate groups of children. Miller et al. (1995) focused on language as the causal factor in differences between groups but did not provide evidence about factors other than language, such as parents' education (Davis-Kean, 2005; Entwisle & Alexander, 1990), children's home experiences (Skwarchuk et al., 2014), or the extent of preschool experience (Englund, Luckner, Whaley, & Egeland, 2004). In Xu and LeFevre (2015), we found that the parents of the Chinese-speaking children had higher levels of education than parents of the

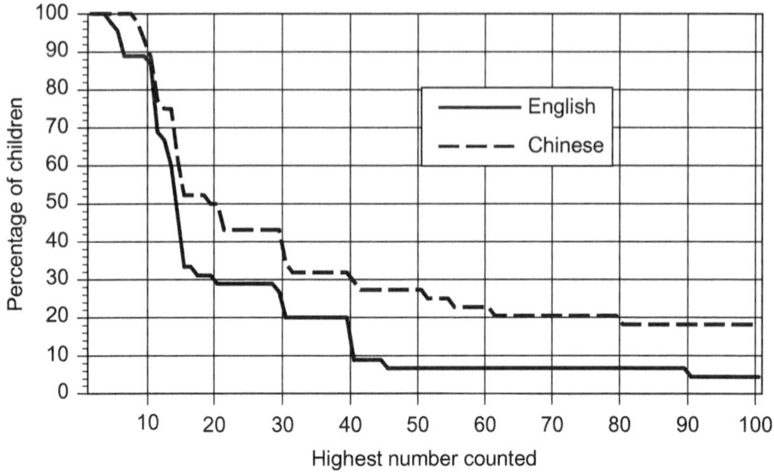

FIG. 1 The percentage of English- and Chinese-speaking children in each language group who successfully counted to each number without making an error. Children were stopped when they reached 100 (Xu & LeFevre, 2015). Note that the steeper decline is between 10 and 20 in the percentage of English- vs Chinese-speaking children who were successful.

English-speaking children and that education mediated the language difference. Davis-Kean (2005) also found that parental education directly predicted children's math and reading achievement and furthermore education was also indirectly related to math and reading performance through home activities and parents' educational expectations. Therefore, an alternative interpretation of the differences between the Chinese- and English-speaking children in Xu and LeFevre (2015) is that the early numeracy experiences of these two groups differed. This interpretation is consistent with the finding that the English-speaking children performed worse than the Chinese-speaking children on a range of numeracy measures, even those for which knowledge of the number words was not required. Hence, in the remaining comparisons in this chapter (summarized in Table 2), we compare children's counting performance and other early numeracy skills across groups, but we also discuss and, if possible, control children's other relevant experiences, with the goal of disentangling the effects of language and experience on early formal numeracy skills.

English- Vs French-Speaking Canadian Children

Counting words from 1 to 20 in French may be more challenging to learn than those in English or other languages. As shown in Table 1, the French counting system comprises distinct words for 1 to 16, whereas from 17 to 19 it switches to be a decade-unit pattern similar to that in Chinese, Spanish, and Turkish.

TABLE 2 Summary of Studies Described in This Chapter: Variables (Language, Country, Culture), Design (Between- or Within-Groups), Method (Correlational Vs Quasiexperimental Training), and Observed Differences in Children's Home Experiences (Parent Education, Numeracy Activities), and Children's Performance (Verbal Counting, Other Numeracy Activities)

					Demographics		Numeracy	
Reference	Description[a]	Design	Method	Country	Parent Education	Numeracy Activities	Verbal Count	Other [b]
Xu and LeFevre (2015)	**Chinese**- vs English-speaking Canadians	Between	Correlational	✠	!!	✠	!!	!!
LeFevre et al. (2002)	French- vs **English**-speaking Canadians	Between	Correlational	✠	✠	!!	!!	!!
Jiménez Lira (2016)	**English**-speaking Canadians vs Spanish-speaking Mexicans	Between	Correlational	!!	!!	✠	!!	!!
Cankaya (2013)	**Chinese**- vs Turkish-speaking Canadians	Between	Correlational	✠	✠	✠	!!	!!
Cankaya (2013)	Turkish vs **Canadian** children	Between	Correlational	!!	!!	✠	✠	!!
Joy (2006)	English-speaking Canadian children learning French	Within	Repeated measures	–	–	–	–	–
Cankaya et al. (2014)	Turkish vs Canadian children	Between	Training	!!	!!	✠	✠	–
Dunbar et al. (2016)	English-speaking Canadian children	Between	Training	✠	✠	–	–	–

✠, not different across the groups; !!, different across the groups.
[a]*Bolded indicates which group had better early numeracy skills.*
[b]*Other numeracy activities; these varied across studies but typically included number naming, object counting, and KeyMath numeration.*

We hypothesized that children learning French might find it even more difficult to learn number words from 13 to 20 than children learning English[2] (LeFevre et al., 2002). To control for socioeconomic and experiential factors across language groups, we compared French- and English-speaking preschoolers living in Canada. The English-speaking children ($n=27$) were from Ottawa, Ontario, and the French-speaking children ($n=23$) were from Gatineau, Quebec. These communities are separated by a river, such that the distance between communities is less than 20 km. The parents of the two groups were equally highly educated, the children were all in daycare (not preschools), and the language at home was the same as the language of the daycare. Both groups of children ranged in age from 3.8 to 5.3 years with means of 4.0 years. Thus, there were fewer cultural and societal factors that may differ between these groups than, for example, between children in China and the United States. Parents were also asked to complete a questionnaire about their expectations for children's knowledge at school entry and the home activities (numeracy and literacy) that they provided for their children.

The pattern of verbal counting performance between the French- and English-speaking children is shown in Fig. 2. All of the children in both groups counted successfully to 10. Overall, the same percentage of English-speaking children and French-speaking children counted to 15; however, at 16, the performance of French-speaking children declined dramatically, such

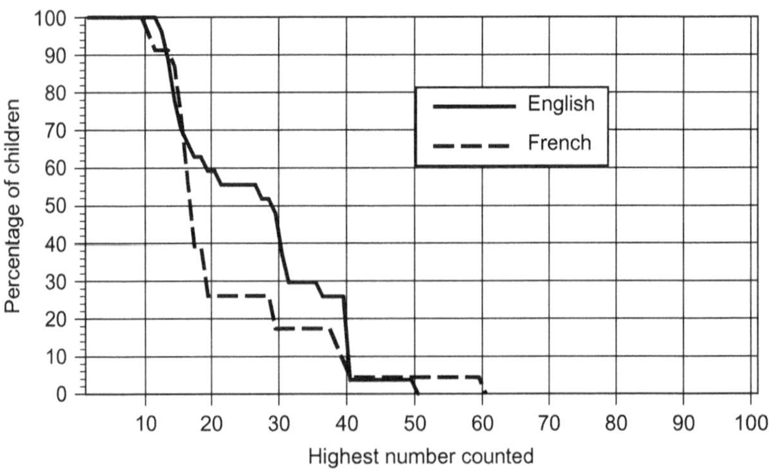

FIG. 2 The percentages of French- and English-speaking children who successfully counted to each number without making an error (LeFevre et al., 2002).

2. Although French is also more complicated after 60 than English because there are base-20 elements (see Table 1), the preschoolers in LeFevre et al. (2002) could not count beyond 60. We address the higher numbers in a later study.

that only about 25% counted to 20, compared with 60% of English-speaking children. After 20, both groups showed the typical scalloped pattern, such that declines occurred at decade boundaries.

Do these results for French- and English-speaking children represent language differences? As shown in Table 1, the way teen number names in French are composed for 17–19 differs from those for 11–16; the latter are unique words that are weakly related to comparable numbers from one to six (e.g., from deux to douze, from trois to treize, from quatre to quatorze, from cinq to quinze). In contrast, the numbers 17, 18, and 19 follow a 10-unit pattern (e.g., dix-sept, dix-huit, and dix-neuf), whereas in English, the pattern of unit-teen names continues the unit-teen structure from 13 (i.e., thir-teen, four-teen, fif-teen, six-teen, seven-teen, eigh-teen, and nine-teen). It is possible that the switch in pattern at 17 is confusing for French-speaking children.

Another explanation of the overall poorer counting performance is that French-speaking children may have had different home numeracy experiences and thus had not acquired as much symbolic numeracy knowledge as English-speaking children. Consistent with this possibility, French-speaking children also named fewer digits than the English-speaking children, whereas the two groups had similar cardinality and object-counting skills. The groups also differed in that the French-speaking parents had lower expectations for children's early numeracy skills, reported less frequent home numeracy activities, and had fewer children's books at home. In general, French-speaking parents are less concerned about children's academic experiences prior to schooling, possibly because French is easier to learn to read than English, and this more relaxed attitude may generalize beyond early literacy (Bruck, Genesee, & Caravolas, 1997; Sénéchal, 2000). In LeFevre et al. (2002), multiple regression analyses showed that differences in parents' reports of home numeracy activities mediated the relation between language and verbal counting. Thus, this study suggests that, even for children from the same country, language and culture cannot be easily separated and home experiences rather than the structure of the number words may account for observed differences in early numeracy skills.

English-Speaking Canadian Vs Spanish-Speaking Mexican Children

In this comparison, children aged 3–5 years who were learning to count in English in Canada were compared with children of the same age learning to count in Spanish in Mexico (Jiménez Lira, 2016; Chapter 4 by Okamoto). There is little existing research on the early numeracy skills of Latin American children (Susperreguy, Jiménez Lira, & LeFevre, 2017). There are some number language differences in that Spanish is similar to French in switching from unique words to a 10-unit structure at 16, but perhaps, more importantly, the groups also differed on a variety of other factors, and these might be more

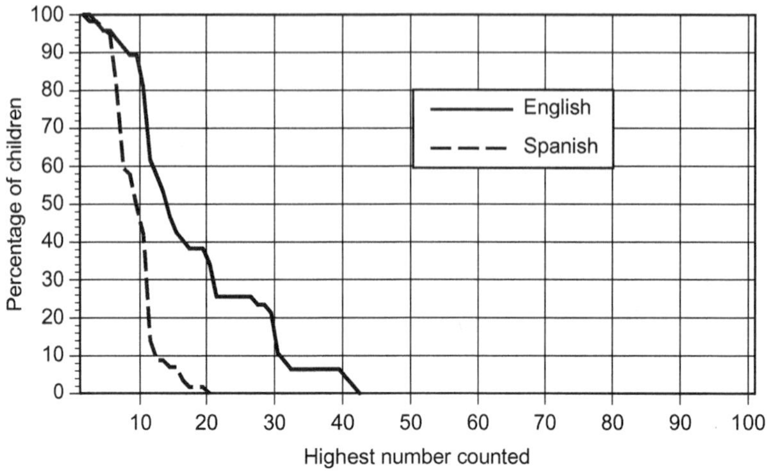

FIG. 3 The percentages of Spanish- ($n=57$) and English-speaking children ($n=45$) who successfully counted to each number without making an error (Jiménez Lira, 2016).

important than language differences. Fig. 3 shows that the Mexican children knew considerably less of the verbal counting string than the Canadian children, despite being the same age, on average, and having more formal schooling. Most children in both groups could count to 5. Although the Canadian children showed strong performance to 10, Mexican children showed steady declines from 6 onward, and none could count past 19.

What factors might explain the performance differences between the groups? The Mexican children were attending mandatory preschools, whereas the Canadian children were recruited from daycares that do not have explicit curricula for early numeracy. The Mexican parents had considerably less education than the Canadian parents. Moreover, despite reporting similar levels of home numeracy activities as the Canadian parents, for Mexican parents, there was no relation between these reports and their children's performance, whereas for the Canadian families, the frequency of home numeracy activities reported by parents was correlated with children's performance (see Susperreguy et al. 2017 for a detailed comparison). Accordingly, we concluded that the Mexican children had less numeracy experience than the Canadian children, despite being matched on age. Other research suggests that Latin American parents believe that academic skills are the purview of the school (Reese & Gallimore, 2000). These data show that substantial differences in counting skills can be obtained for groups of children that may be only weakly related to the regularity of a language; instead, these differences may reflect variability across groups in experiences at home or in school.

Turkish- and Chinese-Speaking Canadian Children

Similar to Chinese, Turkish number words from 11 to 19 consist of the word for 10 plus the unit word, as shown in Table 1. If the regularity of the number word structure is helpful for learning to count, then we might expect Turkish-speaking children to show benefits of the language structure. In comparing Turkish- to Chinese-speaking children, we also explored cultural differences related to the role of home experiences and parents' attitudes. Participants included 20 Turkish- and 22 Chinese-speaking children, all living in Canada (see Cankaya, 2013, Chapter 3 by LeFevre et al. for detailed information), whose parents were immigrants from Turkey or China and who were equally highly educated. We hypothesized that children would show similar counting performance between 10 and 20 if the regularity of the number system is an important factor in learning to count. As shown in Fig. 4, however, the Chinese-speaking children performed better than the Turkish-speaking children on verbal counting, with Turkish-speaking children showing the typical dramatic decline in performance between 10 and 20.

In addition to verbal counting, Chinese-speaking children performed better than Turkish-speaking children on all language-related numeracy skills including digit naming, object counting, and the KeyMath numeration subtest (Connolly, 2000), even though the groups did not differ on spatial span or nonsymbolic arithmetic tasks. Chinese-speaking children also counted higher in English than did Turkish-speaking children. These results indicate that factors other than language influence children's early numeracy performance.

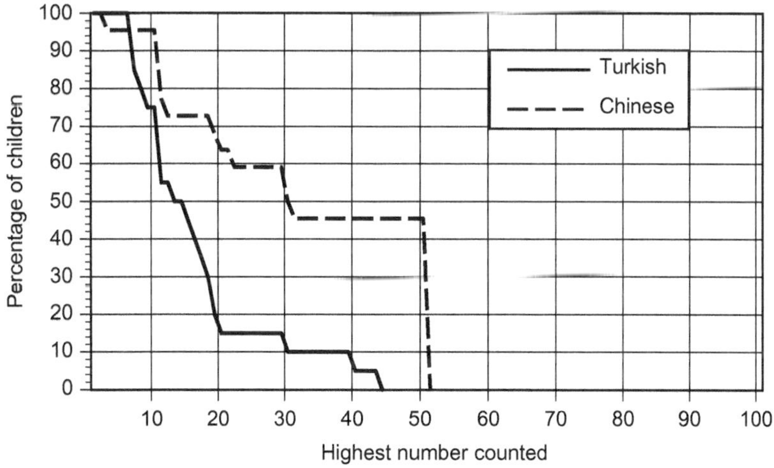

FIG. 4 The percentage of Turkish- ($n=20$) and Chinese-speaking ($n=22$) Canadian children who successfully counted to each number without making an error. Children were stopped when they reached 50.

Although parents' education did not differ between these groups, the parents of Chinese-speaking children had higher expectations for their children's academic skill development, both in literacy and numeracy, than did Turkish parents. The two groups reported similar overall frequencies of formal home numeracy activities, but there was a trend for Turkish parents to report more games and informal activities (e.g., cooking). In contrast, Chinese children were more comfortable with the early numeracy tasks involving number symbols, suggesting that they had more experience with activities preparatory for school learning. These differences are generally consistent with other research on patterns of parent-child interactions for Chinese children (Huntsinger et al., 1997), but there is no equivalent research with Turkish children.

In summary, we found no evidence that the simple structure of the Turkish number words between 11 and 19 provided an advantage to these young children, presumably because they had less exposure to counting than the Chinese-speaking children. We further addressed this issue below in the context of a training study. Exploration of the home environments in which these two groups of children were learning about numbers suggested that experiential factors such as providing exposure to academic activities and higher parental expectations about academic skills played a role in the children's early numeracy accomplishments.

Correlation Vs Causation in Comparisons Across Languages

The results for the comparisons of groups of children speaking different languages suggest that cultural and experiential differences may be important sources of variability in early numeracy skills, even within the same country. Why is it so difficult to draw clear conclusions about the role of language in early numeracy? In essence, comparing groups of children on any specific attribute (such as language) and finding a difference is equivalent to reporting a correlation. Correlations mean that the two variables are related (in this case, language and verbal counting), but they do not indicate a causal relationship. Moreover, in comparisons of children who speak different languages, the groups may differ on other factors that could contribute to early numeracy differences. Accordingly, researchers need to collect data about other potentially causal factors that might predict early numeracy skill and control for as many of these factors as possible, either statistically or by devising alternative ways to assess language differences.

OTHER APPROACHES TO EXPLORING LANGUAGE EFFECTS

In this section, we describe three studies in which we attempted to control or manipulate factors other than language that may be potential sources of early numeracy differences across groups of children. First, we studied English-speaking children learning French, a within-group comparison that removes the influences of parental education and home experiences in relation to language differences. Second, we conducted a training study in which we provided children with practice counting, comparing across languages (Turkish and English). Third, we used a training study where we manipulated English-speaking children's experiences to more directly evaluate whether supported practice could potentially overcome language limitations.

English-Speaking Children Learning to Speak French

As part of a longitudinal study of children in Early French Immersion (Joy, 2006; Sénéchal & LeFevre, 2014[3]), we also collected data on children's counting and number transcoding, that is, their ability to translate between spoken and written Arabic numbers. These English-speaking children were taught all subjects in French in kindergarten (age 5, half-day program) and grade 1, although in grade 1 they were also taught to read in English. English was the first language of the majority of children, and all spoke English fluently, although some parents also spoke French or other languages with them at home.

As noted earlier, the number language from 70 to 100 in French uses a base-20 structure for the decade words and thus differs dramatically from English and the other languages shown in Table 1. The word for seventy is the equivalent of sixty-ten, 80 is quatre-vingts (literally, four-twenties), and ninety is quatre-vingt-dix (four-twenty and ten). The unit names in the seventies and the nineties, therefore, correspond to the teen words (e.g., 96 is quatre-vingt-*quinze* or four-twenty-*sixteen*). Other languages, such as Basque (Colomé, Laka, & Sebastián-Gallés, 2010), use a consistent base-20 structure from 20 to 100; French is interesting because this structure applies only from 60 to 100. Although variants of French, for example, the version that is used in Belgium, use a base-10 structure for the decade words from 70 to 90 (i.e., septante, octante, and nonante; Seron & Fayol, 1994), the French number words used in Canada follow the base-20 structure used in France. We hypothesized that English-speaking children who were learning to speak French might find this structure difficult to learn, even if they were proficient at verbal counting in English, because it violates the rules they have learned for generating verbal numbers in this range.

Fig. 5 shows the counting performance of the children in Early French Immersion nearing the end of their grade 1 year (Joy, 2006). Note that 85% of the children counted successfully to 100 in English. In contrast, in French, the number words starting at 70 were especially difficult. Before 70, the percentage of children who were successful in French declined gradually,

3. Sénéchal and LeFevre (2014) analyzed the relations between children's English literacy skills and their home numeracy experiences; the data for counting and number transcoding are described by Joy (2006).

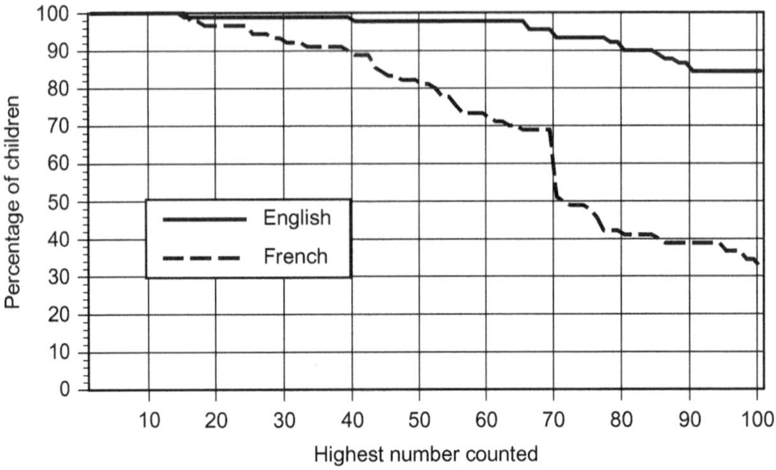

FIG. 5 The percentage of children in Early French Immersion (near the end of first grade) who successfully counted to each number without making an error ($n=89$; mean age 6 years and 10 months).

whereas at 70, abruptly, an additional 20% of the children were unable to continue counting successfully. This pattern supports the view that children may experience difficulty learning number words in their second language when the structure of the number language is more complex than that of their first language, even when they are skilled verbal counters in their first language.

In further support of the view that it was number language, and not the general concept of number words that was problematic, these children also showed difficulties transcoding from verbal to written forms of French numerals for numbers above 70, despite excellent performance in English. Children heard the number words for 15, 23, 46, 72, 84, and 97 in English and French and were asked to write the corresponding Arabic digits. They made almost no errors when transcoding from English to Arabic digits and were successful over 90% of the time for 15, 23, and 46 in French. However, transcoding from French to Arabic digits was much more difficult for 72, 84, and 97, such that 39%, 20%, and 51% of children, respectively, made errors on these numbers. The types of errors reflected the complexity of the number language. For example, when they heard quatre-vingt-dix-sept (i.e., 97), 36% of the children wrote 87 (quatre-vingt-sept), and 6% wrote 47 (quarante-sept); other errors included 40, 417, 48, 99, 717, 917, and 4067. Errors on 72 similarly reflected the complexity of the base-20 number word structure. The children found it similarly difficult to name the Arabic numbers 76, 84, and 90 in French, despite perfect performance in English.

This situation of children learning a second language who have already mastered the number words and the mappings from number words to digits in their first language is important for testing the question that language complexity affects early numeracy skills. Despite being taught solely in French in school, these English-speaking children had excellent number word knowledge in English as they neared the end of first grade, knowledge they had presumably acquired at home or from sources other than school. They had not yet mastered the French number language to 100, however, and their errors reflected difficulties with the base-20 structure of French and some confusion with similar sounding words. Evidence for a connection between learning the number words and the development of their numeracy skills was found: their performance on verbal addition and subtraction problems in English was correlated with their number naming and writing performance in French, suggesting that the language effects were related to their overall numeracy skill.

Other studies with young bilingual children suggest that language may not be the only factor that influences children's learning of the counting sequence (Chapter 9 by Sarnecka et al.). Some research has shown that the amount of practice in each language is also predictive of children's counting performance. For example, Rasmussen et al. (2006) reported that the verbal counting performance of bilingual English-Chinese children was related to children's relative proficiency in each language. These results with bilingual children also suggest that it is important to equate the amount of experience and practice with counting across different languages.

Training Study I: Guided Practice Counting for English- and Turkish-Speaking Children

In Cankaya, LeFevre, and Dunbar (2014), we compared the performance of Canadian children learning number words in English to that of Turkish children learning number words in Turkish. To provide guided practice, children participated in a number game that involved counting in the range of 1–20. If the regularity of the number words from 11 to 19 in Turkish is helpful, then we expected the Turkish children to benefit more from an equivalent amount of counting practice than the Canadian children. This study is a major departure from the previous ones we have discussed because we used a training paradigm in an attempt to capitalize on the differential structures of the two languages.

In a pretest session, Turkish ($n=89$; mean age 46 months) and Canadian ($n=59$; mean age 48 months) children completed a variety of early numeracy tasks (see Cankaya, 2013 and Cankaya et al., 2014 for details of all the pre- and posttest measures). The two groups did not differ on spatial span, nonverbal arithmetic, or subitizing tasks, nor did they differ on verbal counting. However, on all other measures of early numeracy skills, including counting objects, the KeyMath numeration subtest, number recognition, and number-line performance, the Canadian children had higher scores than the Turkish children. Because the children were quite young, few children in either group counted successfully to 20. The lack of differences in counting, despite the overall better numeracy skills of the Canadian children, suggests that the more regular counting system is helpful for Turkish children's counting but does not provide an advantage on other numeracy measures.

As expected, given the substantial cultural differences between Canada and Turkey, there were many factors other than language that differed between the Turkish and Canadian children, including educational experiences in the daycare or kindergarten, parents' academic expectations for children's preschool learning, home experiences (as reported by parents), and the cultural milieu more generally. Rather than trying to control these differences statistically, we used a quasiexperimental intervention to provide equivalent practice to the two groups of children. If the Turkish language affords an advantage when children are learning to count, we expected that Turkish children would benefit more from the same amount of practice compared with the Canadian children, who were learning to count in English. The alternative possibility was that the overall better early numeracy skills of the Canadian children might mediate the effects of the guided practice. Thus, this design provides a relatively strong test of the hypothesis that number language, specifically, provides an advantage when children are learning to count.

Training Phase

As described in detail in Cankaya et al. (2014), children in each language group were randomly assigned to one of two game conditions. The games were modeled on those used by Siegler and Ramani (2008, 2009). In the number game condition, children moved game pieces along a linear path that included the numbers from 1 to 20. In the color game condition, the squares were colored but not numbered. On each turn, children used a spinner to determine whether they moved one square or two (in the number game) or which of four colors they were to move to next (in the color game). As in Siegler and Ramani (2008, 2009), in the number game, children had to count on from their current position. So if their game piece was on the number 10 and they spun the number 2, then they had to count on from 10—saying 11, 12 as they moved their game piece. Hence, the game allowed them to practice the counting words in order. The experimenter helped out as needed so that each turn provided correct practice of the number sequence. Children played the game in four sessions of about 15 min each. At posttest, they completed all of the same numeracy measures as at pretest. They also counted as high as possible at the beginning of each game session.

Results

The results of the training study supported the hypothesis that the Turkish number language was easier to learn than English. Although both the Turkish and Canadian children benefitted more from the number game experience than from the color game experience, more of the Turkish children had improved counting performance after the number game experience than the Canadian children (Cankaya et al., 2014). Neither group showed improved performance on the other early numeracy measures as a function of game condition, however, presumably because the other tasks did not involve numbers greater than 10 (cf. Siegler & Ramani, 2008, 2009). Note that the children in this study were younger than those in previous work using similar number games and, because the sequence was extended from 10 to 20, all the children received only half as much practice on the numbers below 10.

The findings of this study support the conclusion that the regularity of the Turkish number words from 11 to 19 can help Turkish children learn to count in this range. However, this advantage did not generalize to other aspects of their early numeracy learning, and the effect of the intervention was modest, suggesting that the advantage of a regular number language may be limited. Overall, Canadian children had better early numeracy skills (e.g., cardinality, number naming, and KeyMath numeration) than Turkish children, despite similar verbal counting and nonverbal arithmetic scores (before the intervention). Factors such as parents' education and their reports of home numeracy activities were related to children's performance, but did not mediate the differences between groups on the early numeracy skills (see Cankaya, 2013 for details). There were no clear differences in the overall frequency with which parents reported numeracy experiences at home, although Turkish parents reported less frequent early literacy activities. It is possible that other factors in the children's environments such as preschool quality might differentiate their early experiences. Further research on the specific types of home and preschool experiences of children in these cultures is necessary to understand the sources of the differences in early numeracy performance.

More generally, these findings for the comparison between Turkish and Canadian children suggest that research on language effects should also report on other factors that may contribute to children's early numeracy learning. Many of the comparisons that have been used as evidence of language differences include Asian children. In those studies, it seems likely that cultural, educational, and parental factors differ substantially in Asian families from those in families from other cultures in a direction that predicts superior mathematical skills, independently of language differences. Of course, Asian languages are more transparent beyond 20, as compared with Turkish or English—our work with French-speaking children suggests that irregular or more complex number languages may continue to influence children's learning of the number system. It is possible that the combination of all of these factors may interact to accelerate Chinese children's early learning of number. Further research is needed to determine when or how these differences might influence children's early numeracy development.

Despite all the limitations of comparing English-speaking children's counting performance to that of children speaking other languages, overall, our results are consistent with the view that English-speaking children find it difficult to learn to count from 11 to 20; more specifically, we might conclude that children speaking any of the languages with variable number names in that range (i.e., also French and Spanish) will find the learning experience more difficult than Chinese- or Turkish-speaking children. Hence, the goal of the final study we describe here was to determine whether extra practice with the counting words in the difficult "teen" stretch could help English-speaking children learn to count.

Training Study II: Structured Counting Practice for English-Speaking Children

Dunbar, Ridha, Cankaya, Jiménez Lira, and LeFevre (2016) modified the number game paradigm to test whether counting practice that emphasized the base-10 structure of the number system would help English-speaking children acquire the counting words. We studied 3-year olds because most children of this age in Canada have not yet learned to count beyond 10. Children participated in one of three training conditions. In the control condition, they either played the color game or simply counted aloud for the experimenter. In the linear number game, they moved their game pieces along a board with a linear arrangement of the numbers from 1 to 20, as in Cankaya et al. (2014). In the structured number game, numbers from 1 to 10 were presented in a row, and numbers from 11 to 20 were arranged in another row above, with the two parts of the game connected by a "magic bridge." When the children counted beyond 10, they moved their game piece along the bridge to 11 and continued from there. We hypothesized that the structured game would help children to organize their number knowledge by pairing the consistent structure shown in the written Arabic digits to the less consistent number words in the counting sequence.

Children counted as high as possible six times, at pretest, before each of four game-playing sessions, and at posttest. Counting performance did not differ among children in the control, linear game, and structured game conditions at pretest or at the first two game sessions. However, after the children had played the games twice, those in the structured game condition counted to a higher number than those in the other two conditions. Their performance continued to improve on the last two counting sessions, whereas that of the children in the control and color game conditions did not (see Dunbar et al., 2016 for details). Thus, providing children with counting practice that emphasizes the structure of the number system can ameliorate the specific limitations of the English number words from 11 to 19. Although we do not know exactly how the modified game supported children's learning of the counting sequence in this range, the results support the view that providing extra practice and emphasizing the base-10 structure of the Arabic digits may facilitate children's learning of the verbal sequence.

CONCLUSIONS

In this chapter, we explored young children's counting performance across five different languages. Table 2 lists the studies discussed in this chapter and indicates whether or not the groups being compared in each study differed on various factors that might be relevant to early numeracy skills, specifically, language, country, parents' education, and home numeracy experiences; the table also summarizes the research. Children's early numeracy skills differed in all of the group comparisons, with one exception. Turkish-speaking children had counting skills similar to those of English-speaking children but performed more poorly on other early numeracy measures. Across the various groups, differences in early numeracy skills were sometimes correlated with differences in parent education, and home experiences were related to performance (except for Mexican children), but neither parent education nor home experiences consistently mediated group differences (cf. LeFevre et al., 2002). Hence, the picture is complex for these correlational comparisons. In contrast, the results of the final three studies provided more direct support for relations between language complexity and counting performance. In sum, we conclude that children's early numeracy performance varies according to many different factors; language is not the only plausible explanation for overall differences in mathematical performance even though it accounts for specific differences in some tasks (see also Chapter 10 by Göbel; Paik & Mix, 2015). Similar conclusions have been reached by Dowker and Roberts (2015) in comparisons of children learning Welsh or English, by Dowker, Bala, and Lloyd (2008) for Tamil versus English and by Mark and Dowker (2015) for Chinese versus English.

Moreover, although Chinese-speaking children were more efficient counters than English-speaking children, they also counted higher than Turkish-speaking children and performed very well on other early numeracy skills, suggesting that factors such as parents' beliefs about the importance of math and early practice also contribute to the Chinese advantage (Huntsinger et al., 1997). The perspective that parental beliefs and home experiences are important was supported by the comparison between Mexican children and Canadian children. Despite earlier entry to formal instruction, Mexican children had less advanced skills than Canadian children. Parents' academic expectations of their reports of home numeracy activities were related to early numeracy performance for the Canadian but not for the Mexican children. Further, Turkish children had less developed numeracy skills than Canadian children, with the exception of verbal counting that was similar to that of Canadian children. This pattern suggested that the structure of the counting sequence in 11–20 range helped Turkish children with counting, but did not support the development of their other early numeracy skills, none of which required knowledge of the number system beyond 10.

In all of the comparisons between groups of children, cultural and experiential differences were also related to children's performance. Parents' education differed across groups in some cases, and parents' academic expectations and their reports of the home numeracy activities they engaged in with their children also predicted the differences between English-speaking children and both the Turkish- and French-speaking groups. Within groups, parents' reports of home numeracy activities were correlated with children's performance in all except Mexican children; these results are consistent with a growing body of research on the role of parent numeracy activities and parents' use of number talk with preschool children in the United States and Canada (Chapter 9 by Sarnecka et al.; Skwarchuk et al., 2016).

In this chapter, we showed that all of the comparisons across groups of children speaking different languages were consistent with the predictions based on language regularity; however, experiential, cultural, and other parental factors were confounded with the language comparisons. We conclude, therefore, that studies that explore differences in early numeracy skills across languages must also measure and, ideally, control for these other factors. These comparisons also support the importance of using other methods to test for language-related effects in early numeracy learning. The first training study showed that it was easier for Turkish children to learn the counting sequence from 11 to 20 than English-speaking children, given equivalent practice (Cankaya et al., 2014) and the second showed that English-speaking children's performance improved when counting practice was structured to reflect the base-10 arrangement of the number sequence from 11 to 19 (Dunbar et al., 2016). These studies do not address the larger issue of how delays or advantages in learning the number words might influence children's subsequent learning; nevertheless, they support the view that the structure of the number language can influence children's acquisition of counting.

Learning the number words is an important part of early numeracy acquisition and may mediate the more general requirement to translate among the various number representations (i.e., quantities, number words, and Arabic digits; Wiese, 2003; Jiménz Lira, Carver, Douglas, & LeFevre, 2017). Although there is insufficient evidence to link specific number languages to children's learning of more complex number system knowledge, such as place value, knowledge of the number system does predict children's acquisition of other mathematical skills (Geary, 2011; LeFevre, Polyzoi, et al., 2010; Purpura et al., 2013), and the complexity of the number language continues to challenge older children (Chapter 10 by Göbel; Göbel, Moeller, Pixner, Kaufman, & Nuerk, 2014; Imbo, Vanden Bulcke, De Brauwer, & Fias, 2014; Seron, & Fayol, 1994). Learning the number word sequence continues beyond kindergarten and grade 1, with new rules and additional vocabulary words being acquired between grades 1 and 5 (Skwarchuk & Anglin, 2002).

Our findings suggest that children whose languages support the mapping process have an advantage that, combined with other factors, may contribute to a more rapid acquisition of the formal number system. Because this knowledge is critical to further learning (especially from Grade 1 on), children who acquire it earlier may have an advantage in acquiring advanced skills. The studies described in the current chapter focused on preschool children; research exploring the longitudinal effects of number languages is necessary to address the question of whether there are predictive or reciprocal relations between language complexity and mathematical performance. Nevertheless, we also note that number words (i.e., the verbal number system) are only one aspect of early numeracy learning; written symbolic representations may quickly dominate the representational landscape (Knudsen, Fischer, Henning, & Aschersleben, 2015; Purpura et al., 2013). Thus, our work contributes to a more general view in which number language is just one of the many factors that influence the development of children's numeracy skills.

ACKNOWLEDGMENTS

This research was support by several Discovery Grants from the Natural Sciences and Engineering Council of Canada to J. LeFevre and by a Standard Operating Grant from the Social Sciences and Humanities Research Council of Canada to M. Sénéchal and J. LeFevre (French Immersion Study). We thank Jenna Herdman for helpful comments on an earlier version of this chapter. This research has greatly benefitted from the comments and insights of many students from a wide range of linguistic and cultural backgrounds. In particular, we recognize the contributions of Kristina Dunbar, Aala Ridha, Heather Douglas, and Erin Joy. Their enthusiasm and willingness to collaborate on these projects was invaluable. Finally, we thank the many children who counted for us over the last 15 years.

REFERENCES

Aunio, P., & Niemivirta, M. (2010). Predicting children's mathematical performance in grade one by early numeracy. *Learning and Individual Differences*, *20*(5), 427–435. https://doi.org/10.1016/j.lindif.2010.06.003.

Aunola, K., Leskinen, E., Lerkkanen, M.-K., & Nurmi, J.-E. (2004). Developmental dynamics of math performance from preschool to grade 2. *Journal of Educational Psychology*, *96*(4), 699–713. https://doi.org/10.1037/0022-0663.96.4.699.

Bruck, M., Genesee, F., & Caravolas, M. (1997). A cross-linguistic study of early literacy acquisition. In B. Blachman (Ed.), *Foundations of reading acquisition and dyslexia* (pp. 145–162). Hillsdale, NJ: Erlbaum.

Cankaya, O. (2013). *The role of number naming systems and numeracy experience in the development of early numeracy*. [Unpublished Ph.D. Thesis]. Carleton University.

Cankaya, O., & LeFevre, J.-A. (2016). The home numeracy environment: What do cross-cultural comparisons tell us about how to scaffold young children's mathematical skills? In B. Blevins-Knabe & A. Berghout Austin (Eds.), *Early childhood mathematics skill development in the home environment* (pp. 87–104). Switzerland: Springer International Publishing.

Cankaya, O., LeFevre, J.-A., & Dunbar, K. (2014). The role of number naming systems and numeracy experiences in children's rote counting: Evidence from Turkish and Canadian children. *Learning and Individual Differences*, *32*, 238–245.

Chan, W. W. L. (2014). Understanding and processing numbers among Chinese children. *Psychology and Neuroscience*, *7*(4), 583–591. https://doi.org/10.3922/j.psns.2014.4.18.

Colomé, A., Laka, I., & Sebastián-Gallés, N. (2010). Language effects in addition: How you say it counts. *Quarterly Journal of Experimental Psychology, 63*(5), 965–983. https://doi.org/10.1080/17470210903134377.

Connolly, A. J. (2000). *KeyMath-revised/updated Canadian norms*. Richmond Hill, ON: Psycan.

Davis-Kean, P. E. (2005). The influence of parent education and family income on child achievement: The indirect role of parental expectations and the home environment. *Journal of Family Psychology, 19*(2), 294–304. https://doi.org/10.1037/0893-3200.19.2.294.

Dowker, A., Bala, S., & Lloyd, D. (2008). Linguistic influences on mathematical development: How important is the transparency of the counting system? *Philosophical Psychology, 21*(4), 523–538. https://doi.org/10.1080/09515080802285511.

Dowker, A., & Roberts, M. (2015). Does the transparency of the counting system affect children's numerical abilities? *Frontiers in Psychology, 6*(July), 1–7. https://doi.org/10.3389/fpsyg.2015.00945.

Dunbar, K., Ridha, A., Cankaya, O., Jiménez Lira, C., & LeFevre, J.-A. (2016). Learning to count: Structured practice with spatial cues supports the development of counting sequence knowledge in 3-year-old english-speaking children. *Early Education and Development, 9289*(August), 1–15. https://doi.org/10.1080/10409289.2016.1210458.

Duncan, G. J., Dowsett, C. J., Claessens, A., Magnuson, K., Huston, A. C., Klebanov, P., et al. (2007). School readiness and later achievement. *Developmental Psychology, 43*(6), 1428–1446. https://doi.org/10.1037/0012-1649.43.6.1428.

Englund, M. M., Luckner, A. E., Whaley, G. J. L., & Egeland, B. (2004). Children's achievement in early elementary school: Longitudinal effects of parental involvement, expectations, and quality of assistance. *Journal of Educational Psychology, 96*(4), 723–730. https://doi.org/10.1037/0022-0663.96.4.723.

Entwisle, D. R., & Alexander, K. L. (1990). Beginning school math competence: Minority and majority comparisons. *Child Development, 61*(2), 454–471.

Geary, D. C. (2011). Cognitive predictors of achievement growth in mathematics: A 5-year longitudinal study. *Developmental Psychology, 47*(6), 1539–1552. https://doi.org/10.1037/a0025510.

Göbel, S. M., Moeller, K., Pixner, S., Kaufmann, L., & Nuerk, H.-C. (2014). Language affects symbolic arithmetic in children: the case of number word inversion. *Journal of Experimental Child Psychology, 119*, 17–25. https://doi.org/10.1016/j.jecp.2013.10.001.

Huntsinger, C. S., Jose, P. E., Liaw, F.-R., & Ching, W.-D. (1997). Cultural differences in early mathematics learning: A comparison of Euro-American, Chinese-American, and Taiwan-Chinese families. *International Journal of Behavioral Development, 21*(2), 371–388.

Huntsinger, C. S., Jose, P. E., & Luo, Z. (2016). Parental facilitation of early mathematics and reading skills and knowledge through encouragement of home-based activities. *Early Childhood Research Quarterly, 37*, 1–15. https://doi.org/10.1016/j.ecresq.2016.02.005.

Imbo, I., Vanden Bulcke, C., De Brauwer, J., & Fias, W. (2014). Sixty-four or four-and-sixty? The influence of language and working memory on children's number transcoding. *Frontiers in Psychology, 5*(April), 1–10. https://doi.org/10.3389/fpsyg.2014.00313.

Jiménez Lira, C. (2016). *Children's acquisition of the mappings among the number representations*. [Unpublished Ph.D. Thesis]. Carleton University.

Jiménz Lira, C., Carver, M., Douglas, H., & LeFevre, J.-A. (2017). The integration of symbolic and non-symbolic representations of exact quantity predicts number comparison performance in preschool children. *Cognition, 166*, 382–397. https://doi.org/10.1016/j.cognition.2017.05.033.

Joy, E. (2006). *Number knowledge and language as a function of numerical production: Children in early French immersion* [Unpublished B.A. Thesis]. Carleton University.

Kleemans, T., Peeters, M., Segers, E., & Verhoeven, L. (2012). Child and home predictors of early numeracy skills in kindergarten. *Early Childhood Research Quarterly, 27*(3), 471–477. https://doi.org/10.1016/j.ecresq.2011.12.004.

Knudsen, B., Fischer, M. H., Henning, A., & Aschersleben, G. (2015). The development of Arabic digit knowledge in 4- to 7-year-old children. *Journal of Numerical Cognition, 1*(1), 21–37. https://doi.org/10.5964/jnc.v1i1.4.

LeFevre, J.-A., Clarke, T., & Stringer, A. P. (2002). Influences of language and parental involvement on the development of counting skills: Comparisons of French- and English-speaking Canadian children. *Early Child Development and Care, 172*(3), 283–300. https://doi.org/10.1080/03004430212127.

LeFevre, J.-A., Polyzoi, E., Skwarchuk, S., Fast, L., & Sowinski, C. (2010). Do home numeracy and literacy practices of Greek and Canadian parents predict the numeracy skills of kindergarten children? *International Journal of Early Years Education, 18*(1), 55–70. https://doi.org/10.1080/09669761003693926.

LeFevre, J.-A., Skwarchuk, S.-L., Smith-Chant, B. L., Fast, L., Kamawar, D., & Bisanz, J. (2009). Home numeracy experiences and children's math performance in the early school years. *Canadian Journal of Behavioural Science/Revue Canadienne Des Sciences Du Comportement, 41*(2), 55–66. https://doi.org/10.1037/a0014532.

Luo, Z., Jose, P. E., Huntsinger, C. S., & Pigott, T. D. (2007). Fine motor skills and mathematics achievement in East Asian American and European American kindergartners and first graders. *British Journal of Developmental Psychology, 25*(4), 595–614. https://doi.org/10.1348/026151007X185329.

Mark, W., & Dowker, A. (2015). Linguistic influence on mathematical development is specific rather than pervasive: Revisiting the Chinese Number Advantage in Chinese and English children. *Frontiers in Psychology, 6*(February), 1–9. https://doi.org/10.3389/fpsyg.2015.00203.

Miller, K. F., Smith, C. M., Zhu, J., & Zhang, H. (1995). Preschool origins of cross-national differences in mathematical competence: The role of number-naming systems. *Psychological Science, 6*(1), 56–60.

Muira, I. T., Okamoto, Y., Kim, C., Steere, M., & Fayol, M. (1993). First graders' cognitive representation of number and understanding of place value: Cross-national comparisons—France, Japan, Korea, Sweden, and the United States. *Journal of Educational Psychology, 85*(1), 24–30.

Paik, J. H., & Mix, K. S. (2015). U. S. and Korean children's comprehension of fraction names: A reexamination of cross-national differences. *Child Development, 74*(1), 144–154.

Purpura, D. J., Baroody, A. J., & Lonigan, C. J. (2013). The transition from informal to formal mathematical knowledge: Mediation by numeral knowledge. *Journal of Educational Psychology, 105*, 453–464. https://doi.org/10.1037/a0031753.

Rasmussen, C., Ho, E., Nicoladis, E., Leung, J., & Bisanz, J. (2006). Is the Chinese number-naming system transparent? Evidence from Chinese-English bilingual children. *Canadian Journal of Experimental Psychology/Revue Canadienne de Psychologie Expérimentale, 60*(1), 60–67. https://doi.org/10.1037/cjep2006007.

Reese, L., & Gallimore, R. (2000). Immigrant Latinos' cultural model of literacy development: An evolving perspective on home-school discontinuities. *American Journal of Education, 108*(2), 103–134.

Segers, E., Kleemans, T., & Verhoeven, L. (2015). Role of Parent Literacy and Numeracy Expectations and Activities in Predicting Early Numeracy Skills. *Mathematical Thinking and Learning, 17*(2–3), 219–236. https://doi.org/10.1080/10986065.2015.1016819.

Sénéchal, M. (2000). Examen du lien entre la lecture de livres et le développement du vocabulaire chez l'enfant préscolaire. *Enfance, 53*, 169–186. https://doi.org/10.3406/enfan.2000.3175.

Sénéchal, M., & LeFevre, J.-A. (2014). Continuity and change in the home literacy environment as predictors of growth in vocabulary and reading. *Child Development*, *85*(4), 1552–1568. https://doi.org/10.1111/cdev.12222.

Seron, X., & Fayol, M. (1994). Number transcoding in children: A functional analysis. *British Journal of Developmental Psychology*, *12*, 281–300.

Siegler, R. S., & Mu, Y. (2008). Chinese children excel on novel mathematics problems even before elementary school. *Psychological Science*, *19*(8), 759–763.

Siegler, R. S., & Ramani, G. B. (2008). Playing linear numerical board games promotes low-income children's numerical development. *Developmental Science*, *5*, 655–661. https://doi.org/10.1111/j.1467-7687.2008.00714.x.

Siegler, R. S., & Ramani, G. B. (2009). Playing linear number board games—But not circular ones—Improves low-income preschoolers' numerical understanding. *Journal of Educational Psychology*, *101*(3), 545–560. https://doi.org/10.1037/a0014239.

Siegler, R. S., & Robinson, M. (1982). The development of numerical understandings. *Advances in Child Development and Behavior*, *16*(C), 241–312.

Skwarchuk, S.-L. (2009). How do parents support preschoolers' numeracy learning experiences at home? *Early Childhood Education Journal*, *37*(30), 189–197.

Skwarchuk, S., & Anglin, J. M. (2002). Children's Acquisition of the English cardinal number words: A special case of vocabulary development. *Journal of Educational Psychology*, *94*(1), 107–125. https://doi.org/10.1037/0022-0663.94.1.107.

Skwarchuk, S.-L., Sowinski, C., & LeFevre, J.-A. (2014). Formal and informal home learning activities in relation to children's early numeracy and literacy skills: The development of a home numeracy model. *Journal of Experimental Child Psychology*, *121*, 63–84. https://doi.org/10.1016/j.jecp.2013.11.006.

Skwarchuk, S., Vandermaas-Peeler, M., & LeFevre, J.-A. (2016). Optimizing the home numeracy environments of three- to six-year-old children in the United States and Canada. In B. Blevins-Knabe & A. M. Berghout Austin (Eds.), *Early childhood mathematics skill development in the home environment* (pp. 1–36). Switzerland: Springer International Publishing.

Susperreguy, M., Jiménez Lira, C., & LeFevre, J. (2017). *Cross-cultural Comparisons of Formal Home Numeracy Practices: Canada, Mexico, and Chile*. Manuscript under review.

Wang, J., & Lin, E. (2009). A meta-analysis of comparative studies on Chinese and US students' mathematics performance: Implications for mathematics education reform and research. *Educational Research Review*, *4*(3), 177–195. https://doi.org/10.1016/j.edurev.2009.06.003.

Wiese, H. (2003). *Numbers, language, and the human mind*. Cambridge: Cambridge University Press.

Xu, C., & LeFevre, J.-A. (2015). *Effects of a midpoint reference on young children's number line performance*. [unpublished manuscript].

FURTHER READING

LeFevre, J.-A., Fast, L., Skwarchuk, S.-L., Smith-Chant, B. L., Bisanz, J., Kamawar, D., et al. (2010). Pathways to mathematics: Longitudinal predictors of performance. *Child Development*, *81*(6), 1753–1767. https://doi.org/10.1111/j.1467-8624.2010.01508.x.

Purpura, D. J., & Reid, E. E. (2016). Mathematics and language: Individual and group differences in mathematical language skills in young children. *Early Childhood Research Quarterly*, *36*, 1–10. https://doi.org/10.1016/j.ecresq.2015.12.020.

Chapter 4

Effects of Mathematics Language on Children's Mathematics Achievement and Central Conceptual Knowledge

Yukari Okamoto
University of California, Santa Barbara, CA, United States

The primary goal of this chapter is to consider the relation among mathematics achievement, mathematics language, and central numerical knowledge. My colleagues and I have been examining how culturally bound factors such as characteristics of numerical language impact children's learning of school mathematics content (e.g., Miura & Okamoto, 2003). At the same time, from a neo-Piagetian perspective, my colleagues and I have examined the development of central numerical knowledge across cultures (e.g., Okamoto, Case, Bleiker, & Henderson, 1996). The former line of work suggests that mathematics achievement is impacted by cultural factors. The latter suggests that despite achievement differences, children develop numerical knowledge at similar rates. These two suggestions appear contradictory. How do I reconcile these seemingly contradictory suggestions?

My current view is that children growing up in cultures as different as Japan and the United States, for example, develop foundational numerical concepts at about the same rate. However, achievement gaps in mathematics reflect different problem-solving experiences children encounter in and out of school that are valued by particular cultures (see also Chapter 3 by LeFevre et al. and Chapter 5 by Opfer et al.). Differences in numerical language characteristics may provide qualitatively different problem-solving experiences. This is particularly so for young children who are just beginning formal schooling. For these children, differences in numerical language characteristics impact both their learning of school mathematics content and development of their central numerical knowledge.

Mathematical Cognition and Learning, Vol. 4. https://doi.org/10.1016/B978-0-12-812574-8.00004-3

As such, this chapter is organized into two broad sections. The first focuses on mathematics achievement differences and factors that may explain these differences. Evidence is presented in favor of numerical/mathematical language as a potential factor to explain the achievement gap. In the second section, I describe a series of cross-cultural studies that examined the relation between the development of central conceptual knowledge and the attainment of specific skills in particular domains. With the exception of one anomaly, the data point to similar rates of development of central conceptual knowledge. The anomaly was found in Japanese 6-year olds' numerical knowledge. These children showed the development of numerical knowledge 1–2 years ahead of their US peers. This anomaly will be explained in terms of different characteristics of numerical language.

MATHEMATICS LANGUAGE AND ACHIEVEMENT

Achievement gaps between US and East Asian students have been documented since the publication of the results from the First International Mathematics Study (FIMS; Husén, 1967). Although US students have shown an upward trend in international standings over the last five cycles of the Trends in International Mathematics and Science Study (TIMSS), their performance remains substantially lower than their East Asian counterparts (Beaton et al., 1997; Mullis & Martin, 2008; Mullis et al., 1998, 2000; Mullis, Martin, Foy, & Arora, 2012; Mullis, Martin, Gonzalez, & Chrostowski, 2005). In fact, the most recent TIMSS results showed that at the fourth-grade level, US students were 10–14 percentage points behind their East Asian peers (Mullis et al., 2012). Similar results have been obtained in the Program for International Student Assessment. The most recent data from the 2012 study show that only 9% of US 15-year-old students reached the highest category of performance (i.e., scores 607 or above out of 1000 maximum) in mathematics literacy whereas 55% of Chinese students in Shanghai did so (retrieved from http://nces.ed.gov/surveys/pisa/pisa2012/index.asp).

In both these large-scale international comparisons, East Asian students show strengths in the attainment of mathematics content deemed important by the experts, including those who constructed the measures. These results have inspired many researchers to attempt to explain why such differences exist (see Chapter 5 by Opfer et al.).

Early research comparing characteristics of schools in Japan, Taiwan, and the United States documented that Chinese and Japanese students spent more time in school in general—and, in particular, studying mathematics—than did their US counterparts (e.g., Stigler, Lee, & Stevenson, 1987). Subsequent research began to focus on teaching practices. This line of work is most clearly represented in Stigler and Hiebert's (1999) book, *The Teaching Gap*. They argued that differences in teaching methods—not teachers' ability to teach—led to varying learning opportunities for US and Japanese students. For example, the video analysis of teaching practices in the United States,

Germany, and Japan (TIMSS videotape study; Stigler, Gonzales, Kawanaka, Knoll, & Serrano, 1999) showed a wide disparity in the types of mathematical activities in which students were engaged: during seatwork, Japanese students spent far more time inventing, analyzing, and proving than did their US and German peers, who, in turn, spent almost their entire time practicing routine procedures. Japanese practices documented in the TIMSS videotape study are consistent with the practices recommended by the National Council of Teachers of Mathematics (2000).

It is hardly surprising that variations in school experiences, in particular mathematics classroom experiences, account for achievement differences. However, marked differences in mathematics performance have been found as early as in kindergarten when the impact of teaching effectiveness and other school-related factors is minimal (e.g., Stevenson, Lee, & Stigler, 1986). This finding calls for consideration of factors other than teaching practices to account for early differences in mathematics achievement (see Chapter 5 by Opfer et al.).

Mathematics Language

The factor of central interest in explaining achievement differences is variations in how mathematical terms are expressed in different language groups. Our focus has been to examine the impact of numerical language characteristics between East Asian (e.g., Japanese) and non-East Asian (e.g., English) languages (e.g., Miura & Okamoto, 2003). Although written numerical symbols (i.e., Arabic numerals) are practically universal across cultures, spoken words associated with written symbols differ from one language to another (e.g., 1 is spoken as "yi," "ichi," and "il" in Chinese, Japanese, and Korean, respectively). As in any language, different words are used to distinguish the base sequence of numbers from 1 to 10. The number names above 10, however, show interesting variations.

The number naming systems of East Asian languages such as Chinese, Japanese, and Korean have their roots in ancient Chinese. In these languages, the number names above 10 are congruent with the traditional base-10 numeration system (see Table 1). That is, a number word for any given two-digit number can be generated from a set of base-10 rules and a base sequence of number names. For example, 11 is spoken as "ju-ichi" (i.e., ten-one) in Japanese as opposed to "eleven" in English. The latter is a new word for children to memorize. Similarly, the teen words in English (e.g., fourteen) have the single-digit number word first, followed by the word that signals "ten." In Korean, 14 is "shib-sah," which means "ten-four." The numbers 20 and above can also be generated by applying the base-10 rules in these East Asian languages. In Chinese, 55 is spoken as "wu-shi-wu," which is "five-tens-five." In English, however, it is "fifty-five," which implies "fifty" as a *chunk*, not five tens. It should be clear from these examples that Chinese, Japanese, and Korean children need to memorize only the number names for 1–10,

TABLE 1 Number words in English, Japanese, Chinese, and Korean

Number	English	Japanese	Chinese	Korean
1	One	Ichi	Yi	Il
2	Two	Ni	Er	Ee
3	Three	San	San	Sam
4	Four	Shi	Si	Sah
5	Five	Go	Wu	Oh
6	Six	Roku	Liu	Yook
7	Seven	Shichi	Qi	Chil
8	Eight	Hachi	Ba	Pal
9	Nine	Kyuu	Jiu	Goo
10	Ten	Juu	Shi	Shib
11	Eleven	Juu-ichi	Shi-yi	Shib-il
12	Twelve	Juu-ni	Shi-er	Shib-ee
13	Thirteen	Juu-san	Shi-san	Shib-sam
14	Fourteen	Juu-shi	Shi-si	Shib-sah
15	Fifteen	Juu-go	Shi-wu	Shib-oh
16	Sixteen	Juu-roku	Shi-liu	Shib-yook
17	Seventeen	Juu-shichi	Shi-qi	Shib-chil
18	Eighteen	Juu-hachi	Shi-ba	Shib-pal
19	Nineteen	Juu-kyuu	Shi-jiu	Shib-goo
20	Twenty	Ni-juu	Er-shi	Ee-shib
22	Twenty-two	Ni-juu-ni	Er-shi-er	Ee-shib-ee
33	Thirty-three	San-juu-san	San-shi-san	Sam-shib-sam
44	Forty-four	Shi-juu-shi	Si-shi-si	Sah-shib-sah
55	Fifty-five	Go-juu-go	Wu-shi-wu	Oh-shib-oh
66	Sixty-six	Roku-juu-roku	Liu-shi-lie	Yook-shib-yook
77	Seventy-seven	Shichi-juu-shichi	Qi-shi-qi	Chil-shib-chil
88	Eighty-eight	Hachi-juu-hachi	Ba-shi-ba	Pal-shib-pal
99	Ninety-nine	Kyuu-juu-kyuu	Jiu-shi-jiu	Goo-shib-goo
100	One hundred	Hyaku	Bai	Baek

and the rest can be generated by using these names and applying the base-10 rules.[1]

On a side note, English is not the only language that does not follow the base-10 rules. Spanish is somewhat closer to English in that new number names are used to express numbers 11 through 15 and decade names do not suggest multiples of tens. There are number names in Spanish that are similar to those of East Asian languages: the numbers 16–19 are spoken as ten and single-digit number (e.g., 17 is "diecisiete," which is ten-seven). Other interesting number names include 43 in German (which means three and forty), 80 in French (four times twenty), and 50 in Danish ((two and a half) times twenty). (See Chapter 3 by LeFevre et al. for related information about Turkish as well.)

Cognitive Organization of Numbers

Given a variation in characteristics of numerical language, it is plausible to expect that such a variation may have differentiated effects on how children mentally organize numbers. In a series of studies, Miura and colleagues (e.g., Miura, 1987; Miura & Okamoto, 1989) pursued this very question. That is, they wondered if children speaking regular or irregular number words develop different mental representations of two-digit numbers.

The method used in these studies involved commercially available base-10 blocks (see Fig. 1). Children were shown small cubes (one-blocks) and

FIG. 1 One-blocks (right) and ten-blocks (left).

1. Although children's acquisition of two-digit numbers is the focus of this chapter, the number names in these East Asian languages remain regular beyond 99. In Japanese, for example, the numbers up to 99,999 can be generated by applying the base-10 rules.

rectangular prisms (ten-blocks). They then saw a demonstration in which ten small cubes (10 one-blocks) were equivalent to one rectangular prism (1 ten-block). They were next asked to construct two-digit numbers using one- and ten-blocks. This was considered as children's *spontaneous* preference to represent the number. Once children complete the task, they were shown their own construction of each number and asked if they could think of another way to show the same number. In the initial study conducted by Miura (1987), Japanese-speaking first graders[2] spontaneously used the precise combinations of ten- and one-blocks to show two-digit numbers (e.g., 1 ten-block and 3 one-blocks for 13). In contrast, English-speaking first graders rarely chose to use ten-blocks to represent two-digit numbers. Instead, they only used one-blocks (e.g., 13 one-blocks for 13).

Miura's (1987) initial study sparked interest among those who study the relation between language and thought in general and mathematics in particular. Replications, however, were necessary to determine if Miura's initial findings would hold up. In later replication studies, Miura and colleagues recruited first graders, who had not been taught two-digit numbers, from a wider range of nations. From East Asia, they recruited children from China, Japan, and Korea. From non-East Asia, children from France and Sweden, in addition to the United States, were included. The Swedish number naming system is almost identical to the English system. The French system shares some characteristics similar to English but has its own unique number names as mentioned earlier. The results of these studies showed remarkable similarity to Miura's initial findings. For example, when asked to construct 28, the overwhelming majority of Chinese-, Japanese-, and Korean-speaking first graders preferred to use 2 ten-blocks and 8 one-blocks (i.e., the base-10 construction), whereas almost all English-speaking counterparts selected 28 one-blocks as their first construction (Miura, Kim, Chang, & Okamoto, 1988). French- and Swedish-speaking first graders were found to be much like English-speaking participants in that they also selected one-blocks to show two-digit numbers during the first attempt (Miura, Okamoto, Kim, Steere, & Fayol, 1993).

It is possible that children's initial construction does not tell the whole story about their cognitive representation of number. Miura and colleagues (Miura et al., 1988) thus asked children to come up with another way to show the same number. Considering the two constructions children made, they found large differences in how English- and East Asian-speaking children conceived of two-digit numbers: 76%, 79%, and 98% of Chinese, Japanese, and Korean first graders, respectively, made base-10 constructions for all of the numbers in either the first or second attempt or both, whereas only 13% of English-speaking first graders did so. Furthermore, about half of the English-speaking children did not use the base-10 construction at all.

2. The participants resided in California, attending a Saturday Japanese school. All of them spoke Japanese fluently, and their home language was Japanese.

Somewhat similar results were found in the 1993 study: 67% and 100% of Japanese and Korean children, respectively, made base-10 constructions for all of the numbers in either the first or second attempt or both. In contrast, 35%, 65%, and 8% of French, Swedish, and US children did so (Miura et al., 1993). The Swedish result was close to that of Japanese. When the first attempt was considered, however, over 98% of the Swedish first graders chose to use one-blocks only, whereas over 72% of the Japanese counterparts made base-10 constructions.

Across the three studies, common findings are that East Asian speakers differed significantly from non-East Asian speakers in the kinds of constructions they made for two-digit numbers. From these results, Miura and colleagues inferred that ancient Chinese-based number systems influenced the way that young children mentally represent two-digit numbers.

A major critique of these studies is that the results could be due to cultural factors other than linguistic differences (cf., Towse, Muldoon, & Simms, 2015; Towse & Saxton, 1997). Because the participants came from different cultures, children's cultural and educational experiences could be confounding factors. Dowker, Bala, and Lloyd (2008) tested this possibility with children who lived in the same region of Wales. Children in Wales speak Welsh or English or both. The number system in Welsh is congruent with the base-10 system. Dowker et al. found that Welsh-speaking children understood two-digit numbers more accurately than their English-speaking counterparts. Thus, the results from their study provide support for the idea that there is a causal link between numerical language and children's numerical thinking.

Numerical Language and Estimation of Numerical Magnitudes

More recently, several cross-cultural studies were conducted to examine the link between numerical language and children's understanding of numerical magnitudes. Siegler and Mu (2008), for example, asked Chinese and US kindergartners to estimate single- and two-digit numbers on a 0–100 number line. Children were told the locations of 0 (far left) and 100 (far right) on the number line and asked to show where each of the target numbers would go. The literature on this topic has shown that kindergartners in the United States typically overestimate smaller numbers and underestimate larger numbers, resulting in an estimation pattern that is best described as logarithmic (e.g., Siegler & Booth, 2004). By second grade, children gain better understanding of the correspondence between numerals and their magnitudes. Thus, their estimation pattern is best described as linear. Siegler and Mu's study found that Chinese kindergartners already showed a linear pattern of numerical estimation. US kindergartners' estimation, however, was logarithmic as has been found in other studies (see Chapter 5 by Opfer et al. for a thorough analysis of number-line estimation and cross-cultural differences in this skill).

Although Siegler and Mu (2008) did not attribute their findings to regular and irregular counting systems of Chinese and English, respectively, other studies have made an explicit link to the linguistic factor. Dowker and Roberts (2015) examined this question with Welsh- and English-speaking 6-year-old children who lived in the same region of Wales. Similar to Siegler and Mu, they asked children to mark where a particular number would go on the 0–100 number line. They found that overall, Welsh-speaking children were more accurate in their estimation than their English-speaking counterparts. Their further analyses revealed interesting findings. For the numbers under 10, Welsh- and English-speaking children did not differ in their estimation. However, for numbers above 10 and in particular above 20, Welsh-speaking children were significantly more accurate than their English-speaking counterparts. Similar findings were reported between German- and Italian-speaking children (Helmreich et al., 2011). The German counting system includes inversion properties (e.g., 48 is spoken as "eight and forty"), whereas no such properties appear in Italian. Helmreich et al. found that Italian-speaking children were more accurate in number-line estimation than their German-speaking counterparts. Thus, the cross-cultural comparisons of numerical estimation further provide support for the link between numerical language and children's numerical thinking.

Relation of Numerical Language to Mathematics Achievement

The studies discussed above provided evidence that numerical language characteristics influence the way children understand two-digit numbers. The question then is in what ways, if any, differences in cognitive representation of number might be related to mathematics performance. This question was addressed in Miura and Okamoto's (1989) study, albeit indirectly. They assessed English- and Japanese-speaking children's mathematics achievement and their cognitive representation of number. The achievement measures were not the same across cultures. In the United States, it was the Educational Records Bureau's standardized achievement test. In Japan, it was based on teacher ratings. Because the two measures were drastically different, analyses were carried out only within each nation. Miura and Okamoto reported that being able to show two-digit numbers in two different ways (e.g., the correct combination of ten- and one-blocks vs one-blocks only) was significantly correlated with mathematics achievement ($r=0.49$ and $r=0.37$ for the Japanese and United States participants, respectively).

Place-Value Understanding

Place value refers to the idea that the value of a digit depends on its relative position in a number. That is, the value of a given digit in a multidigit numeral depends on the face value of the digit (0–9) and on its position in the numeral, with the value of each position increasing by powers of 10 from right to left. In other words, to say that a child has an understanding of place value requires that the child understands the rules of the base-10 numeration system (e.g., the "2" in

28 stands for 20). It follows then that children who think of numbers as consisting of tens and ones are likely to show strong place value understanding.

Miura and colleagues tested this possibility (Miura & Okamoto, 1989; Miura et al., 1993). In addition to constructing two-digit numbers using base-10 blocks, they asked a series of questions about children's understanding of place value. A relatively easy question involved showing children a card with a two-digit number written on it and asking them to point to the numeral in the one's and ten's position. A slightly more difficult question entailed presenting children with 3 ten-blocks and 12 one-blocks and asking them to write the number the blocks represented. A more difficult question was posed when children were shown 3 clear plastic cups and 13 one-blocks. They were then asked to put four one-blocks in each cup. This resulted in *3* cups (with four one-blocks in each cup) and *1* one-block remaining. Thus, children saw *1* cube and *3* cups. The experimenter then showed a card with 13 written on it and pointed to the "1" and the "3" in turn and asked what each meant in relation to the cups and cubes in front of them. Children with weak or no understanding of place value were often distracted by the visual display and said that the 1 in 13 referred to 1 cube and the 3 in 13 referred to 3 cups.

When Miura and Okamoto (1989) compared US and Japanese first graders' place value understanding, they found a statistically significant difference in favor of Japanese children. In fact, one-half of English-speaking first graders were unable to respond to any of the place value questions, whereas all of the Japanese counterparts responded to at least one problem correctly, and 42% of them solved all of the problems correctly. Due, in part, to the small sample, no statistically significant correlations between cognitive representation of number and place value understanding were found in either group. In another study in which first graders from a greater number of nations were recruited, Miura et al. (1993) reported that East Asian-speaking children from Japan and Korea performed significantly better on the place value task than those from France, Sweden, and the United States. To examine the relation between cognitive representation of number and place value understanding, all children were combined for a regression analysis. The results showed that 58% of the variance in place value performance was explained by children's initial base-10 constructions as opposed to their second construction.

Although the evidence is correlational, the results across the two studies suggest that children's spontaneous preference to represent two-digit numbers as consisting of tens and ones (as opposed to only ones) facilitates the acquisition of place value of two-digit numbers, which, in theory, should provide a strong basis for generalizing the rules of base-10 for two-digit numbers to multidigit numbers. This should, in turn, facilitate children's understanding of later arithmetic performance, including multidigit column addition and subtraction. In fact, Moeller, Pixner, Zuber, Kaufmann, and Nuerk (2011) found that early mastery of base-10 knowledge in first grade was a reliable predictor of later arithmetic performance in third grade. The opposite has also been found to be true, in that children who struggle to develop place value understanding early on struggle to do well on multidigit addition and subtraction computations with regrouping (e.g., Ho & Cheng, 1997). Taken together, it seems reasonable to conclude that the compatibility of the spoken number words with the base-10 rules influences the way young children organize two-digit numbers, which, in turn, has a significant impact on the rate at which children form an understanding of place value and multidigit number arithmetic (Alsawaie, 2004; Dowker et al., 2008; Ho & Fuson, 1998).

Numerical Language and Instructional Methods

Multidigit column addition and subtraction are difficult for children who have yet to understand place value. These children tend to treat digits in different columns as if they are independent. When asked what the "2" in 28 means, English-speaking children without place value understanding would say "two" not "twenty." When faced with column addition with regrouping (i.e., carrying and borrowing) such as 34 – 18, the lack of place value understanding results in answers such as 24 ("smaller-from-larger bug"; see Brown & Burton, 1978). Teachers do their best to teach children how to borrow and carry so that they will not make this sort of error. Miura and Okamoto (1999) pointed out that different teaching methods could result in part from characteristics of numerical language. US children are typically taught 34 – 18 in the following manner (see Table 2a):

> *You can't take the 8 from the 4. So you borrow 10 from the 30 to make 14 (fourteen). Now you can take the 8 from the 14 and you get 6.*

This method is also feasible in Japan. In addition to this method, however, Japanese children understand the method that takes advantage of its base-10 numerical language (see Table 2b):

> *You can't take the 8 from the 4. So you borrow 10 from the 30 to make 14 (ten-four). 10 take away 8 is 2. You add the 4 to the 2 and you get 6.*

The Japanese method naturally arises from the base-10 language characteristics. Teachers would not have to explain why they break 14 into 10 and 4. Because 14 is "ten-four," it makes sense to think of 14 as consisting of 1

TABLE 2 Subtraction strategies

(a) Strategy based on *fourteen*		(b) Strategy based on *ten-four*	
2 10		2 10	
34	14 – 8 = 6	34	10 – 8 = 2
–18		–18	2 + 4 = 6
6		6	

ten and 4 ones. Young children have more practice with number combinations up to 10 and adding two single-digit numbers. Thus, the Japanese method is less prone to errors. The first method, however, requires that children know number facts beyond 10—which is more advanced—or use other strategies such as "counting up from" (e.g., 14 – 8 is accomplished by counting up from 8 to 14), which is more prone to counting errors. These examples suggest that differences in how number words are spoken have implications for how arithmetic computation is taught.

Fraction Terms

Many children—and adults for that matter—find fractions difficult (Ma, 1999; Moseley & Okamoto, 2008; Moseley, Okamoto, & Ishida, 2007; Siegler et al., 2010). It has been reported that only 13% of US fifth graders reached fraction proficiency (Princiotta, Flanagan, & Germino Hausken, 2006). This is of serious concern as understanding of fractions has been identified as a strong predictor of later success in mathematics in the United States and the United Kingdom (Siegler et al., 2012). When children are first learning fractions, they often apply what they know about whole number arithmetic to fraction arithmetic. This is known in the literature as the "whole number bias" (e.g., Ni & Zhou, 2005; Vosniadou, Vamvakoussi, & Skopeliti, 2008). For one thing, children do not see a fraction notation as referring to a particular quantity. Instead, they treat the numerator and denominator as two separate whole numbers. Thus, 2/3 + 1/3 could result in an answer 3/6 (e.g., Mack, 1990). English fraction terms are not particularly helpful in overcoming this misconception. For example, 2/3 is spoken as "two-thirds," which does not readily convey any fraction meaning for children who are learning fractions for the first time. In contrast, in East Asian languages, 2/3 is spoken as "of (the whole) divided into three, (take) two." This conveys the notion of part-whole. Of course, part-whole is just one interpretation of rational numbers (Moss & Case, 1999), and full-fledged understanding of rational numbers requires expanded and multiple interpretations (Behr, Harel, Post, & Lesh, 1992; Kieren, 1993). Nonetheless, part-whole is one of the meanings that fractional notations convey, and children need to develop this understanding as well.

Miura, Okamoto, Vlahovic-Stetic, Kim, and Han (1999) hypothesized that East Asian children who hear and speak fraction terms in which the part-whole relation is embedded might find it easier to identify the part-whole interpretation of fractions than their non-East Asian counterparts. They tested this hypothesis in a cross-sectional study of first and second graders from Croatia, Korea, and the United States. Data were collected from different age groups of children in each nation: the first group in the middle of the first grade, the second group at the end of first grade, and the third group at the beginning of second grade. Children saw and heard each fraction name in their native language and were asked to select one from four pictorial representations that corresponded to the target fraction (see Fig. 2). All four

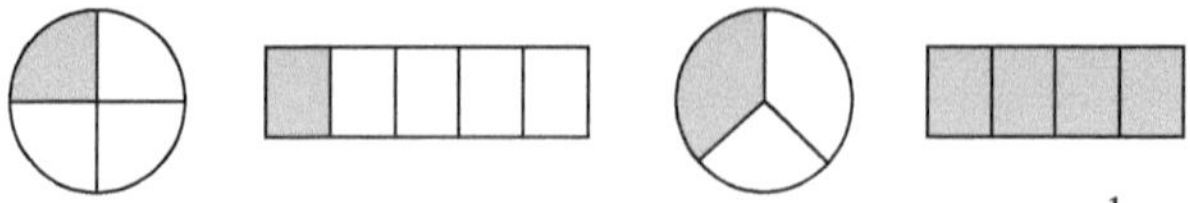

FIG. 2 Four pictorial representations from which to select one that matches $\frac{1}{4}$.

representations conveyed the part-whole meaning of fractions (e.g., a bar or a circle divided into a number of parts) with only one option matching the target fraction. The results showed that Korean children at the end of first grade already outperformed those in Croatia and the United States at the beginning of second grade. The results suggest that East Asian fraction words may facilitate children's interpretation of part-whole fractions by making a stronger association between the spoken words and corresponding part-whole visual representations. It should be noted that these results await replication (cf., Paik & Mix, 2003).

Terms for Geometric Shapes

English terms for geometric shapes are derived from Greek and Latin words. As shown in Table 3, these terms tend to be long and complicated. For young children learning shape names, these are all new words with little hint embedded in the names to link to visual representations. This is particularly true for shapes with four or more angles. In contrast, Chinese-based geometric terms are relatively straightforward. For example, pentagon is a five-angle shape (5角形) in Japanese. Aside from rectangle, all standard 2-D shapes that children see in the shape chart take a form of "(the number of angles)-angle shape."[3] That is, the meaning of each of these shapes is embedded in its term. Children learning these simple terms should have an easier time linking words to visuals and their properties than those who must memorize complex terms. To the best of my knowledge, no cross-cultural studies have examined this relation. The only cross-cultural studies in this general area of spatial terms examined young English- and Korean-speaking children's categorization of spatial words such as "in" and "on" (Choi & Bowerman, 1991; Choi, McDonough, Bowerman, & Mandler, 1999). Using a preferential looking paradigm, Choi et al. (1999) found that two groups of children as young as 18 months old showed sensitivity to language-specific spatial categories. Coupled with recent findings that spatial language supports children's encoding of spatial properties such as big, tall, and curvy and predicts their later performance on spatial reasoning (Pruden, Levine, & Huttenlocher, 2011), it is conceivable that differences in geometric shape names between English and Japanese could result in different rates of acquisition of spatial properties.

3. 3-D shape names also express property meanings more clearly in Japanese than in English (e.g., six-angle pillar vs hexagonal prism).

TABLE 3 2D geometric shape names in English and Japanese

Shape	English	Japanese
	Triangle	Three-angle shape 三角形
	Square	True (correct) four-angle shape 正四角形
	Rectangle	Long-side shape 長方形
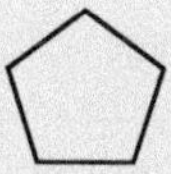	Pentagon	Five-angle shape 五角形
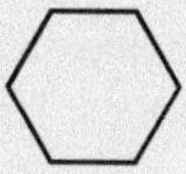	Hexagon	Six-angle shape 六角形
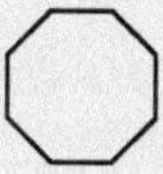	Octagon	Eight-angle shape 八角形

Note: Square in Japanese is also expressed as true (correct) shape (正方形).

In summary, number names, fraction terms, and geometric shape names in East Asian languages facilitate children's learning of difficult mathematics concepts. Geometric shapes are given names that refer to the number of angles. Fractions are spoken in a way to convey the part-whole meaning. Finally, number names for two-digit numbers are congruent with the base-10 numeration system, which not only helps children to develop place value understanding with relative ease but also allows for the teaching of multidigit column computations to take advantage of the number naming system. Taken together, characteristics of East Asian languages make mathematics accessible to the speakers of such languages.

In contrast, children speaking non-East Asian languages are tasked with developing a link between number names and their meaning on their own or through explicit instruction, overcoming whole number bias with fractions, and learning shape names that provide no information about the properties of the shape. East Asian speakers, such as Chinese, Japanese, and Korea children, bypass many of the steps that are necessary for their non-East Asian-speaking children to take. Achievement differences are thus expected to appear as early as when children are first learning number names larger than 10 (and possibly shape names). The initial achievement gap may be small. As shown in large-scale comparison studies, however, it becomes a bigger problem as children learn more complex mathematics.

MATHEMATICS ACHIEVEMENT, CENTRAL CONCEPTUAL KNOWLEDGE, AND NUMERICAL LANGUAGE

Although explaining sources of achievement differences is important, so is considering what these differences tell us about children's understanding of number systems (e.g., whole numbers and rational numbers). Case and colleagues (e.g., Case, 1998; Case & Okamoto, 1996) distinguished mathematics achievement from foundational conceptual knowledge of number. This latter knowledge is a domain-specific conceptual structure, often referred to as a *central conceptual structure* in a specific domain (e.g., Case & Okamoto, 1996). In elementary school years, mathematics achievement in the number domain is a reflection of children's mastery of school arithmetic content. In contrast, children's conceptual knowledge of whole numbers has its root in the kind of core knowledge that develops during infancy (see Okamoto, 2010, for a review). This core knowledge has been found to develop not only in human infants but also in macaque monkeys (e.g., Hauser & Carey, 2003). Furthermore, adults who grew up in cultures without demands for enumeration skills also possess this core knowledge (e.g., Pica, Lemer, Izard, & Dehaene, 2004).

If core knowledge provides foundations for children's central conceptual knowledge, it is possible that central numerical knowledge develops at similar rates in cultures where enumeration skills are valued. However, the particular skills children are encouraged to attain could differ, which might explain large differences in achievement. In a series of studies, we demonstrated that this possibility holds up in general. Evidence counter to this general trend was found among 6-year-old children in Japan who exhibited numerical knowledge 1–2 years ahead of their US peers who, otherwise, showed age-appropriate conceptual development. Among many competing explanations, I have come to conclude that variations in numerical language characteristics play a major role in young Japanese children's acquisition of central numerical knowledge.

Development of Core Conceptual Knowledge

Infants may come into this world with the ability to detect numerical features of the world. However, it takes a few more years before children are able to count objects reliably (e.g., Gelman, 1978) or compare quantities (e.g., Starkey, 1992). As Gelman and Gallistel (1978) have shown, the act of counting requires children to acquire several important principles. For example, children must apply number names in the same order every time they count, count one item at a time without skipping any (and not counting the same item more than once), and know that the last number name spoken represents the cardinal value of the set. In other words, children must learn to adhere to cultural conventions (including number names) associated with the act of counting. Building on Gelman and Gallistel's work, Case (1992a) proposed

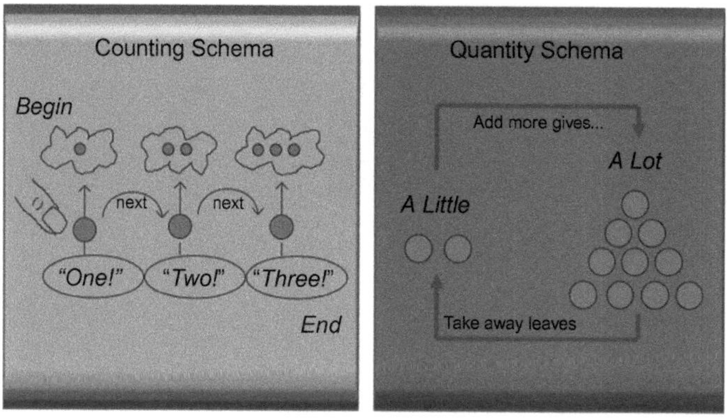

FIG. 3 Four-year olds' counting schema (left) and quantity schema (right).

that by 4 years of age, children develop a schema that allows them to count a small number of items without error (Fig. 3(left)). At the same time, children develop a global quantity schema (Fig. 3(right)). This schema allows preschool-age children to compare quantities and understand the effects of quantity transformations such as addition and subtraction of small numbers (Starkey, 1992; Wynn, 1992). These two schemas provide foundations for preschoolers to interpret the quantitative world around them.

As children acquire more experience working with numbers and quantities in a broad range of situations, they learn to use counting as part of quantitative reasoning. For example, when asked which side of a balance scale (each holding different numbers of weights) would tilt, children by kindergarten or first grade realize that counting is a useful way to make quantitative judgments. In contrast, children in preschool years often fail to count the number of weights but instead visually inspect which pile of weights looks higher (Case, Okamoto, Henderson, McKeough, & Bleiker, 1996). Observations such as these led Case and colleagues to conclude that by around 6 years of age, children typically develop central numerical knowledge that results from the integration of the two earlier schemas (e.g., Case & Okamoto, 1996; see Fig. 4). This knowledge structure is often referred to as a "mental number line" (Case, 1996a).

As shown in Fig. 4, to develop a functional understanding of a mental number line requires that children understand the correspondence between Arabic numerals (written and spoken) and their magnitudes. This point was recently elaborated by Siegler, Thompson, and Schneider (2011). They reviewed studies of numerical magnitude estimation and concluded that it is not until 5 or 6 years of age that children develop a mental number line for single-digit numbers. They further articulated that the acquisition of this initial mental number line provides a foundation for children's understanding of the whole number system and the rational number system. (See also Chapter 5

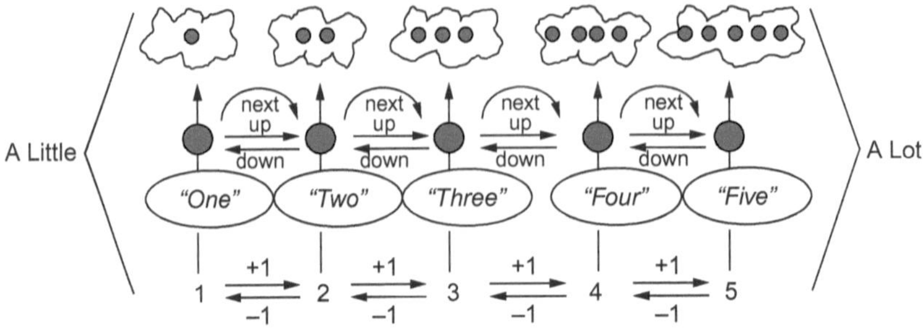

FIG. 4 Six-year olds' central conceptual knowledge (mental number line).

by Opfer et al. for a theory that the cross-national gap in mathematics achievement may be attributable to a difference in the representation of symbolic numerical magnitude that is primarily influenced by early instructional inputs.)

Case and Okamoto (1996) hypothesized that the acquisition of a mental number line provides a conceptual foundation through which children interpret the numerical attributes of the world. That is, it serves as a tool to create new knowledge necessary to interpret cultural concepts of time telling, currency system, and distribution of resources, and to take advantage of instruction in school. It should be emphasized that children's experiences are culturally bound.

By about 8 years of age, children's mastery of the mental number line becomes sufficiently fluent, allowing them to focus on and tentatively begin relating two mental number lines. As a result, new properties of numerical systems such as the base-10 system become part of their understanding. Children's understanding at this level, however, is limited to interpreting the relation between ones and tens. That is, 8-year olds have a difficult time generalizing the rules of base-10 to numbers beyond 99. By about age 10, children begin to generalize the base-10 rules to three-digit numbers and possibly beyond. They are also able to develop the notion that the numerical rules and operations developed earlier can be treated as objects of manipulations to carry out complex arithmetic reasoning. For example, in comparing the numerical magnitudes of $(8-3)$ and $(9-5)$ mentally, children of this age are able to treat the result of one computation *5* $(8-3=5)$ as a new object to compare with the result of another computation *4* $(9-5=4)$. Fig. 5 shows a schematic sketch of the developmental progression.

Cultural Comparisons of Central Conceptual Knowledge and Specific Skills

Central numerical knowledge that develops during elementary school years is shaped by both biologically based factors (e.g., neural development) and children's everyday cultural experiences (e.g., mastering currency). Case (1996b)

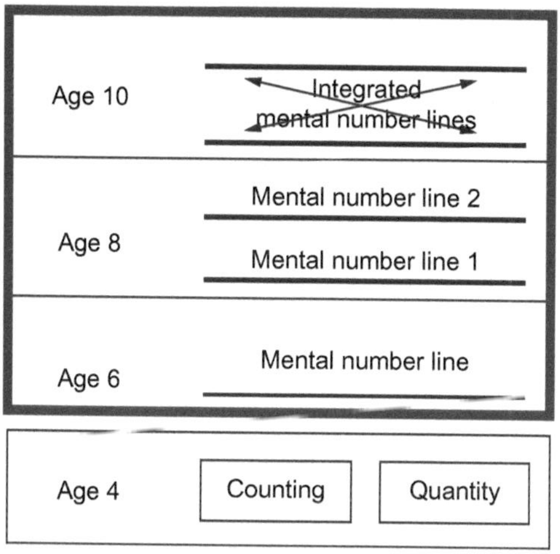

FIG. 5 A schematic representation of the development of central numerical knowledge that results from an integration of the counting and quantity schemas.

suggested that, as long as children's numerical experience includes a broad set of skills in a domain and they follow a normal course of maturation, it is conceivable that central conceptual knowledge takes a similar developmental progression across cultures. If one culture values a particular skill over others and puts more resources in helping children acquire it, children in that culture should excel in the skill emphasized. This in itself, however, does not advance central conceptual development. It is only when children have multiple opportunities to experience whole numbers in different contexts, reflect on them, and come to see numerical connections among them that these specific skills contribute to the development of central numerical knowledge.

To test this possibility, a series of cross-cultural studies were conducted. It should be noted that in Case's theory, parallel developmental progression of central conceptual knowledge takes place in different domains, such as whole numbers, space, and social/narrative knowledge (Case & Okamoto, 1996). As discussed earlier, central conceptual knowledge has its root in the core systems of knowledge on which human cognition is founded. These core systems of knowledge enable human infants to represent objects, number, space, actions, and social interactions (Spelke & Kinzler, 2007). The studies reported below focus on two such domains, space and number.

Central Spatial Knowledge and Drawing Skills

In two separate studies, children's drawings were assessed to examine their central spatial knowledge and drawing skills (Okamoto et al., 1996). In the first study, 4-, 6-, 8-, and 10-year-old Chinese and Canadian children's drawings were compared. In the second study, 6-, 8-, and 10-year-old Japanese and US children's drawings were compared. In both these studies, there were large differences in drawing skills in favor of Asian children. However, there were no statistically significant differences in how children spatially organized objects to be drawn. That is, children in all four nations developed age-appropriate central spatial knowledge. Differences only appeared in children's specific skills of drawing. In both China and Japan, drawing is viewed as an important cultural activity and included in weekly school schedules. In Canada and the United States, drawing is overshadowed by the "three R's" and in some cases completely omitted from the school schedule. In sum, cultural emphasis on drawing was seen in children's drawing skills but not in central spatial knowledge.

Central Numerical Knowledge and Mathematics Achievement

Two studies are reported that examined the relation between central numerical knowledge and mathematics achievement. The first study compared 6-, 8-, and 10-year-old children in Japan and the United States (Okamoto et al., 1996). To test Case's theory of whole number development, children were individually tested on the number knowledge and balance scale tests. The number knowledge test included items that assessed elements described in Figs. 3–5. One important feature was to develop questions that children typically do not encounter in school. All three age groups took this test. As for the balance scale test, the 6- and 10-year olds in each nation took this test. They were asked which side of the scale would tilt and why, as the number of weights and distance from the fulcrum were manipulated. The items for the balance scale test were once again designed to assess their level of central numerical knowledge. An achievement test was group administered to the 10-year-old children in each nation. The items were drawn from the fifth-grade test developed by Stevenson and colleagues (e.g., Stevenson et al., 1986).

As expected, there was a large and statistically significant achievement difference in favor of Japanese 10-year-old children. Despite this difference, no significant differences were found for the 6- or 10-year-old children in their performance on the balance scale test. As for the number knowledge test, no differences were found for the 8- and 10-year-old children. However, Japanese 6-year-old children outperformed their US counterparts. For the most part, these results show the same trend as in the drawing studies. That is, mathematics achievement (specific knowledge) valued by Japanese culture (as described in Stevenson et al., 1986, and Stigler & Hiebert, 1999) showed a substantial difference. In contrast, all age groups of children in both nations—with the exception of the Japanese 6-year olds on the number knowledge test—showed age-appropriate central numerical knowledge.

This cross-cultural study recruited students from middle-income families in each nation. It could be that if one tested US children who are from high-income families who attended a school well-known for its rigorous curriculum and high achievement in mathematics, the results could be different.

TABLE 4 Means (standard deviations) for the number knowledge test by national group, achievement level, and age

The United States Achievement						Japan Achievement		
Middle			*High*			*Middle*		
Age	N	*Mean*	*Age*	N	*Mean*	*Age*	N	*Mean*
6−4	22	1.82 (0.31)	6−0	25	1.69 (0.43)	6−5	21	2.52 (0.73)
8−1	23	3.04 (0.31)	8−1	26	3.12 (0.53)	8−5	26	3.33 (0.77)
10−5	24	4.23 (0.68)	10−2	20	3.79 (0.55)	10−5	20	4.05 (0.76)

Note: Hypothesized age-appropriate mean scores are 2, 3, and 4 for 6-, 8-, and 10-year-old groups, respectively.

Okamoto, Curtis, Chen, Kim, and Karayan (1997) tested this possibility. Children in Okamoto et al.'s study were given the same number knowledge test as above. As shown in Table 4, high-achieving US children were no different from their US and Japanese counterparts from middle-income families in their central numerical knowledge with the exception of the Japanese 6-year olds.[4]

Okamoto et al. (1996) suggested that central numerical knowledge is necessary for the mastery of a range of numerical skills and concepts, including those emphasized in mathematics instruction. In order to contribute to the development of central numerical knowledge, however, a broad range of experiences in the numerical domain is necessary, not just mastery of school arithmetic. Just as drawing was considered to be a specific skill in the spatial domain, mathematics achievement is a specific skill in the numerical domain. Specific skills reflect children's experiences in their cultural and socioeconomic contexts.

Taken together, findings from these cross-national studies suggest that children growing up in nations as different as the United States and Japan develop age-appropriate central conceptual knowledge in spatial and whole number domains. The only exception to these general findings was the performance by Japanese 6-year olds on the number knowledge test.

4. Although children from high- and middle-income families did not differ in their central conceptual knowledge, children from low-income families who seldom experience activities with numerical content tend to be behind their peers in central conceptual knowledge (e.g., Griffin and Case, 1996) and estimation of numerical magnitudes (e.g., Ramani & Siegler, 2008). These researchers also showed that theory-based training significantly improved children's numerical knowledge.

Effects of Numerical Language

The anomaly found in the cross-cultural studies was significantly stronger performance by Japanese 6-year olds than their US counterparts on the number knowledge test. US 6-year olds, however, showed age-appropriate performance on the number knowledge and balance scale tests. Japanese counterparts also showed age-appropriate performance on the balance scale test. Thus, it was only on the number knowledge test that Japanese 6-year olds excelled. In fact, several of these Japanese 6-year olds successfully answered the 8-year-old level items designed to require understanding of two mental numbers or the meaning of tens and ones (see Fig. 5). What do these results mean? Among many competing explanations, I favor the numerical language explanation. As the data from Miura and colleagues' studies showed, East Asian speakers, including Japanese children, develop an understanding of the meaning of tens and ones earlier than their non-East Asian peers. The number knowledge test at the 8-year-old level included items such as "which is bigger, 69 or 71?" Japanese 6-year-old children who hear "six-tens nine" and "seven-tens one" could easily distinguish tens and ones and select the correct answer. In contrast, English-speaking 6-year olds do not have this language advantage. They often mistakenly attend to the number words in the one's column to respond to this sort of question. In contrast, the variables in the balance scale (i.e., the number and location of weights) used numbers less than 10. Thus, there was no advantage for Japanese 6-year olds on this test.

As number words are used to solve many numerical situations in everyday life, it is conceivable that quality and quantity of experience to develop base-10 knowledge helped advance Japanese 6-year olds' central numerical knowledge. The data presented in Case et al. (1996) provide a clue as to if this is a possibility. To verify the existence of the central numerical knowledge, Case et al. developed a battery of six tests in the numerical domain. The 8-year-old-level items on the three of the tests (including the number knowledge test) used two-digit numbers, whereas the three other tests (including the balance scale test) used only single-digit numbers. When this battery was administered to 6-year-old children in the United States, Case et al. found a single latent factor underlying children's numerical thought. It is conceivable that two latent factors might result if this battery was administered to Japanese 6-year olds. Only by administering the entire battery cross-culturally, would we know whether numerical language differentially influences the development of core numerical knowledge. It might confirm that Japanese 6-year olds' performance on the balance scale test *was* the anomaly in the data collected. Given the pattern of performance reflected in the available data, my tentative conclusion is that variations in numerical language characteristics played a major role in young Japanese children's development of central numerical knowledge.

CONCLUSIONS

In this chapter, I examined the role of mathematics language in explaining children's mathematics achievement and the development of central numerical knowledge. A particular focus was placed on children who are just beginning formal schooling. I presented evidence that Chinese, Japanese, and Korean children of this age have an easier time mastering place value because place value is embedded in their spoken language. In contrast, children who speak non-East Asian languages such as English, French, and Swedish do not have this sort of language support. Thus, they must make an explicit link among Arabic numerals, spoken words, and place value. Because the notion of place value is important in carrying out multidigit arithmetic, it follows that the early advantage of these East Asian speakers may facilitate later learning of mathematics. Of course, international comparisons of mathematics achievement go beyond just testing whole number understanding. I presented preliminary evidence that differences in mathematical language characteristics might contribute to achievement differences in other areas such as fractions and geometry.

My primary argument is that language characteristics influence specific aspects of mathematics performance. I am not proposing that these East Asian languages support all aspects of mathematics learning. In fact, when children are first learning single-digit number names, Japanese 2- to 3-year olds were found to be behind their English- and Russian-speaking peers in mastering the cardinal meaning of one, two, and three (Sarnecka, Kamenskaya, Yamana, Ogura, & Yudovina, 2007). Another study also confirmed that Japanese 2-year olds were behind their English-speaking peers in small number comprehension (Barner, Libenson, Cheung, & Takasaki, 2009). Japanese toddlers' difficulty understanding single-digit numbers is due, in part, to the dual counting systems of Japanese number words 1–10. There are two sets of counting words for these numbers in Japanese (see Okamoto, 2015, for a review). It is understandable if Japanese toddlers are delayed because in everyday life they hear two different number words, say "ichi" and "hi" for 1.[5] These initial difficulties, however, are quickly overcome by 4-year olds, and Japanese preschoolers master counting principles just like their English-speaking peers do.

About the time children begin formal schooling, important conceptual development takes place (Case, 1996a). As described in this chapter, by about 6 years of age, children across cultures with demands for enumeration skills develop central numerical knowledge that allows them to mentally reason about quantitative relations with numbers. Although 6-year olds' mental

5. Although number words in Table 1 are more commonly used in everyday life and in learning mathematics, there exist indigenous Japanese counting words for 1–10. Japanese toddlers are exposed to these two different ways of counting from early on (e.g., "ichi-mai" for one sheet of paper and "hi-tori" for one person).

arithmetic is typically limited to single-digit numbers, the data from cross-cultural studies pointed to the possibility that Japanese 6-year olds were beginning to show mental manipulations of two-digit numbers. The cross-cultural studies provided evidence that despite large achievement differences, central conceptual knowledge develops at similar rates across cultures with the exception of the just-mentioned Japanese 6-year olds.

As pointed out in previous studies, children in Japan spend more time studying mathematics than their US peers (e.g., Stevenson et al., 1986), and mathematics teaching in Japan is qualitatively different from teaching in the United States (Stigler & Hiebert, 1999). Furthermore, children's experience with numbers in and out of school is different, due in part to numerical language characteristics. Take, for example, children's acquisition of the currency system. Children in all cultures with trade learn to use their currency to determine how much to pay and how much change to receive. However, the currency systems themselves differ. The coins in the United States, for example, include a penny, a dime, and a quarter, and a 20-dollar bill is commonly used, whereas in Japan, they are one yen, five yen, ten yen, five-tens yen, one-hundred yen, and five-hundreds yen, and there is no bill equivalence to the US $20 bill. The Japanese currency system thus reinforces children's acquisition of tens and ones in everyday experience of shopping.

Cooking is also an activity practiced in all cultures. Japanese children learn to use the metric system using base-10 spoken number words. This is not the case in the United States. These are just some of the everyday experiences in which children speak regular or irregular number words. Japanese 6-year olds are developing specific numerical skills (e.g., money knowledge and metric system), all of which use the regular base-10 number words. It is thus reasonable to expect that Japanese 6-year olds' acquisition of specific skills contributes to the development of central numerical knowledge. Until further data show otherwise, I conclude that spoken Japanese number words that include clear markers for place value influence not only mathematics achievement but also the development of central numerical knowledge for this age group.

REFERENCES

Alsawaie, O. N. (2004). Language influence on children's cognitive representation of number. *School Science and Mathematics*, *104*, 105–111.

Barner, D., Libenson, A., Cheung, P., & Takasaki, M. (2009). Cross-linguistics relations between quantifiers and numerals in language acquisition: Evidence from Japanese. *Journal of Experimental Child Psychology*, *103*, 421–440.

Beaton, A. E., Mullis, I. V. S., Martin, M. O., Gonzalez, E. J., Kelly, D. L., & Smith, T. A. (1997). *Mathematics achievement in the middle School Years: IEA's third international mathematics and science study (TIMSS)*. Chestnut Hill, MA: TIMSS International Study Center, Boston College.

Behr, M., Harel, G., Post, T., & Lesh, R. (1992). Rational number, ratio and proportion. In D. Grouws (Ed.), *Handbook of research on mathematics teaching and learning* (pp. 296–333). New York, NY: Macmillan.

Brown, J. S., & Burton, R. R. (1978). Diagnostic models for procedural bugs in basic mathematical skills. *Cognitive Science*, *2*, 155–192.

Case, R. (1992). *The mind's staircase: Exploring the conceptual underpinnings of children's thought and knowledge*. Hillsdale, NJ: Lawrence Erlbaum Associates.

Case, R. (1996a). Introduction: Reconceptualizing the nature of children's conceptual structures and their development in middle childhood. In R. Case & Y. Okamoto (Eds.), *Monographs of the Society for research in child development: Vol. 61 (1-2). The role of central conceptual structures in the development of children's thought* (pp. 1–26) [Serial No. 246].

Case, R. (1996b). Modeling the dynamic interplay between general and specific change in children's conceptual understanding. In R. Case & Y. Okamoto (Eds.), *Monographs of the Society for research in child development: Vol. 61 (1-2). The role of central conceptual structures in the development of children's thought* (pp. 156–188) [Serial No. 246].

Case, R. (1998). The role of central conceptual structures in the development of children's scientific and mathematical thought. In A. Demetriou, R. Shayer, & A. Elfklides (Eds.), *Neo-Piagetian theories of cognitive development: Implications and applications for education* (pp. 52–64). London: Routledge.

Case, R., & Okamoto, Y. (1996). The role of central conceptual structures in the development of children's thought. *Monographs of the Society for research in child development: Vol. 61 (1-2)*. [Serial No. 246].

Case, R., Okamoto, Y., Henderson, B., McKeough, A., & Bleiker, C. (1996). Exploring the macrostructure of children's central conceptual structures for number and narrative. In R. Case & Y. Okamoto (Eds.), *Monographs of the Society for research in child development: Vol. 61 (1-2). The role of central conceptual structures in the development of children's thought* (pp. 59–82) [Serial No. 246].

Choi, S., & Bowerman, M. (1991). Learning to express motion events in English and Korea: The influence of language-specific lexication patterns. *Cognition*, *41*, 83–121.

Choi, S., McDonough, L., Bowerman, M., & Mandler, J. (1999). Early sensitivity to language-specific spatial categories in English and Korean. *Cognitive Development*, *14*, 241–268. https://doi.org/10.1016/s0885-2014(99)00004-0.

Dowker, A., Bala, S., & Lloyd, D. (2008). Linguistics influence of mathematical development: How important is the transparency of the counting system? *Philosophical Psychology*, *21*, 523–538.

Dowker, A., & Roberts, M. (2015). Does the transparency of the counting system affect children's numerical abilities? *Frontiers in Psychology*, *6*, 945. https://doi.org/10.3389/fpsyg.2015.00945.

Gelman, R. (1978). Counting in the preschoolers: What does and does not develop. In R. Siegler (Ed.), *Children's thinking: What develops?* New Jersey: Lawrence Erlbaum Associates.

Gelman, R., & Gallistel, C. R. (1978). *The child's understanding of number*. Cambridge, MA: Harvard University Press.

Hauser, M. D., & Carey, S. (2003). Spontaneous representations of small numbers of objects by rhesus macaques: Examinations of content and format. *Cognitive Psychology*, *47*, 367–401.

Helmreich, I., Zuber, J., Pixner, S., Kaufmann, L., Nuerk, H.-C., & Moeller, K. (2011). Language effects on children's mental number line: How cross-cultural differences in number word systems affect spatial mappings of numbers in a non-verbal task. *Journal of Cross Cultural Psychology*, *42*, 598–613.

Ho, C. S., & Cheng, F. S. (1997). Training in place-value concepts improves children's addition skills. *Contemporary Educational psychology*, *22*, 495–506.

Ho, C. S., & Fuson, K. C. (1998). Children's knowledge of teen quantities as tens and ones: Comparisons of Chinese, British, and American kindergartners. *Journal of Educational Psychology*, *90*, 536–544.

Husén, T. (Ed.), (1967). *International study of achievement in mathematics: A comparison of twelve countries*. Stockholm: Almqvist & Wiksell. Vols. 1–2.

Kieren, T. E. (1993). Rational and fractional number: From quotient fields to recursive understanding. In T. P. Carpenter, E. Fennema, & T. A. Romberg (Eds.), *Rational numbers: An integration of research* (pp. 261–288). Hillsdale, NJ: Erlbaum.

Ma, L. (1999). *Knowing and teaching elementary mathematics: Teachers' understanding of fundamental mathematics in China and the United State*. Mahwah, NJ: Lawrence Erlbaum Associates.

Mack, N. K. (1990). Learning fractions with understanding: Building on informal knowledge. *Journal for Research in Mathematics Education*, *21*, 16–32.

Miura, I. T. (1987). Mathematics achievement as a function of language. *Journal of Educational Psychology*, *79*, 79–82.

Miura, I. T., Kim, C. C., Chang, C.-M., & Okamoto, Y. (1988). Effects of language characteristics on children's cognitive representation of number: Cross-national comparisons. *Child Development*, *59*, 1445–1450.

Miura, I. T., & Okamoto, Y. (1989). Comparisons of U.S. and Japanese first graders' cognitive representation of number and understanding of place value. *Journal of Educational Psychology*, *81*, 109–113.

Miura, I. T., & Okamoto, Y. (1999). Counting in Chinese, Japanese, and Korean: Support for number understanding. In C. A. Edwards (Ed.), *Perspectives on Asian Americans and Pacific Islanders* (pp. 29-36) of *Changing the Faces of Mathematics*. In W. G. Secada, (Series Ed.). Reston, VA: National Council of Teachers of Mathematics.

Miura, I. T., & Okamoto, Y. (2003). Language supports for mathematics understanding and performance. In A. J. Baroody & A. Dowker (Eds.), *The development of arithmetic concepts and skills: The construction of adaptive expertise* (pp. 229–242). Mahwah, NJ: Lawrence Erlbaum Associates.

Miura, I. T., Okamoto, Y., Kim, C. C., Steere, M., & Fayol, M. (1993). First graders' cognitive representation of number and understanding of place value: Cross-national comparisons—France, Japan, Korea, Sweden, and the United States. *Journal of Educational Psychology*, *85*(1), 24–30.

Miura, I. T., Okamoto, Y., Vlahovic-Stetic, V., Kim, C. C., & Han, J. H. (1999). Language supports for children's intuitive understanding of numerical fractions: Cross-national comparisons. *Journal of Experimental Child Psychology*, *74*, 356–365.

Moeller, K., Pixner, S., Zuber, J., Kaufmann, L., & Nuerk, H. C. (2011). Early place-value understanding as a precursor for later arithmetic performance—A longitudinal study on numerical development. *Research in Developmental Disabilities*, *32*(5), 1837–1851.

Moseley, B., & Okamoto, Y. (2008). Identifying fourth graders' linking of rational number representation: A mixed method approach. *School Science and Mathematics Journal*, *108*, 238–250.

Moseley, B., Okamoto, Y., & Ishida, J. (2007). Comparing U.S. and Japanese elementary school teachers' facility for linking rational number representations. *International Journal of Science and Mathematics Education*, *5*(1), 165–185.

Moss, J., & Case, R. (1999). Developing children's understanding of the rational numbers: A new model and an experimental curriculum. *Journal for Research in Mathematics Education*, *30*, 122–147.

Mullis, I. V. S., Martin, M. O., & Foy, P. (with Olson, J. F., Preuschoff, C., Erberber, E., Arora, A., & Galia, J.). (2008). TIMSS 2007 international mathematics report: Findings from IEA's trends in international mathematics and science study at the fourth and eighth grades. Chestnut Hill, MA: TIMSS & PIRLS International Study Center, Boston College. Retrieved from http://timss.bc.edu/TIMSS2007/mathreport.html.

Mullis, I. V. S., Martin, M. O., Beaton, A. E., Gonzalez, E. J., Kelly, D. L., & Smith, T. A. (1998). *Mathematics achievement in the primary school years: IEA's Third International Mathematics and Science Study (TIMSS)*. Chestnut Hill, MA: TIMSS International Study Center, Boston College.

Mullis, I. V. S., Martin, M. O., Foy, P., & Arora, A. (2012). *TIMSS 2011 international results in mathematics*. Chestnut Hill, MA: TIMSS & PIRLS International Study Center, Boston College Retrieved from: http://timss.bc.edu/timss2011/international-results-mathematics.html.

Mullis, I. V. S., Martin, M. O., Gonzalez, E. J., & Chrostowski, S. J. (2005). *TIMSS 2003 international mathematics report: Findings from IEA's Trends in International Mathematics and Science Study at the Fourth and Eighth Grades*. Chestnut Hill, MA: TIMSS and PIRLS International Study Center, Boston College Retrieved from: http://timss.bc.edu/timss2003i/scienceD.html.

Mullis, I. V. S., Martin, M. O., Gonzalez, E. J., Gregory, K. D., Garden, R. A., O'Connor, K. M., … Smith, T. A. (2000). *TIMSS 1999 international mathematics report: Findings from IEA's repeat of the third international mathematics and science study at the eighth grade*. Chestnut Hill, MA: TIMSS and PIRLS International Study Center, Boston College. Retrieved from: http://timss.bc.edu/timmathbench_report/mathbench_report/t99b_math_report.html.

National Council of Teachers of Mathematics. (2000). *Principles and standards for school mathematics*. Reston, VA: National Council of Teachers of Mathematics.

Ni, Y., & Zhou, Y. D. (2005). Teaching and learning fraction and rational numbers: The origins and implications of whole number bias. *Educational Psychologist, 40*, 27–52.

Okamoto, Y. (2010). Children's developing understanding of number: Mind, brain, and culture. In M. Ferrari & L. Vuletic (Eds.), *Developmental relations among mind, brain, and education: Essays in honor of Robbie Case* (pp. 129–148). New York, NY: Springer.

Okamoto, Y. (2015). Mathematics learning in the USA and East Asia: Influences of language. In R. Kadosh & A. Dowker (Eds.), *Oxford handbook of numerical cognition* (pp. 415–429). Oxford: Oxford University Press.

Okamoto, Y., Case, R., Bleiker, C., & Henderson, B. (1996). Cross cultural investigations. In R. Case & Y. Okamoto (Eds.), *Monographs of the Society for research in child development: Vol. 61 (1-2). The role of central conceptual structures in the development of children's thought* (pp. 131–155) [Serial No. 246].

Okamoto, Y., Curtis, R., Chen, P.-C., Kim, S., & Karayan, S. (1997). *A cross-national comparison of children's intuitive number sense. Its development and relation to formal mathematics learning: Final report* [submitted to Spencer Foundation].

Paik, J. H., & Mix, K. S. (2003). U.S. and Korean children's comprehension of fraction names: A reexamination of cross-national differences. *Child Development, 74*, 144–154.

Pica, P., Lemer, C., Izard, V., & Dehaene, S. (2004). Exact and approximate arithmetic in an Amazonian indigene group. *Science, 306*, 499–503.

Princiotta, D., Flanagan, K. D., & Germino Hausken, E. (2006). *Fifth grade: Findings from the fifth-grade follow-up of the early childhood longitudinal study, kindergarten class of 1998-99 (ECLS-K)*. [NCES 2006-038]. Washington, DC: National Center for Education Statistics, U.S. Department of Education.

Pruden, S. M., Levine, S. C., & Huttenlocher, J. (2011). Children's spatial thinking: Does talk about the spatial world matter? *Developmental Science*, *14*, 1417–1430. https://doi.org/10.1111/j.1467-7687.2011.01088.x.

Ramani, G. B., & Siegler, R. S. (2008). Promoting broad and stable improvements in low-income children's numerical knowledge through playing number board games. *Child Development*, *79*, 375–394. https://doi.org/10.1111/j.1467-8624.2007.01131.x.

Sarnecka, B. W., Kamenskaya, V. G., Yamana, Y., Ogura, T., & Yudovina, Y. B. (2007). From grammatical numbers to exact numbers: Early meanings of 'one', 'two', and 'three' in English, Russian, and Japanese. *Cognitive Psychology*, *55*, 136–168.

Siegler, R. S., & Booth, J. L. (2004). Development of numerical estimation in young students. *Child Development*, *75*(2), 428–444.

Siegler, R. S., Carpenter, T., Fennell, F., Geary, D., Lewis, J., Okamoto, Y., … Wray, J. (2010). *Developing effective fractions instruction for kindergarten through 8th grade: A practice guide (NCEE #2010-4039).* Washington, DC: National Center for Education Evaluation and Regional Assistance, Institute of Education Sciences, U.S. Department of Education. Retrieved from: http://ies.ed.gov/ncee/wwc/PracticeGuide/15.

Siegler, R. S., Duncan, G. J., David-Kean, P. E., Duckworth, K., Claessens, A., Engel, M., … Chen, M. (2012). Early predictors of high school mathematics achievement. *Psychological Science*, *23*, 691–697.

Siegler, R. S., & Mu, Y. (2008). Chinese children excel on novel mathematics. *Psychological Science*, *19*, 759–763.

Siegler, R. S., Thompson, C. A., & Schneider, M. (2011). An integrated theory of whole number and fractions development. *Cognitive Psychology*, *62*, 273–296.

Spelke, E. S., & Kinzler, K. D. (2007). Core knowledge. *Developmental Science*, *10*, 89–96.

Starkey, P. (1992). The early development of numerical reasoning. *Cognition*, *43*, 93–126.

Stevenson, H. W., Lee, S.-Y., & Stigler, J. W. (1986). Mathematics achievement of Chinese, Japanese, and American children. *Science*, *231*, 693–699.

Stigler, J. W., Gonzales, P., Kawanaka, T., Knoll, S., & Serrano, A. (1999). *The TIMSS Videotape Classroom study: Methods and findings from an exploratory research project on eighth-grade mathematics instruction in Germany, Japan, and the United States.* Washington, DC: U.S. Department of Education.

Stigler, J. W., & Hiebert, J. (1999). *The teaching gap: Best ideas from the world's teachers for improving education in the classroom.* New York, NY: The Free Press.

Stigler, J. W., Lee, S.-Y., & Stevenson, H. W. (1987). Mathematics classrooms in Japan, Taiwan, and the United States. *Child Development*, *58*, 1272–1285.

Towse, J. N., Muldoon, K., & Simms, V. (2015). Figuring out children's number representations: Lessons from cross cultural work. In A. Dowker & R. C. Kadosh (Eds.), *Oxford Handbook of Numerical Cognition* (pp. 402–414). New York, NY: Oxford University Press.

Towse, J., & Saxton, M. (1997). Linguistic influences on children's number concepts: Methodological and theoretical considerations. *Journal of Experimental Child Psychology*, *66*, 362–375.

Vosniadou, S., Vamvakoussi, X., & Skopeliti, I. (2008). The framework theory approach to conceptual change. In S. Vosniadou (Ed.), *International handbook of research on conceptual change* (pp. 3–34). Mahwah, NJ: Erlbaum.

Wynn, K. (1992). Children's acquisition of number words and the counting system. *Cognitive Psychology*, *24*, 220–251.

Chapter 5

How Does the "Learning Gap" Open? A Cognitive Theory of Nation Effects on Mathematics Proficiency

John E. Opfer, Dan Kim, and Jike Qin
The Ohio State University, Columbus, OH, United States

The mathematics proficiency of citizens plays a large role in the economic growth of nations (Hanushek & Woessmann, 2008), and individual differences in math proficiency predict a wide range of personal economic and educational outcomes. Among these are likelihood of completing high school, college grades, college graduation, starting salary, and salary growth (Murnane, Willett, & Levy, 1995; National Mathematics Advisory Panel, 2008; Sadler & Tai, 2007). Indeed, the dividends of high math proficiency have increased over time, due largely to secular increases in the wage premium for college attainment (Tyler, 2004).

Mathematics proficiency also differs greatly among nations (Gonzales et al., 2008; Husén, 1967; OECD, 2010). In two of the most widely cited cross-national comparisons, the Program for International Student Assessment (PISA) and the Trends in International Mathematics and Science Study (TIMSS), the difference in math scores between the top-ranking nations and the bottom-ranking nations has been about three standard deviations. To grasp this difference intuitively, if men in nation A were three standard deviations taller than U.S. men (average height = 5 ft, 10 in.), then men in nation A would have an average height of 6 ft, 6 in. tall. Thus, the difference in math scores between top-ranking nations and bottom-ranking nations is like the difference between the average height of U.S. men and the average height of a professional basketball player. Of course, in economic terms, the effect of math proficiency on income is much greater than the effect of height (Mankiw & Weinzierl, 2010; Rose & Betts, 2004)—and thus a good reason to care about these very large national differences in math proficiency.

Mathematical Cognition and Learning, Vol. 4. https://doi.org/10.1016/B978-0-12-812574-8.00005-5

Although the top spots vary somewhat from year to year, there has been a very large and consistent difference in the mathematical performance of U.S. and East Asian students. The difference between U.S. and East Asian students came to the broad attention of scientists largely due to the efforts of Harold Stevenson and his colleagues, first in an article that appeared in *Science* (Stevenson, Lee, & Stigler, 1986) and more broadly publicized in his book *The Learning Gap* (Stevenson & Stigler, 1992). Not much has changed over this period of time. In the 25 years since the publication of *The Learning Gap*, East Asian students have routinely outperformed their American peers across a range of math assessments (Bailey et al., 2015; Stevenson, Chen, & Lee, 1993), with the magnitude of this "learning gap" typically appearing before the first year of formal schooling and growing larger with each successive year (Geary, Bow-Thomas, Liu, & Siegler, 1996).

In this chapter, we review two approaches for understanding these national differences in math abilities. The first approach attempts to link cross-national differences in educational input to cross-national differences in test scores. A tacit assumption of this approach is that children's performance on tests of mathematics is a direct effect of input. The second approach attempts to link cross-national differences in educational input to cross-national differences in test scores by means of indirect effects through more basic cognitive abilities. The assumption of this approach is that the specific educational inputs that differ among nations do not make much difference unless they have an effect on basic cognitive abilities that improve learning for all students. We argue that this approach has two related virtues. Because it is mechanistic, it greatly constrains the search for potential causes. Also, because it is mechanistic, it offers a clearer path to closing the learning gap. We then examine our theory in light of existing cross-national research on math-related cognitive abilities and preview results of a study (under review) that tested our idea.

THE TRADITIONAL APPROACH: DIFFERENCES IN INPUT, DIFFERENCES IN OUTPUT

Early work on cross-national differences in educational outcomes involved intensive analyses of how nations differ in their educational practices (Husén, 1967). Data on this topic came from a wide range of sources, including surveys and interviews with parents and teachers, systematic video recordings of classroom lessons, and experimental studies seeking to uncover the implicit attitudes of students living in different nations. Many potentially important national differences in educational input were uncovered in this early work, including cross-national differences in the quantity (Stevenson & Stigler, 1992) and quality of math instruction (Richland, Zur, & Holyoak, 2007; Stigler & Hiebert, 1999), quality of teacher's math knowledge (Ma, 1999), and cultural differences in math attitudes (Chen & Uttal, 1988; Stevenson & Stigler, 1992).

Quantity of Math Instruction

One of the most powerful predictors of academic achievement generally is "time on task," which is partly a function of the time that a student is available to learn (Carroll, 1989). This quantity of time differs greatly between East Asian and American schools, where the former has longer school days and more days in the school year. As a result, Chinese and Japanese children finish the sixth grade having spent between 1500 and 3000 more hours (equivalent to 1–2 more years) in school than their American peers (Stevenson & Stigler, 1992). This difference might not be as important as it seems when one considers that Chinese and Japanese children spend more time at recess and in lunch than American children. Alternatively, one must also recall that Chinese and Japanese children spend more time in teacher-directed activities, more time allocated for mathematics instruction, more time within math lessons involving teacher activity, and less time within math lessons involving "off-task" activities (Stevenson & Stigler, 1992). Thus, compared with their American counterparts, East Asian students spend more days of the year in school, more hours of the day in school, more time of the day in math class, and more time in math class encountering math instruction.

Quality of Math Instruction

All other things being equal, more time with math instruction would be expected to lead to greater math learning. Many studies, however, have shown that these potential gains may be compounded by higher quality math instruction in East Asian schools. Measuring instruction quality is a notoriously difficult task. However, experimental studies on the features of high-quality math lessons typically point to the importance of using real-world problems, connecting symbolic to concrete representations of quantity, using students' answers in discussion, and reflecting on students' errors for insight (National Research Council (NRC), 2005). These are also much more likely to be features of East Asian math classes than American ones according to observational studies in Sendai, Taipei, Beijing, and Chicago (Stevenson & Stigler, 1992; for recent reviews, see Geary, Berch, Ochsendorf, & Mann Koepke, 2017).

Another feature of high-quality instruction—one that we have used in our own studies teaching number line estimation (Thompson & Opfer, 2008)—involves providing cognitive support for connecting novel problems to familiar ones (Richland, Stigler, & Holyoak, 2012; Richland et al., 2007). Despite the potential of these cognitive supports to help young children, their use is by no means ubiquitous in the mathematics classroom—at least not in American classrooms. Support for this conclusion comes from a secondary analysis performed by Richland et al. (2007) on the video portion of the Trends in International Mathematics and Science Study (Stigler, Gonzales, Kwanaka, Knoll, & Serrano, 1999), a large-scale international comparison of mathematics instruction.

To study how often teachers provided cognitive support for analogies between familiar and novel problems, Richland et al. (2007) analyzed how analogies were used in mathematics lessons in the United States and two regions in Asia—Hong Kong and Japan—where students consistently outperform U.S. students in mathematics achievement and where the gap between high and low achievers is substantially smaller. Richland et al. specifically coded use of six cognitive supports for analogy that structure-mapping theory had identified as potentially helpful: (1) using a familiar source analog to compare with the target analog being taught, (2) presenting the source analog visually, (3) keeping the source visible to learners during comparison with the target, (4) using spatial cues to highlight the alignment between corresponding elements of the source and target (e.g., diagramming a scale below the equal sign of an equation), (5) using hand or arm gestures that signaled an intended comparison (e.g., pointing back and forth between a scale and an equation), and (6) using mental imagery or visualizations (e.g., "picture a scale when you balance an equation"). For every one of these six cognitive supports coded, teachers in the U.S. sample were less likely to promote relational learning than were teachers in Hong Kong and Japan. Thus, in addition to providing more math instruction, East Asian teachers also appear to provide better math instruction.

Quality of Teacher Knowledge

Teachers' ability to provide high-quality math instruction partly depends on their own fluency with mathematical content, including their ability to envision real-world math problems, to link the elements of the problem to mathematical language and to discuss the mathematical ideas implied by their students' errors. Unfortunately, American elementary school teachers struggle with mathematical concepts much more than their East Asian peers (e.g., Lo & Luo, 2012; Ma, 1999). For example, in one cross-cultural study of American and Chinese early mathematics teachers, nearly all of the 23 American teachers who were interviewed struggled to accurately solve and generate word problems about fraction division (Ma, 1999). In contrast, not only all of the Chinese teachers could accurately solve fraction division problems, but also they were able to describe multiple ways of conceptualizing fraction division and multiple strategies for approaching this topic with their students.

Cultural Differences in Math Attitudes

A final difference in educational input concerns students' willingness to spend the time required to learn math, that is, to expend effort. Stevenson and Stigler (1992) have documented many ways that attitudes toward effort differ greatly between American and East Asian parents, teachers, and students. American mothers were found to rate the importance of effort for academic achievement as lower than did their East Asian counterparts, and American mothers rated the importance of ability much higher than East Asian mothers. Also, American students were found to rate the importance of ability as higher than did Chinese students. Finally, when given a difficult math problem, American students spent much less time trying to solve it than their East Asian counterparts.

Beyond Habits of Highly Successful Educational Systems

The traditional approach of documenting differences in educational input has an obvious and intuitive appeal; for children in a given nation to be doing better at mathematics, those children must have been exposed to something in their environment that would lead to better outcomes. Thus, the scientist's job is simply to find what those environmental factors are.

At first glance, the only alternative approach to the traditional one would posit that children in different nations have different intrinsic math ability (e.g., Lynn, 1982; Rushton, 1992). Such an essentialist approach, however, does not appear to be supported by the evidence. First, the "learning gap" seems to be a recent phenomenon. That is, although contemporary Chinese students outperform American students in arithmetic, elderly Chinese individuals perform no better than elderly Americans (Geary et al., 1996). This would not be expected if Chinese students were intrinsically better at math than Americans. Yet, another reason to reject the essentialist approach is that within-nation differences in many basic cognitive abilities—such as IQ, perceptual speed, working memory, and spatial skill—are as large or larger than between-nation differences, at least when good translations of the tests of these abilities are used (Geary et al., 1996; Stevenson et al., 1985). Thus, with the essentialist approach having been disconfirmed, the traditional approach seems rather sensible.

The traditional approach, however, has two major problems. The first is that—in the absence of a good theory of how math proficiency develops—it is tempting to consider any and every national difference in child socialization as potentially causing national differences in math proficiency. The problem is not simply a matter of inferring causation from correlations (often a reasonable source for our hypotheses). Rather, the problem is chasing down one dependent variable (math scores) with too many independent variables, which is bound to lead to some spurious correlations and wrong-headed educational reforms.

Another limitation of the traditional approach is that it is nonmechanistic. That is, even if national differences in the quantity or quality of math instruction do reliably lead to national differences in math ability, we would want to know *how*. Does increasing the quantity of math instruction lead directly or indirectly to greater math ability? Distinguishing between direct and indirect effects is important both theoretically and practically. The issue is important theoretically because it allows us to identify the mechanisms by which nations produce better math students. The issue is also important practically because it allows us to identify novel targets for interventions.

To illustrate, consider a problem with many parallels to the "learning gap" that Stevenson identified: the gap between children of high and low socioeconomic status. Like the difference between East Asian and American children, children in low-income families achieve lower math scores than peers in higher-income families (Case, 1975; Jordan, Huttenlocher, & Levine, 1992; Saxe, Guberman, & Gearhart, 1989), despite going to schools with a similar number of days in the school year, a similar number of hours in the school day, a similar cultural beliefs about effort and ability, and so on. Intriguingly, children from low-income families were also found to lag considerably behind their middle-income peers in their knowledge of fundamental aspects of number itself, thereby impairing their ability to compare the size of Arabic numbers, to order a group of four numbers from smallest to largest, or to determine which of two numbers was closer to a target number (Case & Griffin, 1990; Case & Sandieson, 1987; Griffin, Case, & Sandieson, 1992). Put another way, the performance of low-income children suggested they lacked a "mental number line" that associated ordered numbers with a linearly increasing series of magnitudes. Hypothesizing that this "central conceptual structure" might explain much of the learning gap, Griffin, Case, and Siegler (1994) developed a program to improve low-income children's numerical magnitude judgments, such as having them play games where they had to locate numbers on a number line. The results were quite dramatic, with the children in the experimental training group catching up to their middle-income peers in tests of arithmetic ability. Thus, rather than simply documenting all the ways that children from low- and high-income families differ, the investigators made progress in closing the learning gap by identifying a crucial *cognitive* difference between the groups and targeting it for intervention.

A COGNITIVE THEORY ON HOW THE LEARNING GAP OPENS

Our theory is that national differences in early math proficiency emerge for the same reason that socioeconomic differences lead to differences in math proficiency. That is, what mediates the effect of socioeconomic and national differences on early math proficiency is the same cognitive feature—the quality of numerical magnitude judgments, especially symbolic numerical magnitude judgments.

As we will argue, numerical magnitude judgments are important for mathematical thinking at any given point in time—a point that is consistent with

Siegler, Thompson, and Schneider's (2011) theory about fraction learning and recent work on math achievement in at-risk preschoolers (Geary & vanMarle, 2016). Additionally, numerical magnitude judgments help to foster *learning* from mathematical instruction (Booth & Siegler, 2008; Siegler & Ramani, 2008). In this way, high-quality numerical magnitude judgments can be thought of as a kind of malleable math aptitude. As this aptitude rises, the amount of time that a student needs in math instruction to reach a given level of competence will decline. If true, this theory is important because it suggests that East Asian students' math learning is not likely to be matched simply by increasing American students' time in math class, the quality of their math instruction, or their willingness to spend time on mathematics. At the same time, it suggests that the learning gap between East Asian and American children may be closed—perhaps substantially—by boosting American students' numerical magnitude judgments.

Why Numerical Magnitude Judgments?

Math proficiency—and quantitative reasoning more broadly—entails the ability to encode magnitudes, and the development of math proficiency builds and expands on the kinds of magnitudes that can be accurately encoded at any given age (Butterworth, 2005; Dehaene, 1997). What kinds of information-processing capacities must develop to encode these relations among magnitudes? Two kinds—ones that correlate with culturally invariant vs culturally variant abilities—are processing of nonsymbolic and symbolic numeric magnitudes, respectively. Proficiency at processing nonsymbolic numeric magnitudes (or "number sense acuity") (Halberda, Mazzocco, & Feigenson, 2008) involves facility with the ability to represent and compare nonsymbolic quantities accurately, such as comparing the relative quantities of two collections of items. Proficiency at processing symbolic numeric magnitudes (or "numeracy") (Hofstadter, 1982) involves the ability to accurately translate between magnitudes and symbolic numbers (e.g., "9," "nine," "1/9") and to encode quantitative relations among the numbers (e.g., 10 is closer to 1 than 100; 1/10 is closer to 1/100 than 1).

The number acuity/numeracy distinction provides a useful way to think about the varieties of numerical magnitude judgments that might lead to national differences in math proficiency. Numbers were only invented in the last 10,000 years of human history; the human brain is unlikely to have evolved specifically to handle numbers (Geary, 1995). Instead, numbers must be learned by modifying the brain through socialization and other interactions with the environment. As we will see, number acuity and numeracy differ in many ways that are consistent with this theoretical perspective. Number acuity emerges in infancy; numeracy develops much later. Number acuity is approximate and ratio-dependent; numeracy allows precision and encoding of mathematically important (linear) quantitative relations. Finally, numeracy varies widely among children growing up in different nations, whereas number acuity shows less variation and (where it exists) may be diagnostic of neurocognitive disorders (Piazza et al., 2010). Therefore, just as the distinction between processing of speech sounds (/kæt/) and signs (cat) is important for reading, the distinction between number acuity and numeracy is likely to be an important one for arithmetic generally (Vanbinst, Ansari, Ghesquière, & De Smedt, 2016) and in cross-cultural studies of arithmetic in particular.

Varieties of Numerical Magnitude Judgments

In the following section, we review findings on three major types of numerical magnitude judgments: comparing sets and numbers, estimating the location of a set or number on a number line, and approximately adding two sets or numbers (Fig. 1).

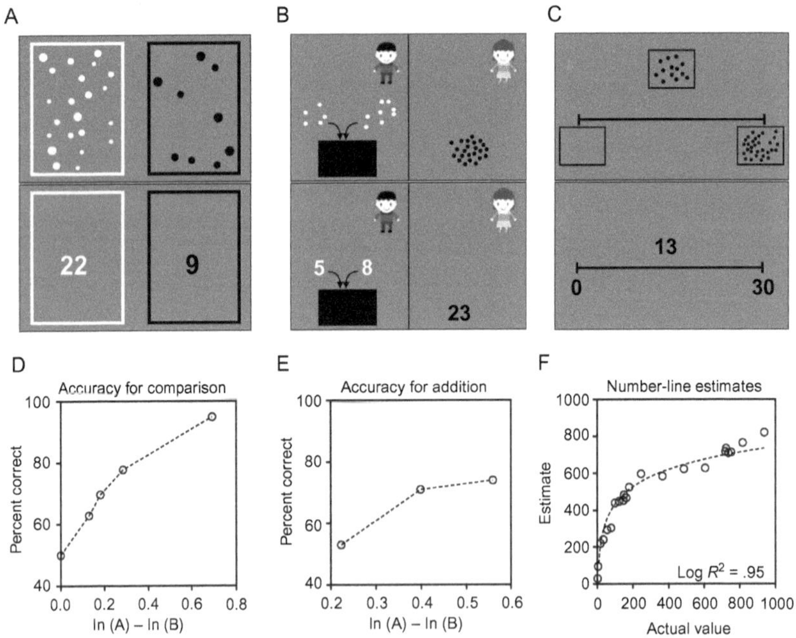

FIG. 1 Illustration of six numerical magnitude tasks and evidence for compressive representation of numbers. (A) Nonsymbolic and symbolic number comparison. (B) Nonsymbolic and symbolic approximate addition. (C) Nonsymbolic and symbolic number line estimation. (D) Ratio effect in comparison (Halberda et al., 2008) (A > B). (E) Ratio effect in approximate addition (Barth, La Mont, Lipton, & Spelke, 2005) (A > B, on half of trials A was the sum of two addends). (F) Logarithmic responses in number line estimates (Opfer & Siegler, 2007).

Comparing Sets and Numbers

A breakthrough in research on the development of mathematical abilities came from developmental psychologists' discovery that human infants notice changes in set sizes, typically sets of 2 vs 3 objects (Starkey & Cooper, 1980; Strauss & Curtis, 1981) and prefer to look at the larger set when the two sets differ numerically, for example, 128 vs 32 elements (Fantz & Fagan, 1975), 32 vs 16 (Xu, Spelke, & Goddard, 2005), or 16 vs 8 (Lipton & Spelke, 2003). In a typical study, investigators repeatedly presented babies with N_1 objects until looking times decreased and then presented N_2 objects. Infants' recovery of attention (suggesting they noticed a difference) to N_2 was evident only when the ratio between N_1 and N_2 was 3:2 or 2:1 (2 vs 3, Strauss & Curtis, 1981; 8 vs 4, Xu, 2003), a finding that appeared even in 21- to 144-hour-old neonates (Antell & Keating, 1983).

Research on infants' and children's ability to compare sets led to three broad generalizations. First, at any given age, the number required to discriminate two sets of large numbers (i.e., 4 or more) does not depend on an absolute difference in number. Rather, as in the Weber-Fechner psychophysical function, *the probability of discrimination is proportional to the difference in the logarithms of the numbers*, where $\ln(32)-\ln(16)=\ln(16)-\ln(8)=\ln(8)-\ln(4)>\ln(12)-\ln(8)=\ln(6)-\ln(4)$. (The term "ln" refers to the natural logarithm.) Second, the difference in logarithms required to discriminate numbers decreases with age. Thus, older infants were found to discriminate small ratios (e.g., 4:6 or 8:10) that younger infants could not (Cordes & Brannon, 2008). Third, for very small numbers (i.e., 3 or less), the probability of discrimination was uniformly high, higher than would be predicted from considering the differences in logarithms alone. Thus, discriminating 2 vs 3 is easier for infants than discriminating 4 vs 6. These three generalizations also hold for a wide range of animals' ability to compare sets (honeybees, Dacke & Srinivasan, 2008; fish, Piffer, Petrazzini, & Agrillo, 2013; pigeons, Scarf, Hayne, & Colombo, 2011; rats, Mechner, 1958; and monkeys, Brannon & Terrace, 2000) making it likely that number acuity is the product of evolution and not schooling.

The time required for older children and adults to compare sets parallels findings with infants. When prevented from counting and asked to select the larger of two sets, the time required to select the larger set is generally proportionate to the difference in logarithms, with the time required to select the larger of 4–6 being less than 6–7 (Birnbaum, 1980; Buckley & Gillman, 1974; Temple & Posner, 1998). Also, the difference in logarithms necessary for discriminating sets decreases with age. That is, a small difference in set size takes longer for younger children to discriminate than it does for older children and longer for older children than for adults (Temple & Posner, 1998). To quantify this development, a useful formalism is the "internal Weber fraction" (Halberda & Feigenson, 2008; Piazza & Izard, 2009;

Piazza, Izard, Pinel, Le Bihan, & Dehaene, 2004). For example, both schooled and unschooled adults can reliably discriminate numbers that are only within 11%–15% of each other (Halberda & Feigenson, 2008; Halberda et al., 2008; Piazza et al., 2010), which would be measured as an internal Weber fraction of 0.11–0.15. Regardless of cultural experiences, this internal Weber fraction changes greatly from infancy to adulthood (Piazza & Izard, 2009; Pica, Lemer, Izard, & Dehaene, 2004), starting from an average of 0.34 for kindergarteners, down to 0.25 for 10-year olds, and 0.15 for adults. Finally, the speed of older children's and adults' discrimination of very small numbers of objects is consistently greater than would be expected from considering the ratio in isolation. For example, the time required to judge the greater of two sets of 3 or fewer objects is uniformly low, much lower than would be expected given their ratios alone (Chi & Klahr, 1975; Revkin, Piazza, Izard, Cohen, & Dehaene, 2008).

Unlike the early ability to compare sets of objects, the ability to compare numbers develops much later. Young children—and some adults—who reliably notice the difference between collections of 3 and 4 objects are hopelessly innumerate when asked to compare the magnitudes of numerals like "3" and "4," "30%" and "40%," or "3/10" and "2/5." Unlike the ability to compare sets of objects, these skills must be learned and show substantial variability. For example, 3-year olds from middle-income backgrounds or 4- and 5-year olds from ow-income backgrounds perform near-chance levels when asked to compare numbers between 1 and 9 (Ramani & Siegler, 2008). By kindergarten, most (but not all) children are generally accurate at comparing numerals, but their accuracy and solution times—like those of older children and adults—are subject to the same Weber-Fechner law that characterizes comparisons of nonsymbolic quantities (Dehaene, 1989; Laski & Siegler, 2007; Moyer & Landauer, 1967; Sekuler & Mierkiewicz, 1977).

The parallels between the ability of babies and other animals to compare sets and the ability of children and other humans to compare numbers suggests that once the meaning of numeric symbols are learned, numeric symbols automatically activates the brain's system for representing *magnitude*, thereby leading to the Weber-Fechner pattern in response times. In support of this idea, presentations of either sets or numerals evoke number-related activation in the intraparietal cortex of educated adults (Naccache & Dehaene, 2001). Also, after adaptation with a set of 17, 18, or 19 dots, blood oxygenation recovery is observed when the number 50 is presented, but not when the number 20 is presented. These results suggest that—at least in educated human adults—populations of neurons in parietal and prefrontal cortex are jointly activated by nonsymbolic and symbolic quantities (Naccache & Dehaene, 2001). This is a powerful result in that it provides a link between the mechanism of magnitude representation and the symbols that evoke that mechanism.

It also illustrates why all human children may possess some mathematical aptitude, but why culture is needed to fully develop that aptitude.

Estimating the Placement of Sets and Numbers on a Number Line

A still more direct method of assessing numerical magnitude judgments asks children to estimate the placement of a number on a number line (Siegler & Opfer, 2003). Like comparisons of sets and numbers, children's early number line estimates follow the Weber-Fechner law, with children—like babies, rats, and pigeons comparing sets—estimating the differences among small numbers to be greater than the differences among large numbers (as on a logarithmic ruler). With education, however, children's estimates typically follow a more linear function.

This "logarithmic-to-linear shift" was first observed on a number line task in which a symbolic number was estimated on a line flanked by another symbolic number at each end and with no intermediate anchors provided. On this task, young children's placement of numbers typically follows an approximately logarithmic function, but this logarithmic pattern changes to a linear one with age and experience, with the timing of the shift coinciding with the magnitude of the number ranges tested and children's experiences with numbers in school (Booth & Siegler, 2006; Kim & Opfer, in press; Laski & Siegler, 2007; Opfer & Martens, 2012; Opfer & Siegler, 2007; Opfer, Siegler, & Young, 2011; Opfer & Thompson, 2008; Siegler & Booth, 2004; Siegler & Opfer, 2003; Siegler, Thompson, & Opfer, 2009; Thompson & Opfer, 2008). The logarithmic-to-linear shift has since been observed across a wide range of other number line tasks, including ones in which a midpoint anchor is provided for children (Booth & Siegler, 2006; Opfer, Thompson, & Kim, 2016; Siegler & Booth, 2004), one in which there is no upper boundary marked (Kim & Opfer, in press), and ones in which the numbers are replaced by sets of dots (Opfer, Qin, & Kim, submitted for publication).

An interesting feature of number line estimation is that the magnitude of the logarithmic-to-linear shift is much greater when subjects are asked to estimate the placement of numbers instead of sets of dots. On this nonsymbolic version of the task, estimates of nonschooled adults and children are more logarithmic than would be expected from their age alone (Dehaene, Izard, Spelke, & Pica, 2008; Sasanguie, De Smedt, Defever, & Reynvoet, 2012; Sella, Berteletti, Lucangeli, & Zorzi, 2015), and the estimates of even educated adults show a substantial degree of logarithmicity (Anobile, Cicchini, & Burr, 2012; Dehaene et al., 2008). These findings suggest that providing linear number line estimates—where the distance between any two successive estimates (e.g., 5 and 6 vs 155 and 156) is equal—depends on schooling and culture, where children gain proficiency with symbolic quantities.

Approximately Adding Sets and Numbers

A mental number line makes basic addition and subtraction trivial: traveling four spaces forward from four registers the sum of four and four, traveling four spaces back from eight registers the difference between eight and four, and so on. Thus, if infants encode the approximate numerical value of a set and possess something akin to a mental number line (for at least nonsymbolic numbers), they should be able to register sums and differences of numeric quantities (at least approximately).

The hypothesis that infants could approximately add sets led Wynn (1992) to conduct a series of surprising experiments on infants' arithmetic abilities. In one condition, Wynn tested babies' ability to compute $1+1$: she recorded infants' looking times as they watched one object appear to be placed behind an opaque screen and then another object added to it behind the screen. When the screen dropped, seeming to reveal the arithmetically impossible event $1+1=1$, infants looked longer than when the screen dropped and revealed an outcome consistent with the arithmetically realistic event $1+1=2$. Additionally, infants' ability to approximately add sets is not limited to numbers less than 4 (McCrink & Wynn, 2004). Consistent with Wynn's interpretation that infants were surprised by the numerical outcome, when 9-month-old babies were confronted with arithmetic transformations of large sets (e.g., $5+5=10$ vs $5+5=5$ or $10-5=5$ vs $10-5=10$), babies also looked longer at the arithmetically impossible events than the arithmetically possible ones. Thus, it seems that nonsymbolic, approximate arithmetic may not require any schooling at all.

How are babies able to add and subtract sets of objects (at least approximately)? One possibility is that nonsymbolic, approximate addition relies on the same logarithmically compressed number line representation that we reviewed above. In this view, adding $n2$ to $n1$ involves traveling $n2$ spaces forward from $n1$ on the mental number line. If this is right, then babies would actually arrive at the position $\log(n1)+\log(n2)$, which would be a considerable overestimation of the actual result. From this perspective, babies would have no doubt found $5+5=5$ surprising because they would have experienced it as $\log(5)+\log(5)$ and thus expected to see $\log(25)$! Conversely, subtraction through traversing a logarithmically spaced number line would yield a considerable underestimation of the actual results.

The idea that the approximate addition relies on a logarithmically spaced number line is consistent with the mistakes that babies make when adding sets. Specifically, babies' expectations do overestimate the results of additive operations and underestimate the results of subtractive operations (McCrink & Wynn, 2009). Thus, when babies were initially shown a sequence of events equivalent to $6+4$, they looked significantly longer when the raised screen revealed 5 objects than when it revealed 20 objects. Similarly, when babies were shown a sequence equivalent to $14-4$, they looked significantly longer when the raised screen revealed 20 objects than when it revealed 5. In the same study, the infants looked equally long at 20 objects as 10 at the end of the first example and looked as long at 5 objects as at 10 at the end of the second example.

Approximately adding sets of objects appears quite similar in infants, young children, and adults (Barth, Beckmann, & Spelke, 2008; Barth et al., 2006; McCrink, Dehaene, & Dehaene-Lambertz, 2007; Park, Bermudez, Roberts, & Brannon, 2016; Pica et al., 2004). In a representative study, McCrink et al. (2007) showed adults several hundred videos of two successive sets of dots and asked them to approximate their sum or difference by choosing one of two sets of dots. As with infants, adults almost always overshot the correct outcomes on addition problems, whereas they almost always undershot the correct outcomes on the subtraction problems. Researchers found that from adults' modal response, the distribution of other responses tapered off as a function of the ratio of the true and alternative quantities, just as would be predicted by Weber-Fechner's law. To illustrate the magnitude of this error, the presented subtraction problem of $32-16=8$ was judged to be correct approximately 60% of the time, which is quite a radical departure from moving along a linearly scaled mental number line.

Before formal arithmetic training, children's ability to add numbers is very similar to adults' ability to add sets of objects. One source of evidence comes from Gilmore, McCarthy, and Spelke's (2007) study of preschoolers' estimates of answers to arithmetic problems that they had not yet encountered in school. The investigators presented 5- and 6-year olds with problems such as "Sarah has 21 candies; she gets 30 more; John has 34 candies—who has more?" To ensure that the preschoolers understood the symbols that were being used, the problems were simultaneously presented both orally—as spoken numerals—and in writing, as Arabic numerals. Despite the fact that the preschoolers had received no training with numbers of that size, they spontaneously performed better than chance. Performance was still approximate, however, and depended on the ratio of the two sums that the children were choosing between, a signature of the Weber-Fechner law.

Connections Between Numerical Magnitude Judgments and Math Proficiency

In the previous section, we reviewed evidence that three types of numerical magnitude judgments—comparison, estimation, and approximate addition—bear striking psychophysical similarities and may rely on highly overlapping cortical resources. In this section, we examine the potential importance of this cognitive feature on math proficiency by reviewing the literature that examines relations between numerical magnitude judgments, math proficiency, and aptitude for math learning.

Comparing Sets and Numbers

Given that arithmetic proficiency is usually measured by participants' speed and accuracy with symbolic problems, it was a real surprise to many investigators to learn that individual differences in the ability to compare sets of objects (thereby tapping the ANS) could (retrospectively) predict individual differences in mathematics ability (Halberda et al., 2008). The finding that number acuity predicts math skill, however, has now been tested repeatedly in preschoolers and kindergartners (Bonny & Lourenco, 2013; Libertus, Feigenson, & Halberda, 2011, 2013), older children (Holloway & Ansari, 2009; Mundy & Gilmore, 2009), and adults (Castronovo & Göbel, 2012; Halberda, Ly, Wilmer, Naiman, & Germine, 2012; Lyons & Beilock, 2011). The breadth of these studies allowed Fazio, Bailey, Thompson, and Siegler (2014) to conduct a metaanalysis of 19 studies examining the relation between number acuity and math proficiency. Their general finding was that there is a small but reliable connection between number acuity and math proficiency, with the ability to compare sets accounting for about 4% of the variation in math scores. One way to make sense of these findings is that math ability partly develops as an elaborated, conceptual understanding of numerosity—the property shared by sets of *N* objects or *N* tones or *N* events (Butterworth, 2005), which is often impaired in a subpopulation of individuals with dyscalculia (Butterworth, 2005).

Children's ability to compare symbolic numbers also has a positive relation to their arithmetic proficiency (Castronovo & Göbel, 2012; De Smedt, Verschaffel, & Ghesquière, 2009; Fazio et al., 2014; Holloway & Ansari, 2009; Vanbinst, Ghesquière, & De Smedt, 2012), with the magnitude of the correlation between number comparison and math scores typically being much larger than the correlation between ability to compare sets and math scores (for a metaanalysis of 45 studies, see Schneider et al., 2016). For example, in the study by Fazio et al. (2014), the speed of number comparison explained 33% of the variation in standardized math scores (compared with 4% from their metaanalysis on number acuity and math proficiency). Additionally, individual differences in the magnitude of the numerical distance effect explain about 11% of the variation in math fluency (Holloway & Ansari, 2009). Thus, if Chinese children have greater facility with comparing numbers than U.S. children, one might expect them to also become more fluent in math and score better on standardized tests.

Estimating the Placement of Sets and Numbers on a Number Line

As with magnitude comparison, estimating the placement of sets and numbers on a number line shows a strong and consistent relation to math skills, with the magnitude of the relation again being larger for symbolic than nonsymbolic number line estimation. For example, in the Fazio et al. (2014) study, accuracy of nonsymbolic whole number line estimation accounted for about

10% of the variation in standardized math scores. In our own studies (Opfer, Qin, et al., submitted for publication), the logarithmicity of nonsymbolic number line estimation explains about the same proportion of variance in addition and subtraction scores after controlling for all other variables in the model (such as age) and experiment-wise error.

The relation between ability to place numbers on a number line and math proficiency is also a very consistent finding. Individual differences in the accuracy (Fazio et al., 2014; LeFevre et al., 2013; Lyons, Price, Vaessen, Blomert, & Ansari, 2014; Muldoon, Towse, Simms, Perra, & Menzies, 2013; Sasanguie, Göbel, Moll, Smets, & Reynvoet, 2013; Schneider, Grabner, & Paetsch, 2009; Siegler & Ramani, 2008, 2009) and logarithmicity of estimates of a number on a number line (Kim & Opfer, in press; Opfer, Qin, et al., submitted for publication) correlates positively and strongly with speeded arithmetic and standardized math scores. In studies that compare placement of numbers with placement of sets, the proportion of variance in math scores explained by symbolic number line estimates is typically around 30% (Fazio et al., 2014; Kim & Opfer, in press; Opfer, Qin, et al., submitted for publication) and roughly twice that of nonsymbolic number line estimation.

Probably, the best evidence that linearity of number line estimates improves math proficiency comes from experimental studies that have studied the effects of number line training (Booth & Siegler, 2008; Siegler & Ramani, 2009). These studies find that presenting randomly chosen school aged children with accurate number line representations of the magnitudes within arithmetic problems improves children's learning of the answers to the problems (Booth & Siegler, 2008). Similarly, playing a linear board game with randomly chosen preschoolers improves their ability to learn answers to arithmetic problems (Siegler & Ramani, 2009). Thus, rather just linking number line estimate accuracy to arithmetic performance, these studies suggest that accurate representations of numerical magnitude make learning math easier for children. This is an important finding because it suggests that Chinese children might gain more from their experiences in school than American children if they have more linear numerical magnitude representations (Fig. 2).

Cross-National Differences in Numerical Magnitude Judgments

Given the early emergence of competence in nonsymbolic numerical magnitude judgments and the relatively late emergence of competence in symbolic numerical magnitude judgments, it seems reasonable to expect a certain pattern to cross-national differences in numerical magnitude judgments. Specifically, because proficiency with symbols depends on experiences in early childhood, cross-national differences with symbolic numbers would be expected to be larger than cross-national differences in proficiency with nonsymbolic numbers. This hypothesis has been tested indirectly with Western

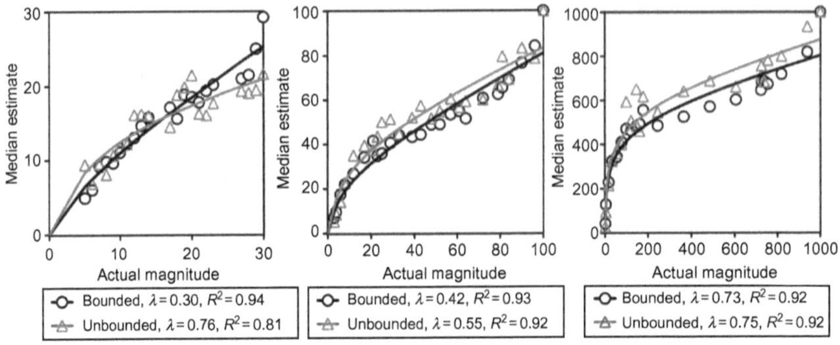

FIG. 2 Median estimates for American kindergartners' estimates on 0–30 number lines, first graders' estimates on 0–100 number lines, and second graders' estimations on 0–1000 number lines. On unbounded versions of the number line task, estimates were generally more logarithmic than on the bounded versions studied in Siegler and Opfer (2003).

and East Asian children in a number of studies (Rodic et al., 2015; Siegler & Mu, 2008; Xu, Chen, Pan, & Li, 2013; Zhou et al., 2007). These studies have examined cross-national differences in comparing sets and numbers and estimating the placement of sets and numbers on a number line. However, no studies have yet examined cross-national differences in approximate addition.

Comparing Sets and Numbers

When comparing two sets of dots for the greater quantity, Chinese children perform somewhat better than their peers in Western countries. For example, Rodic et al. (2015) tested ethnic and national Chinese children and their Western peers on a large test battery, including comparing sets of dots. Chinese children's judgments were reliably more accurate than children growing up in the United Kingdom, but the judgments were typically made just as quickly. This is important because the standard positive relation between nonsymbolic number comparison and math performance typically uses a composite measure of speed and accuracy, suggesting that this component of numerical magnitude judgments may not mediate the effect of nation on math proficiency.

When comparing two numbers for the greater quantity, Chinese children also performed better than their Western peers, and this difference was larger and better supported by a range of evidence than evidence regarding set size comparison (Rodic et al., 2015). More strikingly, Chinese children were found to automatically process symbolic magnitudes as early as 5 years old (Zhou et al., 2007). In their study, Zhou et al. (2007) asked children to estimate the *physical* size of numbers but ignore their magnitude information. In this Stroop-like task, 5-year-old Chinese children showed a significant size congruity effect, suggesting magnitude information of numbers interfered with their physical size comparison. This automatic processing of symbolic magnitude ability is striking because it emerges at a much older age for children in Western nations (Berch, Foley, Hill, & Ryan, 1999; Girelli, Lucangeli, & Butterworth, 2000; Rubinsten, Henik, Berger, & Shahar-Shalev, 2002).

Estimating the Placement of Sets and Numbers on a Number-Line

When estimating the placement of sets on a number line, Chinese kindergartners' estimates were also found to be more accurate and linear than their Western peers. However, both groups showed a similar developmental pattern (a logarithmic-to-linear shift), and both showed a linear pattern of estimates on 0–10 number lines and a logarithmic pattern on 0–100 number lines (Sasanguie et al., 2012; Sella et al., 2015; Xu et al., 2013). Given the small but reliable relation between nonsymbolic number line estimation and math proficiency, generally, these results should lead us to expect that this component of numerical magnitude judgments will mediate a small but reliable effect of nation on math proficiency.

When estimating the placement of numbers on number lines, a somewhat similar pattern emerges. Specifically, Chinese children have shown the same logarithmic-to-linear shift as their Western peers, but they typically show an earlier onset of attaining linear representations (Siegler & Mu, 2008; Xu et al., 2013). For example, Xu et al. (2013) found the estimates of Chinese preschoolers (3- to 7-year olds) were more accurate and linear on the 0–10, 0–100, and 0–1000 number lines than their Western peers. Moreover, the linearity indexes of estimates for Chinese *kindergartners* were similar to those for American *first and second graders* (Booth & Siegler, 2006; Opfer & Siegler, 2007; Siegler & Booth, 2004; Siegler & Opfer, 2003, Xu et al., 2013). Given the previously observed association between symbolic number line estimation and math proficiency and math learning, this is an important finding. It suggests that a given hour of math instruction in China might have a larger effect on children than the same hour of math instruction in the United States because children in the former group have a linear representation of numerical magnitude—and thus a stronger foundation for symbolic magnitude representations than their Western peers.

DO NUMERICAL MAGNITUDE JUDGMENTS MEDIATE NATION EFFECTS ON MATH PROFICIENCY?

In the previous sections, we have reviewed evidence that (1) East Asian children typically show greater math proficiency than their American peers, (2) numerical magnitude judgments (particularly symbolic ones) are generally correlated with greater math proficiency, and (3) East Asian children typically show more accurate numerical magnitude judgments (particularly symbolic ones) than their American peers. Together, these three lines of work suggest a simple model in which the effect of nation on math proficiency—especially early, but possibly all math proficiency (see below)—is mediated by the quality of numerical magnitude judgments (Fig. 3).

A test of this model has recently been conducted by Opfer, Kim, Siegler, Fazio, and Zhou (submitted for publication). Although these results have not been published yet, we can preview a narrative description of their findings. At present, these preliminary results provide the only test of our cognitive theory of nation effects on mathematics proficiency.

In their study, Opfer et al. examined kindergarteners recruited at the end of the school year from four schools in Beijing, China, and Columbus, OH. All children completed a mathematics assessment (Geary et al., 1996), a subtest of the WISC-III with a numeric response code (Stevenson et al., 1985), and six numeric magnitude judgment tasks. The numeric magnitude tasks varied in the format of numeric information (symbolic numerals vs dot arrays) and type of magnitude judgment (comparison, approximate addition, and number line estimation). Time for each task was roughly equivalent; each took less than 5 min.

Consistent with previous reports (Geary et al., 1996), Opfer et al. found that arithmetic scores of Chinese kindergartners were roughly three times higher than those of U.S. kindergartners. Each element of the arithmetic score differed by nationality: U.S. kindergartners provided fewer accurate answers than Chinese kindergartners, and U.S. children provided more wrong answers than Chinese children. Finally, when wrong answers were given, the difference between the child's answer and the correct one was significantly smaller among Chinese than U.S. children.

What might account for these differences in math proficiency? Opfer et al. first reexamined the essentialist hypothesis that East Asian kindergartners are simply smarter than American ones. To do this, they used a coding task (WISC-III; also Digit-Symbol Substitution task, WAIS-III) that is widely used to measure children's and adults' performance IQ (Ryan & Lopez, 2001; Wechsler, 1991). The task is thought to index the overall speed of information processing (Lezak, Howieson, Loring, Hannay, & Fischer, 2004; Stevenson

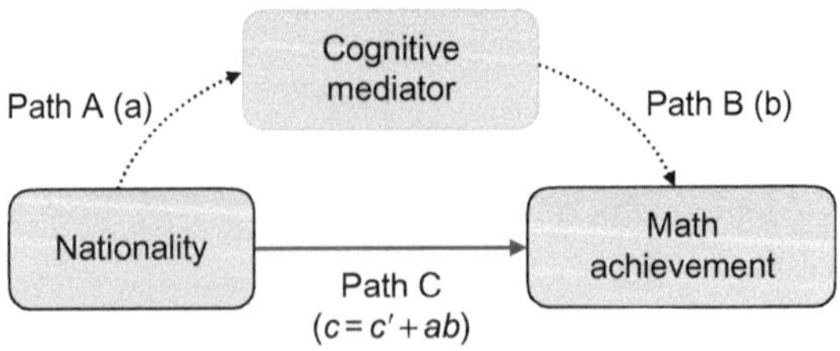

FIG. 3 Mediation model of the effects of nationality on math proficiency. The total effects of nationality (c) are divided into two effects: the direct effect (c') and the indirect effect via a cognitive mediator (ab).

et al., 1985) and correlates more highly with arithmetic performance than any other IQ subscale (Osborne & Lindsey, 1967). Additionally, because the test relies on a numeric response code, it provides a simple test of the idea that Chinese children are simply more familiar with numbers overall.

Consistent with Stevenson et al. (1985) findings with first graders (Geary et al., 1997; Stevenson et al., 1985), Opfer et al. found that average scores on this coding task did not differ between Chinese and U.S. kindergarteners, suggesting their samples were equivalent to previous studies and national norms. Additionally, variability in scores did not differ between the two groups, suggesting the samples were equivalently homogenous. Although IQ scores did correlate highly with mathematics scores, the lack of difference between the American and Chinese children in IQ meant that nationality could not be having an effect on math proficiency via IQ.

Comparing Sets and Numbers: U.S. Vs Chinese children

Opfer et al. next examined the effect of nationality on three measures of nonsymbolic and symbolic number comparison. These measures included overall accuracy (Lourenco, Bonny, Fernandez, & Rao, 2012; Mazzocco, Feigenson, & Halberda, 2011), Weber fraction (*w*; Halberda et al., 2008; Inglis, Attridge, Batchelor, & Gilmore, 2011), and numeric distance effect (Holloway & Ansari, 2009; Price, Palmer, Battista, & Ansari, 2012). Across all three indexes of nonsymbolic number comparison, a remarkably consistent pattern emerged. Overall accuracy scores, the average Weber fraction, and the numeric distance effect were similar for Chinese and U.S. children. As we have argued, comparing sets is likely to be a universal cognitive ability, and the similarity of Chinese and U.S. children should not be surprising. Also, like IQ, accuracy of nonsymbolic number comparison was somewhat correlated with mathematics scores, but the lack of difference between the Chinese and U.S. children's ability to compare sets meant that numerical magnitude judgments could not have mediated the effect of nation on mathematics proficiency.

Chinese and U.S. children's ability to compare numbers was examined similarly. Unlike set comparison, number comparison differed markedly between the U.S. and Chinese children. In comparing numbers, U.S. children provided less accurate answers and required a larger numeric difference to accurately discriminate two numbers. Additionally, because accuracy of symbolic number comparison was correlated with mathematics scores and because U.S. children provided less accurate symbolic number comparisons, symbolic number comparison could partially mediate the effect of nationality on arithmetic performance. Consistent with this conjecture, Opfer et al. found that nearly half the effect of nationality on arithmetic performance was an indirect effect that occurred via symbolic number comparison.

Estimating the Placement of Sets and Numbers on a Number Line: U.S. Vs Chinese Children

Children's ability to estimate the placement of sets and numbers on a number line was also evaluated in terms of overall accuracy and by assessing the degree of logarithmicity in the responses. Chinese children's estimates where a set of dots would fall on the number line were surprisingly more accurate than U.S. children's estimates. Additionally, the estimates of Chinese children were less logarithmic than estimates of U.S. children. Because erroneous nonsymbolic number line estimation was somewhat negatively correlated with mathematics scores and because U.S. children provided less accurate nonsymbolic number line estimates than Chinese children, nonsymbolic number line estimation would be expected to partially mediate the effect of nationality on arithmetic performance. Consistent with this conjecture, Opfer et al. found that almost half of the effect of nationality on arithmetic performance was an indirect effect that occurred via nonsymbolic number line estimation.

Chinese and U.S. children's ability to place numbers on number lines was evaluated and compared similarly (see also Chapter 4 by Okamoto for a related discussion of cross-cultural differences in number line estimation). Consistent with previous reports, accuracy of Chinese children's symbolic number line estimates was greater than accuracy of U.S. children's estimates, and the estimates of Chinese children were less logarithmic than the estimates of U.S. children (Siegler & Mu, 2008). Because erroneous symbolic number line estimation was negatively correlated with mathematics scores and because U.S. children were less accurate at symbolic number line estimation, it was expected that symbolic number line estimation ability might mediate the effect of nationality on arithmetic performance. Consistent with this conjecture, Opfer et al. found that nearly half of the effect of nationality on arithmetic performance was an indirect effect that occurred via symbolic number line estimation.

Approximately Adding Sets and Numbers: U.S. Vs Chinese Children

Children's ability to approximately add sets of dots and numbers (e.g., $a+b>$ or $<c$?) was evaluated in both the U.S. and Chinese samples. When approximately adding sets of dots, Chinese and U.S. children were remarkably similar in accuracy and the average Weber fraction. In both groups, accuracy improved as a function of the difference in the logarithms of the numbers compared, and the magnitude of this distance effect was similar as well. Thus, although accuracy of nonsymbolic approximate addition was somewhat correlated with mathematics scores, there was very little indirect effect of nation on math through this route.

Children's ability to approximately add numbers, somewhat unsurprisingly, did differ between Chinese and U.S. children. Overall accuracy scores were higher for Chinese than for U.S. children, and the average Weber fraction was also lower for Chinese than for U.S. children. Because accuracy of symbolic approximate addition was correlated with mathematics scores and because U.S. children were less accurate at symbolic approximate addition, symbolic approximate addition would be expected to partially mediate the effect of nationality on arithmetic performance. Consistent with this conjecture, Opfer et al. found that a little more than half the effect of nationality on arithmetic performance was an indirect effect that occurred via symbolic approximate addition.

Cognitive Mechanisms of Nation Effects on Mathematics Proficiency

In the previous three sections, independent mediation analyses indicated that the effect of nation on mathematics proficiency was partially mediated by a number of symbolic numerical magnitude judgments. These cognitive abilities, however, were largely correlated with one another. As a result, the total indirect effect of nation on mathematics via these cognitive abilities cannot be a sum of these partial effects (e.g., more than two different mediators accounted for roughly half the indirect effect, which is mathematically impossible). Additionally, each independent mediation analysis is associated with the risk of experiment-wise error. To deal with these issues, Opfer et al. conducted a multiple mediation analysis to measure how much the total effect of nationality on mathematics occurred via an indirect effect on numerical magnitude judgments and which of these were the mechanisms by which nation affected mathematics proficiency (Fig. 4). Overall, they found that only a very small proportion of the total effects of nation on arithmetic performance occurred directly via nationality itself, whereas almost all of the effect occurred through the numerical magnitude judgment mediators. Put another way, if the total effect of nationality on mathematics was the size of the "learning gap," almost all of this learning gap was reduced by accuracy of numerical magnitude judgments.

Although the multiple mediators accounted for almost all of the effects of nation on arithmetic scores, contribution of individual mediators to the mediated effects varied considerably. Generally, indirect effects of nationality through the three nonsymbolic numeric judgments were not significant, whereas the symbolic numerical judgments mediated effects of nationality significantly. In contrast, effects of nationality on arithmetic scores were significantly mediated by all symbolic numeric judgment tasks. These results suggest that the effects of nation on mathematics proficiency are mediated by symbolic numeric performance significantly more than by nonsymbolic numeric judgments. Additionally, of these putative mediators, only number line estimation showed a direct effect of nation controlling for differences in math achievement (Fig. 5).

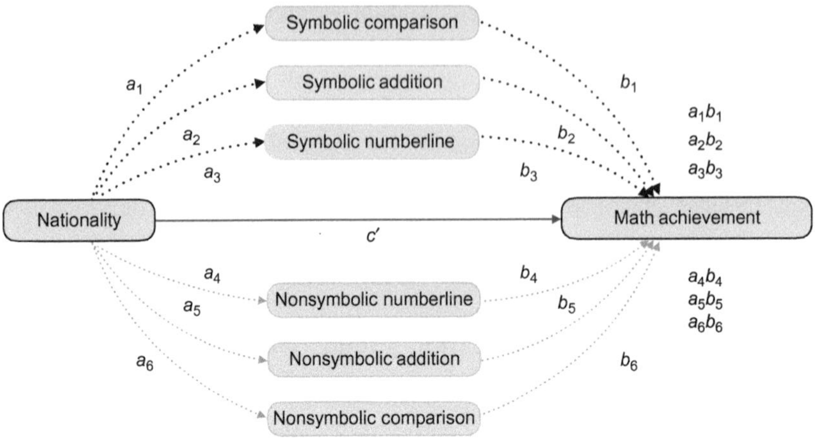

FIG. 4 Multiple mediation analysis. Nearly all of the total effects of nationality occurred via the six tasks (indirect effects, $a_1b_1+a_2b_2+a_3b_3+a_4b_4+a_5b_5+a_6b_6$), whereas nationality itself had no remaining proportion of the total effects (direct effect, c'). The indirect effects through individual symbolic number abilities were significant.

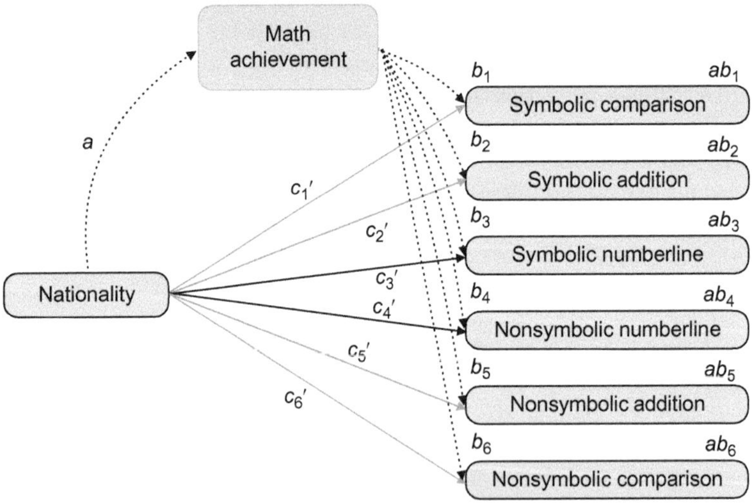

FIG. 5 Mediation analyses with math achievement as a mediator. Only number line estimation presented direct effects of nationality controlling for math achievement (c_3' and c_4').

CONCLUSIONS

For many decades, American children have lagged behind their East Asian peers in math proficiency. Efforts to explain this "learning gap" have traditionally looked for potential differences in input, and these efforts have not come up short: East Asian children have been found to have more opportunities to spend time on math, have math instructors that make their time better spent, and (maybe for these reasons) are more willing to spend time learning math. Emulating the East Asian model seems promising. However, reforms to lengthen the American school day or school year, to improve the quality of American teachers and instruction, and to try to convince American children that math is worthwhile have not narrowed the gap between East Asian and American students' math proficiency in the 25 years since the learning gap first came to broad attention.

In this chapter, we proposed a very different approach to the problem of the "learning gap." Specifically, we hypothesized that the cause of the learning gap may not be simply a matter of differences in schooling, but a *cognitive* difference in representations of numerical magnitude. This cognitive difference is superficially similar to IQ in that representations of numerical magnitude also increases learning from a given unit of input and emerges before formal schooling. Unlike IQ, however, these differences in numerical magnitude representations almost certainly come from differences in informal teaching at home. Thus, the learning gap may result from a kind of aptitude difference (i.e., a greater ability to learn) between East Asian and American children, but this rests on domain-specific knowledge—namely, representations of numeric magnitude—rather than a general cognitive ability, and these representations of numeric magnitude are known to be malleable (Opfer & Siegler, 2007; Opfer & Thompson, 2008; Ramani & Siegler, 2008; Siegler & Ramani, 2008; Thompson & Opfer, 2008).

High-quality numerical magnitude representations—assessed by children's ability to compare, estimate, and approximately add sets and numbers—generally correlate positively with math proficiency and have been shown to increase the ability of children to learn arithmetic. Additionally, evidence suggests that these numerical magnitude representations also differ between East Asian and U.S. children. These facts led us to hypothesize that American children with high-quality numerical magnitude representations might almost be as proficient at mathematics as East Asian children.

Our preliminary results from a study of Beijing and Columbus students provided much support for our cognitive theory. Specifically, we found that accurate numerical magnitude representations were generally associated with greater math proficiency, and we found that accuracy of numerical magnitude representations were generally higher among Chinese students than American ones. Mediation analyses also showed that nearly the entire effect of nation on math proficiency came via the indirect route of improving numerical magnitude representations. Thus, when American children had numerical magnitude representations that were equal to those of their Chinese peers, they were also just as proficient at mathematics—despite being in schools with American-length school years and school days,

American-trained teachers, and presumably the same American attitudes toward mathematics.

These results are broadly consistent with several lines of previous work. First, after controlling for Americans' ethnicity (including Asian and European heritage) and a broad array of general cognitive abilities (such as working memory), symbolic number knowledge in first grade has also been found to predict performance on economically relevant math skills in adolescence (Geary, Hoard, Nugent, & Bailey, 2013). This relation is not unique to early math learning: adolescents who show better symbolic number knowledge for fractions also do better in high school math classes, after controlling for a very wide range of other predictors (Siegler et al., 2012). Thus, rather than symbolic number knowledge simply providing a foundation to early math proficiency, it appears to play a supporting role for proficiency throughout development.

Results from our study of East Asian and American students closely parallel findings about socioeconomic differences in math ability. Like American students, children low in SES also show deficits in numerical magnitude representations and math proficiency. Moreover, experimental lessons that have intervened specifically on low SES children's numerical magnitude representations were also able to eliminate the differences between low and high SES children's math proficiency. Combined with our cross-national findings, previous efforts suggest that much of the learning gap between U.S. and American children could be closed if educational interventions focused on American children's relatively poor-quality numerical magnitude representations.

ACKNOWLEDGMENTS

This chapter was supported in part by IES grant R305A160295. We gratefully acknowledge the help and support of Bob Siegler, Xinlin Zhou, Lisa Fazio, and the Siegler Center for Innovation in Learning. We would also like to thank our Ohio State research assistant Yiwan Wang and the research assistants at Beijing Normal University.

REFERENCES

Anobile, G., Cicchini, G. M., & Burr, D. C. (2012). Linear mapping of numbers onto space requires attention. *Cognition*, *122*(3), 454–459.

Antell, S. E., & Keating, D. P. (1983). Perception of numerical invariance in neonates. *Child Development*, *54*, 695–701.

Bailey, D. H., Zhou, X., Zhang, Y., Cui, J., Fuchs, L. S., Jordan, N. C., et al. (2015). Development of fraction concepts and procedures in US and Chinese children. *Journal of Experimental Child Psychology*, *129*, 68–83.

Barth, H., Beckmann, L., & Spelke, E. S. (2008). Nonsymbolic, approximate arithmetic in children: Abstract addition prior to instruction. *Developmental Psychology*, *44*(5), 1466.

Barth, H., La Mont, K., Lipton, J., & Spelke, E. S. (2005). Abstract number and arithmetic in preschool children. *Proceedings of the National Academy of Sciences of the United States of America*, *102*(39), 14116–14121.

Barth, H., La Mont, K., Lipton, J., Dehaene, S., Kanwisher, N., & Spelke, E. (2006). Nonsymbolic arithmetic in adults and young children. *Cognition*, *98*(3), 199–222.

Berch, D. B., Foley, E. J., Hill, R. J., & Ryan, P. M. (1999). Extracting parity and magnitude from Arabic numerals: Developmental changes in number processing and mental representation. *Journal of Experimental Child Psychology*, *74*, 286–308.

Birnbaum, M. H. (1980). Comparison of two theories of "ratio" and "difference" judgments. *Journal of Experimental Psychology: General*, *109*(3), 304–319.

Bonny, J. W., & Lourenco, S. F. (2013). The approximate number system and its relation to early math achievement: Evidence from the preschool years. *Journal of Experimental Child Psychology*, *114*(3), 375–388.

Booth, J. L., & Siegler, R. S. (2006). Developmental and individual differences in pure numerical estimation. *Developmental Psychology*, *42*(1), 189.

Booth, J. L., & Siegler, R. S. (2008). Numerical magnitude representations influence arithmetic learning. *Child Development*, *79*, 1016–1031.

Brannon, E. M., & Terrace, H. S. (2000). Representation of the numerosities 1–9 by rhesus macaques (Macaca mulatta). *Journal of Experimental Psychology: Animal Behavior Processes*, *26*(1), 31.

Buckley, P. B., & Gillman, C. B. (1974). Comparisons of digits and dot patterns. *Journal of Experimental Psychology*, *103*(6), 1131.

Butterworth, B. (2005). The development of arithmetical abilities. *Journal of Child Psychology and Psychiatry*, *46*(1), 3–18.

Carroll, J. B. (1989). The Carroll Model: A 25-year retrospective and prospective view. *Educational Researcher*, *18*, 26–31.

Case, R. (1975). Social class differences in intellectual development: A Neo-Piagetian investigation. *Canadian Journal of Behavioural Science/Revue canadienne des sciences du comportement*, *7*(3), 244.

Case, R., & Griffin, S. (1990). Child cognitive development: The role of central conceptual structures in the development of scientific and social thought. *Advances in Psychology*, *64*, 193–230.

Case, R., & Sandieson, R. (1987). General developmental constraints on the acquisition of special procedures (and vice versa). In *Paper presented at the annual meetings of the American Educational Research Association, Baltimore, April*.

Castronovo, J., & Göbel, S. M. (2012). Impact of high mathematics education on the number sense. *PLoS ONE*, *7*(4), e33832.

Chen, C., & Uttal, D. H. (1988). Cultural values, parents' beliefs, and children's achievement in the United States and China. *Human Development*, *31*, 351–358.

Chi, M. T., & Klahr, D. (1975). Span and rate of apprehension in children and adults. *Journal of Experimental Child Psychology*, *19*(3), 434–439.

Cordes, S., & Brannon, E. M. (2008). Quantitative competencies in infancy. *Developmental Science*, *11*(6), 803–808.

Dacke, M., & Srinivasan, M. V. (2008). Two odometers in honeybees? *Journal of Experimental Biology*, *211*(20), 3281–3286.

De Smedt, B., Verschaffel, L., & Ghesquière, P. (2009). The predictive value of numerical magnitude comparison for individual differences in mathematics achievement. *Journal of Experimental Child Psychology*, *103*(4), 469–479.

Dehaene, S. (1989). The psychophysics of numerical comparison: A reexamination of apparently incompatible data. *Attention, Perception, & Psychophysics*, *45*(6), 557–566.

Dehaene, S. (1997). *The number sense*. New York, NY: Oxford University Press.

Dehaene, S., Izard, V., Spelke, E., & Pica, P. (2008). Log or linear? Distinct intuitions of the number scale in Western and Amazonian indigene cultures. *Science*, *320*(5880), 1217–1220.

Fantz, R. L., & Fagan, J. F., III. (1975). Visual attention to size and number of pattern details by term and preterm infants during the first six months. *Child Development*, *46*, 3–18.

Fazio, L. K., Bailey, D. H., Thompson, C. A., & Siegler, R. S. (2014). Relations of different types of numerical magnitude representations to each other and to mathematics achievement. *Journal of Experimental Child Psychology*, *123*, 53–72.

Geary, D. C. (1995). Reflections of evolution and culture in children's cognition: Implications for mathematical development and instruction. *American Psychologist*, *50*(1), 24.

Geary, D. C., Berch, D. B., Ochsendorf, R., & Mann Koepke, K. (Eds.), (2017). *Acquiring complex arithmetic skills and higher-order mathematical concepts*. San Diego, CA: Elsevier/ Academic Press.

Geary, D. C., Bow-Thomas, C. C., Liu, F., & Siegler, R. S. (1996). Development of arithmetical competencies in Chinese and American children: Influence of age, language, and schooling. *Child Development*, *67*, 2022–2044.

Geary, D. C., Hamson, C. O., Chen, G. P., Liu, F., Hoard, M. K., & Salthouse, T. A. (1997). Computational and reasoning abilities in arithmetic: Cross-generational change in China and the United States. *Psychonomic Bulletin & Review*, *4*, 425–430.

Geary, D. C., Hoard, M. K., Nugent, L., & Bailey, H. D. (2013). Adolescents' functional numeracy is predicted by their school entry number system knowledge. *PLoS ONE*, *8*(1), e54651.

Geary, D. C., & vanMarle, K. (2016). Young children's core symbolic and non-symbolic quantitative knowledge in the prediction of later mathematics achievement. *Developmental Psychology*, *52*, 2130–2144.

Gilmore, C. K., McCarthy, S. E., & Spelke, E. S. (2007). Symbolic arithmetic knowledge without instruction. *Nature*, *447*(7144), 589–591.

Girelli, L., Lucangeli, D., & Butterworth, B. (2000). The development of automaticity in accessing number magnitude. *Journal of Experimental Child Psychology*, *76*, 104–122.

Gonzales, P., Williams, T., Jocelyn, L., Roey, S., Kastberg, D., & Brenwald, S. (2008). *Highlights from TIMSS 2007: Mathematics and science achievement of US fourth- and eighth-grade students in international context (NCES 2009–001 revised)*. Washington, DC: National Center for Education Statistics.

Griffin, S., Case, R., & Sandieson, R. (1992). *Synchrony and asynchrony in the acquisition of children's everyday mathematical knowledge. The mind's staircase: Exploring the conceptual underpinnings of children's thought and knowledge*. Hillsdale, NY: Lawrence Erlbaum Associates, Inc. pp. 75–97.

Griffin, S. A., Case, R., & Siegler, R. S. (1994). Rightstart: Providing the central conceptual prerequisites for first formal learning of arithmetic to students at risk for school failure. In K. McGilly (Ed.), *Classroom lessons: Integrating cognitive theory and classroom practice* (pp. 25–50). Cambridge, MA: The MIT Press/Bradford Books.

Halberda, J., & Feigenson, L. (2008). Developmental change in the acuity of the "Number Sense": The Approximate Number System in 3-, 4-, 5-, and 6-year-olds and adults. *Developmental Psychology*, *44*(5), 1457.

Halberda, J., Ly, R., Wilmer, J. B., Naiman, D. Q., & Germine, L. (2012). Number sense across the lifespan as revealed by a massive Internet-based sample. *Proceedings of the National Academy of Sciences of the United States of America*, *109*(28), 11116–11120.

Halberda, J., Mazzocco, M. M. M., & Feigenson, L. (2008). Individual differences in non-verbal number acuity correlate with maths achievement. *Nature*, *455*, 665–668.

Hanushek, E. A., & Woessmann, L. (2008). The role of cognitive skills in economic development. *Journal of Economic Literature*, *46*, 607–668.

Hofstadter, D. R. (1982). Number numbness, or why innumeracy may be just as dangerous as illiteracy. *Scientific American*, *246*(5), 20–34.

Holloway, I. D., & Ansari, D. (2009). Mapping numerical magnitudes onto symbols: The numerical distance effect and individual differences in children's mathematics achievement. *Journal of Experimental Child Psychology*, *103*, 17–29.

Husén, T. (1967). *International study of achievement in mathematics: A comparison of twelve countries* (Vols. 1 & 2). New York, NY: Wiley.

Inglis, M., Attridge, N., Batchelor, S., & Gilmore, C. (2011). Non-verbal number acuity correlates with symbolic mathematics achievement: But only in children. *Psychonomic Bulletin & Review*, *18*(6), 1222–1229.

Jordan, N. C., Huttenlocher, J., & Levine, S. C. (1992). Differential calculation abilities in young children from middle-and low-income families. *Developmental Psychology*, *28*(4), 644.

Kim, D., & Opfer, J. (2017). A unified framework for bounded and unbounded numerical estimation. *Developmental Psychology*, *53* (6), 1088.

Laski, E. V., & Siegler, R. S. (2007). Is 27 a big number? Correlational and causal connections among numerical categorization, number line estimation, and numerical magnitude comparison. *Child Development*, *78*(6), 1723–1743.

LeFevre, J. A., Lira, C. J., Sowinski, C., Cankaya, O., Kamawar, D., & Skwarchuk, S. L. (2013). Charting the role of the number line in mathematical development. *Frontiers in Psychology*, *4*, 641.

Lezak, M. D., Howieson, D. B., Loring, D. W., Hannay, J., & Fischer, J. (2004). *Neuropsychological assessment*. New York, NY: Oxford University Press.

Libertus, M. E., Feigenson, L., & Halberda, J. (2011). Preschool acuity of the approximate number system correlates with school math ability. *Developmental Science*, *14*, 1292–1300.

Libertus, M. E., Feigenson, L., & Halberda, J. (2013). Numerical approximation abilities correlate with and predict informal but not formal mathematics abilities. *Journal of Experimental Child Psychology*, *116*(4), 829–838.

Lipton, J. S., & Spelke, E. S. (2003). Origins of number sense: Large-number discrimination in human infants. *Psychological Science*, *14*(5), 396–401.

Lo, J. J., & Luo, F. (2012). Prospective elementary teachers' knowledge of fraction division. *Journal of Mathematics Teacher Education*, *15*(6), 481–500.

Lourenco, S. F., Bonny, J. W., Fernandez, E. P., & Rao, S. (2012). Nonsymbolic number and cumulative area representations contribute shared and unique variance to symbolic math competence. *Proceedings of the National Academy of Sciences of the United States of America*, *109*(46), 18737–18742.

Lynn, R. (1982). IQ in Japan and the United States shows a growing disparity. *Nature*, *297*(5863), 222–223.

Lyons, I. M., & Beilock, S. L. (2011). Numerical ordering ability mediates the relation between number-sense and arithmetic competence. *Cognition*, *121*, 256–261.

Lyons, I. M., Price, G. R., Vaessen, A., Blomert, L., & Ansari, D. (2014). Numerical predictors of arithmetic success in grades 1–6. *Developmental Science*, *17*(5), 714–726.

Ma, L. (1999). *Knowing and teaching elementary mathematics: Teachers' understanding of fundamental mathematics in China and the United States*. Mahwah, NJ: Lawrence Erlbaum Associates.

Mankiw, N. G., & Weinzierl, M. (2010). The optimal taxation of height: A case study of utilitarian income redistribution. *American Economic Journal: Economic Policy*, *2*(1), 155–176.

Mazzocco, M. M., Feigenson, L., & Halberda, J. (2011). Preschoolers' precision of the approximate number system predicts later school mathematics performance. *PLoS ONE*, *6*(9), e23749.

McCrink, K., & Wynn, K. (2004). Large-number addition and subtraction by 9-month-old infants. *Psychological Science*, *15*(11), 776–781.

McCrink, K., & Wynn, K. (2009). Operational momentum in large-number addition and subtraction by 9-month-olds. *Journal of Experimental Child Psychology*, *103*(4), 400–408.

McCrink, K., Dehaene, S., & Dehaene-Lambertz, G. (2007). Moving along the number line: Operational momentum in nonsymbolic arithmetic. *Attention, Perception, & Psychophysics*, *69*(8), 1324–1333.

Mechner, F. (1958). Probability relations within response sequences under ratio reinforcement. *Journal of the Experimental Analysis of Behavior*, *1*(2), 109–121.

Moyer, R. S., & Landauer, T. K. (1967). Time required for judgements of numerical inequality. *Nature*, *215*(5109), 1519–1520.

Muldoon, K., Towse, J., Simms, V., Perra, O., & Menzies, V. (2013). A longitudinal analysis of estimation, counting skills, and mathematical ability across the first school year. *Developmental Psychology*, *49*(2), 250.

Mundy, E., & Gilmore, C. K. (2009). Children's mapping between symbolic and nonsymbolic representations of number. *Journal of Experimental Child Psychology*, *103*(4), 490–502.

Murnane, R. J., Willett, J. B., & Levy, F. (1995). *The growing importance of cognitive skills in wage determination (No. w5076)*. Cambridge, MA: National Bureau of Economic Research.

Naccache, L., & Dehaene, S. (2001). The priming method: Imaging unconscious repetition priming reveals an abstract representation of number in the parietal lobes. *Cerebral Cortex*, *11*(10), 966–974.

National Mathematics Advisory Panel. (2008). *Foundations for success: The final report of the National Mathematics Advisory Panel*. Washington, D.C.: US Department of Education.

National Research Council (NRC). (2005). How students learn: History, mathematics, and science in the classroom. Committee on How People Learn. A targeted report for teachers. In M. S. Donovan & J. D. Bransford (Eds.), Washington, DC: National Academics Press.

OECD. (2010). *PISA 2009 results: Executive summary*. Paris: OECD.

Opfer, J. E., Kim, D., Siegler, R. S., Fazio, L. K., & Zhou, X. (submitted for publication). Cognitive mechanisms of nation effects on mathematics proficiency.

Opfer, J. E., & Martens, M. A. (2012). Learning without representational change: Development of numerical estimation in individuals with Williams syndrome. *Developmental Science*, *15*(6), 863–875.

Opfer, J. E., Qin, J., & Kim, D. (submitted for publication). Varieties of numerical estimation: A unified developmental framework.

Opfer, J. E., & Siegler, R. S. (2007). Representational change and children's numerical estimation. *Cognitive Psychology*, *55*(3), 169–195.

Opfer, J. E., Siegler, R. S., & Young, C. J. (2011). The powers of noise-fitting: Reply to Barth and Paladino. *Developmental Science*, *14*(5), 1194–1204.

Opfer, J. E., & Thompson, C. A. (2008). The trouble with transfer: Insights from microgenetic changes in the representation of numerical magnitude. *Child Development*, *79*(3), 788–804.

Opfer, J. E., Thompson, C. A., & Kim, D. (2016). Free versus anchored numerical estimation: A unified approach. *Cognition*, *149*, 11–17.

Osborne, R. T., & Lindsey, J. M. (1967). A longitudinal investigation of change in the factorial composition of intelligence with age in young school children. *The Journal of Genetic Psychology*, *110*(1), 49–58.

Park, J., Bermudez, V., Roberts, R. C., & Brannon, E. M. (2016). Non-symbolic approximate arithmetic training improves math performance in preschoolers. *Journal of Experimental Child Psychology*, *152*, 278–293.

Piazza, M., & Izard, V. (2009). How humans count: Numerosity and the parietal cortex. *The Neuroscientist*, *15*(3), 261–273.

Piazza, M., Facoetti, A., Trussardi, A. N., Berteletti, I., Conte, S., Lucangeli, D., et al. (2010). Developmental trajectory of number acuity reveals a severe impairment in developmental dyscalculia. *Cognition*, *116*(1), 33–41.

Piazza, M., Izard, V., Pinel, P., Le Bihan, D., & Dehaene, S. (2004). Tuning curves for approximate numerosity in the human intraparietal sulcus. *Neuron*, *44*(3), 547–555.

Pica, P., Lemer, C., Izard, V., & Dehaene, S. (2004). Exact and approximate arithmetic in an Amazonian indigene group. *Science*, *306*(5695), 499–503.

Piffer, L., Petrazzini, M. E. M., & Agrillo, C. (2013). Large number discrimination in newborn fish. *PLoS ONE*, *8*(4), e62466.

Price, G. R., Palmer, D., Battista, C., & Ansari, D. (2012). Nonsymbolic numerical magnitude comparison: Reliability and validity of different task variants and outcome measures, and their relationship to arithmetic achievement in adults. *Acta Psychologica*, *140*(1), 50–57.

Ramani, G. B., & Siegler, R. S. (2008). Promoting broad and stable improvements in low-income children's numerical knowledge through playing number board games. *Child Development*, *79*(2), 375–394.

Revkin, S. K., Piazza, M., Izard, V., Cohen, L., & Dehaene, S. (2008). Does subitizing reflect numerical estimation? *Psychological Science*, *19*(6), 607–614.

Richland, L. E., Stigler, J. W., & Holyoak, K. J. (2012). Teaching the conceptual structure of mathematics. *Educational Psychologist*, *47*(3), 189–203.

Richland, L. E., Zur, O., & Holyoak, K. J. (2007). Cognitive supports for analogies in the mathematics classroom. *Science*, *316*, 1128–1129.

Rodic, M., Zhou, X., Tikhomirova, T., Wei, W., Malykh, S., Ismatulina, V., et al. (2015). Cross-cultural investigation into cognitive underpinnings of individual differences in early arithmetic. *Developmental Science*, *18*, 165–174.

Rose, H., & Betts, J. R. (2004). The effect of high school courses on earnings. *Review of Economics and Statistics*, *86*(2), 497–513.

Rubinsten, O., Henik, A., Berger, A., & Shahar-Shalev, S. (2002). The development of internal representations of magnitude and their association with Arabic numerals. *Journal of Experimental Child Psychology*, *81*, 74–92.

Rushton, J. P. (1992). Cranial capacity related to sex, rank, and race in a stratified random sample of 6,325 U.S. military personnel. *Intelligence*, *16*, 401–413.

Ryan, J. J., & Lopez, S. J. (2001). Wechsler adult intelligence scale-III. *Understanding psychological assessment*. New York City, NY: Springer. pp. 19–42.

Sadler, P. M., & Tai, R. H. (2007). The two high-school pillars supporting college science. *Science*, *317*, 457–458.

Sasanguie, D., De Smedt, B., Defever, E., & Reynvoet, B. (2012). Association between basic numerical abilities and mathematics achievement. *British Journal of Developmental Psychology*, *30*(2), 344–357.

Sasanguie, D., Göbel, S. M., Moll, K., Smets, K., & Reynvoet, B. (2013). Approximate number sense, symbolic number processing, or number–space mappings: What underlies mathematics achievement? *Journal of Experimental Child Psychology*, *114*(3), 418–431.

Saxe, G. B., Guberman, S. R., & Gearhart, M. (1989). *Social processes in early number development. Monographs of the society for research in child development:* Vol. 52 (serial no. 216).

Scarf, D., Hayne, H., & Colombo, M. (2011). Pigeons on par with primates in numerical competence. *Science*, *334*(6063), 1664.

Schneider, M., Beeres, K., Coban, L., Merz, S., Schmidt, S. S., Stricker, J., et al. (2016). Associations of non-symbolic and symbolic numerical magnitude processing with mathematical competence: A meta-analysis. *Developmental Science*, *20*(3), 1–16.

Schneider, M., Grabner, R. H., & Paetsch, J. (2009). Mental number line, number line estimation, and mathematical achievement: Their interrelations in grades 5 and 6. *Journal of Educational Psychology*, *101*(2), 359.

Sekuler, R., & Mierkiewicz, D. (1977). Children's judgments of numerical inequality. *Child Development*, *48*(2), 630–633.

Sella, F., Berteletti, I., Lucangeli, D., & Zorzi, M. (2015). Varieties of quantity estimation in children. *Developmental Psychology*, *51*(6), 758–770.

Siegler, R. S., & Booth, J. L. (2004). Development of numerical estimation in young children. *Child Development*, *75*, 428–444.

Siegler, R. S., Duncan, G. J., Davis-Kean, P. E., Duckworth, K., Claessens, A., Engel, M., et al. (2012). Early predictors of high school mathematics achievement. *Psychological Science*, *23*, 691–697.

Siegler, R. S., & Mu, Y. (2008). Chinese children excel on novel mathematics problems even before elementary school. *Psychological Science*, *19*, 759–763.

Siegler, R. S., & Opfer, J. E. (2003). The development of numerical estimation: Evidence for multiple representations of numerical quantity. *Psychological Science*, *14*(3), 237–250.

Siegler, R. S., & Ramani, G. B. (2008). Playing linear numerical board games promotes low-income children's numerical development. *Developmental Science*, *11*(5), 655–661.

Siegler, R. S., & Ramani, G. B. (2009). Playing linear number board games—but not circular ones—improves low-income preschoolers' numerical understanding. *Journal of Educational Psychology*, *101*(3), 545.

Siegler, R. S., Thompson, C. A., & Opfer, J. E. (2009). The logarithmic-to-linear shift: One learning sequence, many tasks, many time scales. *Mind, Brain, and Education*, *3*(3), 143–150.

Siegler, R. S., Thompson, C. A., & Schneider, M. (2011). An integrated theory of whole number and fractions development. *Cognitive Psychology*, *62*(4), 273–296.

Starkey, P., & Cooper, R. G. (1980). Perception of numbers by human infants. *Science*, *210*(4473), 1033–1035.

Stevenson, H. W., Chen, C., & Lee, S. Y. (1993). Mathematics achievement of Chinese, Japanese, and American children: Ten years later. *Science*, *231*, 693–699.

Stevenson, H. W., Lee, S. Y., & Stigler, J. W. (1986). Mathematics achievement of Chinese, Japanese, and American children. *Science*, *231*, 693–699.

Stevenson, H. W., & Stigler, J. W. (1992). *The learning gap*. New York, NY: Summit.

Stevenson, H. W., Stigler, J. W., Lee, S.-Y., Lucker, G. W., Kitamura, S., & Hsu, C.-C. (1985). Cognitive performance and academic achievement of Japanese, Chinese, and American Children. *Child Development*, *56*, 718–734.

Stigler, J. W., & Hiebert, J. (1999). *Understanding and improving. Comparing Standards Internationally: research and practice in mathematics and beyond.* Oxford, UK: Symposium Books Ltd. 119.

Stigler, J. W., Gonzales, P., Kwanaka, T., Knoll, S., & Serrano, A. (1999). *The TIMSS videotape classroom study: Methods and findings from an exploratory research project on eighth-grade mathematics instruction in Germany, Japan, and the United States*. Washington, DC: U.S. Department of Education, National Center for Education Statistics.

Strauss, M. S., & Curtis, L. E. (1981). Infant perception of numerosity. *Child Development*, *52*(4), 1146–1152.

Temple, E., & Posner, M. I. (1998). Brain mechanisms of quantity are similar in 5-year-old children and adults. *Proceedings of the National Academy of Sciences of the United States of America*, *95*(13), 7836–7841.

Thompson, C. A., & Opfer, J. E. (2008). Costs and benefits of representational change: Effects of context on age and sex differences in symbolic magnitude estimation. *Journal of Experimental Child Psychology*, *101*(1), 20–51.

Tyler, J. H. (2004). Basic skills and the earnings of dropouts. *Economics of Education Review*, *23*(3), 221–235. https://doi.org/10.1016/j.econedurev.2003.04.001.

Vanbinst, K., Ansari, D., Ghesquière, P., & De Smedt, B. (2016). Symbolic numerical magnitude processing is as important to arithmetic as phonological awareness is to reading. *PloS ONE*, *11*(3), e0151045.

Vanbinst, K., Ghesquière, P., & De Smedt, B. (2012). Numerical magnitude representations and individual differences in children's arithmetic strategy use. *Mind, Brain, and Education*, *6*(3), 129–136.

Wechsler, D. (1991). *Manual for the Wechsler intelligence scale for children-(WISC-III)*. San Antonio, TX: Psychological Corporation.

Wynn, K. (1992). Addition and subtraction by human infants. *Nature*, *358*(6389), 749–750.

Xu, F. (2003). Numerosity discrimination in infants: Evidence for two systems of representations. *Cognition*, *89*(1), B15–B25.

Xu, F., Spelke, E. S., & Goddard, S. (2005). Number sense in human infants. *Developmental Science*, *8*(1), 88–101.

Xu, X., Chen, C., Pan, M., & Li, N. (2013). Development of numerical estimation in Chinese preschool children. *Journal of Experimental Child Psychology*, *116*, 351–366.

Zhou, X., Chen, Y., Chen, C., Jiang, T., Zhang, H., & Dong, Q. (2007). Chinese kindergartners' automatic processing of numerical magnitude in Stroop-like tasks. *Memory & Cognition*, *35*, 464–470.

FURTHER READING

Gilmore, C. K., McCarthy, S. E., & Spelke, E. S. (2010). Non-symbolic arithmetic abilities and mathematics achievement in the first year of formal schooling. *Cognition*, *115*, 394–406.

Meck, W. H., & Church, R. M. (1983). A mode control model of counting and timing processes. *Journal of Experimental Psychology: Animal Behavior Processes*, *9*(3), 320.

Nosek, B. A., Smyth, F. L., Sriram, N., Lindner, N. M., Devos, T., Ayala, A., et al. (2009). National differences in gender–science stereotypes predict national sex differences in science and math achievement. *Proceedings of the National Academy of Sciences of the United States of America*, *106*, 10593–10597.

Park, J., & Brannon, E. M. (2013). Training the approximate number system improves math proficiency. *Psychological Science*, *24*, 2013–2019.

Park, J., & Brannon, E. M. (2014). Improving arithmetic performance with number sense training: An investigation of underlying mechanism. *Cognition*, *133*, 188–200.

Paterson, S. J., Girelli, L., Butterworth, B., & Karmiloff-Smith, A. (2006). Are numerical impairments syndrome specific? Evidence from Williams syndrome and Down's syndrome. *Journal of Child Psychology and Psychiatry*, *47*(2), 190–204.

Piazza, M., Mechelli, A., Price, C. J., & Butterworth, B. (2006). Exact and approximate judgements of visual and auditory numerosity: An fMRI study. *Brain Research*, *1106*(1), 177–188.

Piazza, M., Pinel, P., Le Bihan, D., & Dehaene, S. (2007). A magnitude code common to numerosities and number symbols in human intraparietal cortex. *Neuron*, *53*(2), 293–305.

Preacher, K. J., & Hayes, A. F. (2008). Asymptotic and resampling strategies for assessing and comparing indirect effects in multiple mediator models. *Behavior Research Methods*, *40*(3), 879–891.

Rousselle, L., & Noël, M.-P. (2007). Basic numerical skills in children with mathematics learning disabilities: A comparison of symbolic vs non-symbolic number magnitude processing. *Cognition*, *102*, 361–395.

Stevenson, H. W., Lee, S.-Y., Chen, C., Lummis, M., Stigler, J., Fan, L., et al. (1990). Mathematics achievement of children in China and the United States. *Child Development*, *61*, 1053–1066.

Chapter 6

Mathematical Skills of Children With Specific Language Impairments: Testing Developmental Theory

Chris Donlan
University College London, London, United Kingdom

INTRODUCTION

Formal teaching in educational settings relies heavily on spoken language. The learner may be put under linguistic pressure from the outset, frequently required to assimilate new forms of expression and understanding. In the example that follows, we track the conversation between a teacher and a small group of 5-year olds who are struggling to come to terms with simple arithmetic:

> *TEACHER*: What would you do if I asked you to do four add two? What would you do?
> *CHILD A*: Add a one.
> *TEACHER*: Would you?
> *CHILD B*: Add a two.
> *TEACHER*: What would you use to help you?
> *CHILD A*: Plus.
> *TEACHER*: You'd use plus? What does plus mean?
> *CHILD C*: Equals.
> *TEACHER*: Does it? Really?
> *CHILD A*: It's a plus.
> *TEACHER*: Plus means putting things together... putting numbers together. If I put numbers together, do I get more or less?
> *CHILD D*: Yes. Yes. Yes.
> *CHILD A*: More!

Mathematical Cognition and Learning, Vol. 4. https://doi.org/10.1016/B978-0-12-812574-8.00006-7

This short excerpt (K. Mills, personal communication, April 14, 2012) is notable for the misunderstandings it contains. The teacher, who is extremely skilled, is careful to elicit children's understanding of key word meanings. As the conversation develops and in future conversations, she will continue to negotiate these meanings with the children to demonstrate how they operate within sentences, in the context of actions. Such sensitively targeted linguistic support is a privilege of expert small group teaching. Most children will meet the challenges of school-based learning without such individualized input. The extent to which they can use and understand language, under the pressure of new formal demands, will be of critical importance for their future success in education and in later life.

BACKGROUND TO SPECIFIC LANGUAGE IMPAIRMENT

Definition

Specific language impairment (SLI) is a developmental disorder in which significant deficits in expressive or receptive language occur, not due to sensory or environmental factors. Traditionally, SLI is taken to cooccur with nonverbal intelligence within the average range. It is the cooccurrence of significantly impaired language and apparently unimpaired intelligence that justifies the use of the term "specific." Recent debate among researchers, speech and language therapists, and educationists has questioned the specificity criterion (Reilly et al., 2014). A broader category of "language disorder" has been used, though the generality of this term can be seen as disadvantageous (Bishop, 2014). The debate continues (Bishop et al., 2016). What is not in dispute is either the prevalence (a recent UK cohort study (Norbury et al., 2016) is in broad agreement with the 7% US estimate of Tomblin et al., 1997) or the heterogeneity of child language disorders. There may be difficulties with sentence production or comprehension, with grammar or vocabulary, or with pragmatics (language use in context) (Bishop, 1997; Chiat, 2000). Furthermore, although research studies tend to focus on language deficits and aim to exclude children with speech difficulties, their widespread cooccurrence makes it difficult or even artificial to distinguish the two. Educational studies, based on school populations, tend to be more inclusive (Haynes & Naidoo, 1991).

Causes of SLI

Genetic factors are known to be significant in the causation of SLI. Tests of nonword repetition (in which the child hears a nonword such as "bannow" and attempts to repeat it) have been used to assess word learning capacity. Data from these behavioral assessments are then analyzed in order to identify linkage with genetic factors. Convergent evidence from twin studies and

DNA-based analyses indicates that nonword repetition serves as a phenotypic marker for inherited SLI (Bishop, North, & Donlan, 1996; SLI Consortium, 2002, 2004). In a further larger-scale twin study, Bishop, Adams, and Norbury (2006) found evidence of independent genetic transmission of language impairments linked to nonword repetition on the one hand and past-tense production (e.g., producing past tense form "walked" when prompted with present tense "walks") on the other hand. Based on this finding, Bishop et al. (2006) proposed that more than one specialized genetically determined system contributes to language development and that SLI is most likely to occur where genetic variation produces cooccurring deficits in such systems. Bishop (2009) posed the following challenge for future research: to enhance phenotypic identification so as to support differentiation between language deficits whose causation is attributable to (a) rare genetic variants, (b) complex polygenic factors, and (c) environmental causation. As yet, Bishop's challenge of differential diagnosis remains unmet. The heterogeneity of child language disorders is not as yet susceptible to differentiation at the level of etiology.

Mathematical Attainments of Children With SLI

There are numerous findings of low attainment in mathematics by children with SLI. Nelson et al. (2011) reported that 4-year olds with severe language delay scored 1 SD below norms on a mathematics test measuring counting skills and simple addition and subtraction. Significant deficits have been observed throughout the school years, both in the United States and in Europe (Cowan, Donlan, Newton, & Lloyd, 2005; Fazio, 1994; Nys, Content, & Leybaert, 2013). A large-scale longitudinal study by Durkin, Mok, and Conti-Ramsden (2013) found that children with SLI, assessed at age 11, performed as poorly in national tests of mathematics as they did in English. However, alongside these reports of poor mathematics, it is important to acknowledge some more positive findings. Both Fazio (1994) and Donlan, Cowan, Newton, and Lloyd (2007) reported relatively good conceptual understanding, despite procedural limitations in basic arithmetic. Consistent with this pattern, Nys et al. (2013) found that children with SLI were not intrinsically impaired in estimation (identifying approximate arithmetic solutions), though they showed significant deficits in identifying exact addition solutions.

Furthermore, it is clear that mathematical deficits are not inevitable for children with SLI. Additional analyses of the sample on which Cowan et al. (2005) and Donlan et al. (2007) are based showed that 12% of the SLI group performed at levels indistinguishable from their typically developing peers. These intriguing variations raise questions about the particular aspects of SLI that might pose obstacles to mathematical learning and regarding the areas of mathematics that are most severely compromised with SLI. Reporting on an earlier phase of their longitudinal study of children with SLI, Durkin et al. (2013) showed that standard scores on a test of number skills declined from age 7 to 8. In line with these results, longitudinal findings by Fazio (1994, 1996, 1999) indicated a cumulative delay in development of arithmetic procedures, such that any earlier conceptual advantage failed to support overall attainment in the longer term.

How are we to understand these patterns of change? Durkin et al. (2015) looked at individual differences in the decline in math standard scores from 7 to 8 years and found that the decline in language scores over this time period was a significant predictor of math achievement. For Durkin et al. (2015), this association between language deficits and deficits in mathematics indicates a general developmental association between language and mathematics. I will examine this proposal in the next sections, first by considering the literature concerning the influence of language on typical mathematical development, then by looking in detail at research findings from children with SLI.

LINGUISTIC AND NONLINGUISTIC CONTRIBUTIONS TO MATHEMATICAL DEVELOPMENT

The Approximate Number System

The notion that the mental representation of number is modular and depends upon a nonverbal approximate number system (ANS) has been highly influential in studies of arithmetic development (Halberda, Mazzocco, & Feignson, 2008; Piazza & Eger, 2016). Studies with human infants and nonhuman primates (Viswanathan & Nieder, 2013) claim to identify a "primitive" ability to represent numerical magnitudes in an abstract code that is independent of perceptual properties of the input. A much cited paper by Gilmore, McCarthy, Shannon, and Spelke (2007) proposed that nonsymbolic magnitude representations drive early arithmetic skills. This study was based on a small sample of kindergarten children, presenting an argument based primarily on the ability of these children to perform above chance levels on tasks requiring approximate addition and subtraction using Arabic numerals. Participants had not received tuition in symbolic arithmetic; therefore, it was argued they must be using nonsymbolic magnitude representations in order to perform the symbolic operation. Subsequently, Halberda et al. (2008) claimed that "individual differences in achievement in school mathematics are related to individual differences in the acuity of an evolutionarily ancient, unlearned approximate number sense" (Halberda et al., 2008, p. 665).

However, the possible associations between the ANS, symbolic number representation, and arithmetic have proved controversial (Price, Palmer, & Battista, 2012). These associations have been explored quite extensively in research with children with SLI, so I will consider them in some detail.

The task commonly used to tap the ANS is a nonsymbolic magnitude comparison task (e.g., choosing the numerically greater of two dot arrays); this task is derived from the classic symbolic magnitude comparison task (choosing the greater value of two Arabic numerals) introduced by Moyer and Landauer (1967). This classic study showed that response time to choose the greater of two numerals conforms to a quasipsychophysical function whereby latency is inversely related to the difference between stimuli, a phenomenon that became known as the symbolic distance effect (SDE).

This finding has been influential in the development of the theory that the ANS underlies mathematical development, based on the assumption that a common underlying semantic representation supports both symbolic and nonsymbolic comparison. Holloway and Ansari (2009) investigated this assumption and the further assumption that this semantic system underlies the development of arithmetic. In a sample of 87 6- to 8-year olds, they found a moderate but significant prediction of mathematical attainment from the distance effect for symbolic comparison but no such prediction from the distance effect in nonsymbolic comparison. Furthermore, there was no simple correlation between distance effects in symbolic and nonsymbolic comparison. While these findings cast doubt on the persistence of influence of the ANS over development, they leave open the possibility that it provides the early semantic framework on which later symbolic networks are constructed and further suggest that the symbolic comparison skills underpin arithmetic learning. Evidence in support of these suggestions can be found in VanMarle, Chu, and Geary (2014) and Chu, vanMarle, and Geary (2015).

The Importance of Number Symbols

Göbel, Watson, Lervag, and Hulme (2014) set out to evaluate the factor structure underlying symbolic vs nonsymbolic magnitude comparisons and their influence on arithmetic skills. They tested 6-year olds and followed up after 12 months. Using a large sample size ($n = 165$), multiple measurements, and structural modeling of longitudinal outcome, their findings were at odds with Holloway and Ansari (2009) with respect to the relation between symbolic and nonsymbolic comparisons. Göbel et al. (2014) found the two tasks to be part of a single factor that was highly correlated with later arithmetic skills. More important, the inclusion of a measure of number knowledge (matching spoken numerals to the corresponding Arabic forms) produced a radical departure from previous findings. It proved to be the sole (and substantial) unique predictor of later arithmetic (apart from the autocorrelate), suggesting that the children's ability to associate Arabic numerals with their spoken forms is primary in the developmental process (see Chapter 10 by Göbel). It is useful to note that a recent study has replicated these findings in a group of 4-year olds followed up at age 6 (Habermann, Donlan, Göbel, & Hulme, 2017).

The Integration of Symbolic Representations

If the association between spoken numbers and Arabic numerals is of central importance for the development of arithmetic, then the transparency of the relation between the numerical symbol systems (which varies widely across languages) should be important. Children should learn number word-numeral pairings more quickly and learn arithmetic faster, in languages where such mappings are more systematic or lawful (see Chapter 4 by Okamoto and Chapter 13 by Bender and Beller). Early studies comparing Chinese and US children's counting and early mathematics skills identified significant differences between the two language groups (Miura, 1987), but these were claimed to be transient and secondary to wider cultural differences (Geary, 1995). Recent studies, however, have developed more stringent controls and have identified crosslinguistic influences on a range of transcoding and arithmetic tasks (Moeller, Korbinian, Göbel, et al., 2015).

Work by Mix, Prather, Smith, and Stockton (2014) offers a different perspective. They studied English speakers and focused on the child's grasp of basic principles of the Hindu-Arabic system: *The same digit represents more when it is to the left than when it is to the right in a string of digits* (which is more, 123 vs 321?); *numbers are read left to right* (which is two hundred sixty-seven, 267 vs 627?); *the number of digits is related to both magnitude* (which is more, 101 vs 99?); and *identity* (which is four hundred two, 402 vs 42?). Surprisingly, many preschool children showed good understanding of these principles. Furthermore, children were better able to map from spoken numeral inputs to written numerals (symbol to symbol) than to blocks or dot arrays (symbol to concrete object). The researchers also conducted a small-scale intervention study demonstrating that training based on symbols (spoken and written forms) was more effective than training based on blocks. Mix and her colleagues interpreted these findings within the broader context of language learning. The preschoolers in their study had not been explicitly taught to use multidigit numerals, but their exposure to such numerals in everyday life is extensive. In the same way that young children typically extract grammatical rules from regularities in the natural language environment without explicit training (Aslin & Newport, 2012), they may also learn the basic correspondence between spoken and written numerals and the basic principles of number notation. Thus, the learning of number notation may be based initially on linguistic forms, their structures and regularities, and their accessibility to the learner.

Natural Language as a Support for Learning Number Meanings

An intriguing study by Hodent, Bryant, and Houde (2005) was based on the contrast between quantifiers (words expressing number and quantity) in French and English. In French, but not in English, the same word, "un," is used as the indefinite article ("a" in English) and as the first element in the number word sequence (un, deux, trois...). Thus, in English, asked if you have any pets, you might say "I've got a cat"; on the other hand, if you were asked "how many pets have you got?," you might say "I've got one cat."

In French, the two statements would be identical (j'ai un chat). The experimental task used by Hodent et al. (2005) made no explicit use of these quantifiers but required that children watch a series of events on a small stage. Puppets entered were concealed and revealed in such a way that simple arithmetic facts were represented (e.g., 1+1=2); the "show" was set up so that some "impossible" events occurred (e.g., one puppet was joined behind a screen by two more puppets, and then, the screen was removed to reveal four puppets). French, but not English, 2-year olds were happy to accept this series of events (1+2=4). The authors had predicted this particular outcome, arguing that the ways in which 2-year olds represent number may be strongly influenced by the natural language environment. Where there is possible confusion between, for example, the meanings "one car" and "a car," the child may be less likely to represent specific numerical values. Hodent et al. (2005) acknowledged that cultural (as opposed to linguistic) differences between their samples were uncontrolled; nonetheless, these findings alert us to the possible importance of language in shaping early number concepts.

A study by Almoammer et al. (2013) provides further insight into the role of language. The study was motivated by the contrasting ways in which different languages use noun morphology (variations in the form of the noun) to mark number. English uses singular/plural marking ("car" vs "cars"). Some other languages (including Saudi Arabic and Slovenian) make a further distinction for sets of two items. These languages have singular/dual/plural marking ("car" vs "car-dual" vs "cars"). Almoammer et al. (2013) compared number word knowledge in 2- to 4-year olds in Saudi Arabia, Slovenia, and the United States. They found that children whose language uses dual markers are much more likely to be able to respond accurately to the instruction "give me two" than those whose language marks just singular and plural. These findings are particularly striking given the different cultural contexts within which Saudi Arabic and Slovenian children are learning. They emphasize the importance of children's experience of number within the grammar of natural language, beyond the rote-counting routine that we often prioritize. Thus, the broad framework of natural language interaction may serve as a primary vehicle for the development of number knowledge and an important means by which children learn to use the number word sequence as a tool for purposeful counting and a base for the elaboration of number concepts.

Pragmatics

Much of the research reviewed above points to the importance of natural language. If we give consideration to the ways in which number word meanings operate in everyday communication, then we must necessarily broaden our view by looking outside the constraints of rigidly controlled experimental tasks and examine linguistic interaction in more naturalistic settings by taking account of pragmatics (the way context contributes to meaning). Of particular interest is the relation between children's understanding of number words ("one," "two," etc.) and the broader set of quantifiers (e.g., "all," "some," "few," "more," and "less"). A useful distinction is made between the logical interpretation of a quantifier (e.g., taking "some" to mean *at least two and possibly all*) and the pragmatic interpretation, taking account of the speaker's intention (e.g., taking "some" to mean *at least two and fewer than all*). A study by Pouscoulous, Noveck, Politzer, and Bastide (2007) not only replicated the surprising finding that young children are much more likely than older children and adults to make the logical inference (Noveck, 2001) but also showed that varying the task demand and the complexity of the context can facilitate pragmatic inference, even in 4-year olds. Although research of this sort is in its infancy, as our knowledge expands, we may find new ways in which the communication that is at the heart of mathematical learning may be enhanced.

Summary

Despite the widespread commitment of researchers to the ANS as the basic driver of symbolic mathematical development, a growing body of research identifies a key role for language. The association between spoken and written numerals is a dominant predictor of arithmetic learning in the early school years. Preschool children's grasp of the integration of spoken numbers and Arabic numerals (transcoding) indicates a possible sharing of resources with natural language learning. Grammatical systems for representing number can provide support for learning to count objects, and we are starting to understand the role of pragmatics in communication about number and quantity. As we shall see, the literature on mathematical skills of children with SLI is in many ways confirmatory of these findings.

EVIDENCE FROM CHILDREN WITH SPECIFIC LANGUAGE IMPAIRMENTS (SLI)

Challenges Presented by the Number Word Sequence and Transcoding

While there are substantial individual differences in mathematical development within this group, most children with SLI find production of the number word sequence particularly difficult. Fazio (1994) found that 5-year olds with SLI could recite the sequence up to six or so, while their typically developing peers were accurate to 20 and beyond. However, Fazio found that her SLI group showed a relatively strong grasp of logical principles in object counting, in particular the principle that the final count word indicates the value of the set (cardinality). A larger study by Donlan et al. (2007) extended these findings. Forty percent of 7-year olds with SLI were unable to produce the sequence to 20 accurately. Only 4% of typically developing 5- to 6-year olds had problems within this range. Most of the children with SLI also showed significant deficits in basic calculation. In order to test their understanding of arithmetic principles independent of notation and counting skills, Donlan et al. (2007) introduced a scenario in which participants were presented with unfamiliar numerals and asked to take the role of the teacher and mark simple addition and subtraction sums as correct or incorrect. The scenario allowed evaluation of children's understanding of logical principles, for example, the principle that the order of addends in addition may be reversed ($2+3=3+2$), though the same principle (commutativity) does not apply to subtraction. This test of arithmetic principles was the only task on which the performance of children with SLI approximated that of typically developing peers.

A common pattern of conceptual strength and procedural weakness emerges from Fazio's (1994) study of 5-year olds with SLI and from Donlan et al. (2007) study of 7-year olds. Further confirmation of the procedural deficit came from follow-up studies (Fazio, 1996, 1999), in which Fazio's original sample subsequently observed at two-year intervals showed particular deficits in procedural aspects of basic calculation (persistent reliance on counting strategies and errors in number fact retrieval). It is notable that the relative strength of conceptual understanding identified by Fazio (1994) and by Donlan et al. (2007) was found with tasks that required no knowledge of Arabic numerals. Where numeral knowledge was tested, deficits were usually found. Even using a symbolic comparison task, in which participants simply chose the greater of two double- or multidigit Arabic numerals, Donlan et al. (2007) found that children with SLI were significantly less accurate than age-matched peers. Although this was a purely nonverbal task, performance was strongly correlated with participants' ability to produce the spoken number sequence and their ability to associate spoken numbers with Arabic numerals. All three tasks were strong correlates of basic calculation (r-values ranged from 0.507 to 0.697). In this way, studies of children with SLI confirm the pattern observed in typically developing children in the first few years of formal school (Göbel et al., 2014; Habermann et al., 2017), whereby transcoding skills are primary predictors of arithmetic. In the case of children with SLI, the challenge of learning the number word sequence appears to be particularly important. The combined cognitive load of phonological processing and phonological memory is likely to be overwhelming for many of these individuals, setting limitations on the effectiveness with which associations between spoken words and Arabic numerals can be established.

A study with a French-speaking Belgian sample (Nys et al., 2013) included 10-year olds with SLI, addressing a particular theoretical question concerning the distinction between exact and approximate arithmetic. Dehaene and Cohen (1995) proposed that exact arithmetic (e.g., the sort of basic calculation tasks reported in the studies reviewed above) is supported by language, while approximate arithmetic (e.g., estimation of approximate solutions) is not language dependent, but rather is supported by nonverbal representations (the ANS). To test this model in children with SLI, Nys et al. administered tasks of exact and approximate arithmetic and symbolic and nonsymbolic magnitude comparison (pairs of double-digit Arabic numerals and pairs of dot arrays carefully controlled for perceptual vs numerical cues). Results confirmed many previous findings. Verbal counting (assessed as a background measure) and exact calculation were particular areas of difficulty for children with SLI. A strong correlation was found between phonological processing and exact arithmetic. Nonsymbolic tasks showed no deficit. Symbolic comparison showed deficits that were mediated by cognitive factors (verbal counting was included as one of these). In conclusion, the authors emphasize the relative strength of the approximate system while acknowledging the particular challenges posed by the Arabic numeral system.

Effects of Cognitive Factors

An important study by Koponen, Mononen, Rasanen, and Ahonen (2006) found a somewhat different pattern of mathematical skills in 9- to 11-year old Finnish-speaking children with SLI. These researchers did not find severe deficits in arithmetic skills, nor did they find evidence of relative strength in nonverbal tasks. Correlational evidence suggested that rapid naming of colors and objects was associated with calculation ability, and the authors proposed that variation in efficiency of access to long-term memory might account for the heterogeneity of mathematical abilities. Findings similar to those of Koponen et al. were reported by Kleemans, Segers, and Verhoeven (2012) who studied cognitive and linguistic skills of a Dutch sample of 6-year olds with SLI, assessing their basic calculation skills a year later. In a structural equation model of predictors of calculation skill, rapid naming emerged as a strong predictor specifically in the SLI group. Both the Finnish and Dutch studies report prediction (concurrent and longitudinal) of calculation skills from a cognitive factor (rapid naming) outside the domain of numerical cognition. Comparability with the findings of Fazio and Donlan and colleagues is hard to evaluate given differences in sampling, task selection and data analysis, and, not least, language.

CONCLUSIONS

The studies reviewed above concerning challenges presented by the number word sequence and transcoding for children with SLI draw an interesting distinction between procedural deficits and conceptual strengths. Recent studies of typical development (Lipton & Spelke, 2005, 2006) have sought to establish linkage between procedural knowledge (the spoken number sequence) and conceptual understanding (conservation of number), such that familiarity with procedures supports understanding of underlying principles. Viewed from this perspective, the limitation imposed by procedural deficits may well accumulate over time, as suggested by the findings of Durkin et al. (2013). Whether it is the case that the challenges of integrating spoken numbers and Arabic numerals and internalizing the combined number system are persistent over development is as yet untested. None of the studies I have reviewed have followed children with SLI into the later years of education using measures that allow evaluation of underlying mechanisms.

There are some important discrepancies between findings concerning the mathematical skills of children with SLI, most notably when comparing the study by Koponen et al. (2006) with other studies. Numerous explanations are possible. Not least among these arc (a) differences in educational experience and (b) differences in the learning challenges presented by the varying transparency of the relations between spoken numbers and Arabic numerals in different languages. It is striking that the literature makes scant reference to these factors; learning challenges specific to the spoken number systems of different languages are seldom mentioned.

All told, the literature on mathematical skills of children with SLI is sparse. Effects of the natural language context for number learning have yet to be explored. The pragmatic context of communication about number and quantity is the subject of a single study so far. Katsos, Roqueta, Clemente, Rosa, et al. (2011) have shown that children with SLI show pragmatic skills consistent with their language level; further work is needed to evaluate effects on mathematical development. Much of the research that has been published lacks the power and comprehensive measurement needed to meet the challenges of heterogeneity in the population. Nonetheless, it is to be hoped that before long, we are able to mount a comprehensive longitudinal comparative study to examine the effects of linguistic deficits on mathematical development throughout the school years, taking account of variation in language and educational experience.

REFERENCES

Almoammer, A., Sullivan, J., Donlan, C., Marusic, F., Zaucer, R., O'Donnell, T., et al. (2013). Grammatical morphology as a source of early number word meanings. *Proceedings of the National Academy of Sciences of the United States of America*, *110*, 18448–18453. https://doi.org/10.1073/pnas.1313652110.

Aslin, R. N., & Newport, E. L. (2012). Statistical learning; from acquiring specific items to forming general rules. *Current Directions in Psychological Science*, *21*, 170–176.

Bishop, D. (2009). Genes, cognition, and communication insights from neurodevelopmental disorders. *Annals of the New York Academy of Sciences*, *1156*, 1–18.

Bishop, D. V. M. (1997). *Uncommon understanding: Development and disorders of language comprehension in children*. Hove: Psychology Press.

Bishop, D. V. M. (2014). Ten questions about terminology for children with unexplained language problems. *International Journal of Language and Communication Disorders*, *49*, 81–415.

Bishop, D. V. M., Adams, C. V., & Norbury, C. F. (2006). Distinct genetic influences on grammar and phonological short-term memory deficits: Evidence from 6-year-old twins. *Genes Brain and Behavior*, *5*(2), 158–169.

Bishop, D. V. M., North, T., & Donlan, C. (1996). Nonword repetition as a behavioural marker for inherited language impairment: Evidence from a twin study. *Journal of Child Psychology and Psychiatry and Allied Disciplines*, *37*(4), 391–403.

Bishop, D. V. M., Snowling, M. J., Thompson, P. A., Greenhalgh, T., & CATALISE consortium. (2016). CATALISE: A multinational and multidisciplinary delphi consensus study. Identifying language impairments in children. *PLOS One*, *11*(7), e0158753.

Chiat, S. (2000). *Understanding children with language problems*. Oxford: Cambridge University Press.

Chu, F., vanMarle, K., & Geary, D. (2015). Early numerical foundations of young children's mathematical development. *Journal of Experimental Child Psychology*, *132*, 205–215.

Cowan, R., Donlan, C., Newton, E., & Lloyd, D. (2005). Number skills and knowledge in children with specific language impairment. *Journal of Educational Psychology*, *97*(4), 732–744.

Dehaene, S., & Cohen, L. (1995). Towards an anatomical and functional model of number processing. *Mathematical Cognition*, *1*, 83–120.

Donlan, C., Cowan, R., Newton, E. J., & Lloyd, D. (2007). The role of language in mathematical development: Evidence from children with specific language impairments. *Cognition*, *103*, 23–33.

Durkin, K., Mok, P., & Conti-Ramsden, G. (2013). Severity of specific language impairment predicts delayed development in number skills. *Frontiers in Psychology*, *4*, 581.

Fazio, B. B. (1994). The counting abilities of children with specific language impairment: A comparison of oral and gestural tasks. *Journal of Speech and Hearing Research*, *37*, 358.

Fazio, B. B. (1996). Mathematical abilities of children with specific language impairment: A 2-year follow-up. *Journal of Speech and Hearing Science*, *39*, 839–849.

Fazio, B. B. (1999). Arithmetic calculation, short-term memory, and language performance in children with specific language impairment: A 5-year follow-up. *Journal of Speech, Language and Hearing Research*, *42*, 420–431.

Geary, D. C. (1995). Reflections of evolution and culture in children's cognition: Implications for mathematical development and instruction. *American Psychologist*, *50*, 24–37.

Gilmore, C., McCarthy, S., Shannon, E., & Spelke, E. (2007). Symbolic arithmetic knowledge without instruction. *Nature*, *447*, 589–591.

Göbel, S., Watson, S., Lervag, A., & Hulme, C. (2014). Children's arithmetic development; it is number knowledge, not the approximate number sense, that counts. *Psychological Science*, *25*(3), 789–798.

Habermann, S., Donlan, C., Göbel, S., & Hulme, C. (2017). The importance of pre-school transcoding in the development of arithmetic skills. In *Paper presented and the London Meeting of Experimental Psychology Society, Jan 4th 2017*.

Halberda, J., Mazzocco, M. M. M., & Feignson, L. (2008). Individual difference in non-verbal number acuity correlate with math achievement. *Nature*, *455*, 665–668.

Haynes, C., & Naidoo, S. (1991). *Children with specific speech and language impairment*. Oxford: Blackwell.

Hodent, C., Bryant, P., & Houde, O. (2005). Language-specific effects on number computation in toddlers. *Developmental Science*, *8*, 420–423.

Holloway, I., & Ansari, D. (2009). Mapping numerical magnitudes onto symbols: The numerical distance effect and individual differences in children's mathematics achievement. *Journal of Experimental Child Psychology*, *103*(1), 17.

Katsos, N., Roqueta, C., Clemente, E., Rosa, A., et al. (2011). Are children with specific language impairment competent with the pragmatics and logic of quantification? *Cognition, 119*, 43–57.

Kleemans, T., Segers, E., & Verhoeven, L. (2012). Naming speed as a clinical marker in predicting basic calculation skills in children with specific language impairment. *Research in Developmental Disabilities, 33*(3), 882–889.

Koponen, T., Mononen, R., Rasanen, P., & Ahonen, T. (2006). Basic numeracy in children with specific language impairment: Heterogeneity and connections to language. *Journal of Speech Language and Hearing Research, 49*, 58–73.

Lipton, J. S., & Spelke, E. S. (2005). Preschool children's mapping of number words to nonsymbolic numerosities. *Child Development, 76*, 978–988.

Lipton, J. S., & Spelke, E. S. (2006). Preschool children master the logic of number word meanings. *Cognition, 98*, B57–B66.

Miura, I. (1987). Mathematics achievement as a function of language. *Journal of Educational Psychology, 79*(1), 79–82.

Mix, K., Prather, R., Smith, L. B., & Stockton, J. (2014). Young children's interpretation of multidigit number names: From emerging competence to mastery. *Child Development*. https://doi.org/10.1111/cdev.12197.

Moeller, K., Korbinian, S., Göbel, S., et al. (2015). Language influences number processing—A quadrilingual study. *Cognition, 136*, 150–155.

Moyer, R. S., & Landauer, T. K. (1967). Time required for judgements of numerical inequality. *Nature, 215*, 1519–1520.

Nelson, K. E., Welsh, J. A., Vance Trup, E. M., & Greenberg, M. T. (2011). Language delays of impoverished preschool children in relation to early academic and emotion recognition skills. *First Language, 31*(2), 164–194.

Norbury, C. F., Gooch, D., Wray, C., Baird, G., Charman, T., Simonoff, E., et al. (2016). The impact of nonverbal ability on prevalence and clinical presentation of language disorder: Evidence from a population study. *Journal of Child Psychology and Psychiatry, 57*(11), 1247–1257. https://doi.org/10.1111/jcpp.12573 [in press].

Noveck, I. A. (2001). When children are more logical than adults: Experimental investigations of scalar implicature. *Cognition, 78*(2), 165–188.

Nys, J., Content, A., & Leybaert, J. (2013). Impact of language abilities on exact and approximate number skills development: Evidence from children with specific language impairment. *Journal of Speech, Language and Hearing Research, 56*(3), 956–970.

Piazza, M., & Eger, E. (2016). Neural foundations and functional specificity of number representations. *Neuropsychologia, 83*, 257–273.

Pouscoulous, N., Noveck, I., Politzer, G., & Bastide, A. (2007). Processing costs and implicature development. *Language Acquisition, 14*(4), 347–375.

Price, G., Palmer, D., & Battista, C. (2012). Nonsymbolic numerical magnitude comparison: Reliability and validity of different task variants and outcome measures, and their relationship to arithmetic achievement in adults. *Acta Psychologica, 140*(1), 50–57.

Reilly, S., Tomblin, B., Law, J., McKean, C., Mensah, F., Morgan, A., et al. (2014). SLI: A convenient label for whom? *International Journal of Language and Communication Disorders, 49*, 416–451.

SLI Consortium. (2002). A genomewide scan identifies two novel loci involved in specific language impairment. *American Journal of Human Genetics, 70*(2), 384–398.

SLI Consortium. (2004). Highly significant linkage to the *SLI1* locus in an expanded sample of individuals affected by specific language impairment. *American Journal of Human Genetics, 74*(6), 1225–1238.

Tomblin, J. B., Smith, E., & Zhang, X. (1997). Epidemiology of specific language impairment: Prenatal and perinatal risk factors. *Journal of Communication Disorders*, *30*, 325–344.

VanMarle, K., Chu, F., Li, Y., & Geary, D. (2014). Acuity of the approximate number system and pre-schoolers' quantitative development. *Developmental Science*, *17*, 492–505.

Viswanathan, P., & Nieder, A. (2013). Neuronal correlates of a visual "sense of number" in primate parietal and prefrontal cortices. *Proceedings of the National Academy of Sciences of the United States of America*, *110*(27), 11187–11192.

Chapter 7

Arithmetic in the Bilingual Brain

Nicole Y. Wicha*, Danielle S. Dickson*, and Amanda Martinez-Lincoln†
*The University of Texas at San Antonio, San Antonio, TX, United States
†The University of Texas at Austin, Austin, TX, United States

INTRODUCTION

In popular culture, numerical and other arithmetic abilities are often discussed as if they involve cognitive processes that are completely independent of language abilities. A fairer assessment is that although the relative independence of these processes is likely true for many levels of numerical cognition, there are also many aspects of mathematical thinking that are nevertheless facilitated by language abilities. Indeed, even critics of the role of language in mathematical learning, such as those who argue that having number words in your language is not necessary for learning advanced math, agree that having names and labels for number concepts provides important support for counting, calculation, and other manipulations of number information (e.g., Gelman & Butterworth, 2005). Additionally, there is a persistent link between certain measures of language ability in children (e.g., phonological awareness scores) and outcomes on simple arithmetic tasks—a finding that has been the source of substantial interest regarding the precise nature and extent of the relationship between language and math (e.g., De Smedt & Boets, 2010; Fuchs et al., 2006; Hecht, Torgesen, Wagner, & Rashotte, 2001; Kleemans, Segers, & Verhoeven, 2011; LeFevre et al., 2010; Rasmussen & Bisanz, 2005; Simmons, Singleton, & Horne, 2008; Simmons & Singleton, 2007; Vukovic & Lesaux, 2013).

Pertinently, simple multiplication problems (e.g., "$2 \times 4 = 8$") have traditionally been memorized through mental repetition and rote verbal rehearsal, generating a language-based representation of these arithmetic facts, much like other overlearned verbal expressions (De Smedt, Noël, Gilmore, & Ansari, 2013; El Yagoubi, Lemaire, & Besson, 2005). Perhaps unsurprisingly, then, fluent use of verbal retrieval strategies for these problems is associated with higher levels of performance (faster and more accurate responding) for these problems (Prado et al., 2013), although procedural approaches to solving problems are sometimes used (e.g., LeFevre, Sadesky, & Bisanz, 1996).

Mathematical Cognition and Learning, Vol. 4. https://doi.org/10.1016/B978-0-12-812574-8.00007-9

BILINGUALISM AND RETRIEVAL OF MATHEMATICAL FACTS

In this chapter, we address how the implementation of verbal retrieval strategies to efficiently access arithmetic-related information might be affected by the ability to speak more than one language. Bilingualism is growing in the United States, both from changes in the population demographics and the rising popularity of bilingual education in U.S. schools. Approximately 1 in 5 children enters school with a language other than English as their first language, and in parts of the United States, like South Central Texas, bilinguals can comprise over 50% of the population (2010 U.S. Census). Much research has been devoted to determining the effects of bilingualism on aspects of a speaker's language acquisition (e.g., with respect to receptive and productive vocabulary; Hoff et al., 2012; Thordardottir, Rothenberg, Rivard, & Naves, 2006; Umbel, Pearson, Fernández, & Oller, 1992) and the influences of bilingualism on broader cognitive abilities (e.g., executive function, for which evidence is mixed and contentious; Bialystok, Craik, & Luk, 2008; Hilchey & Klein, 2011; Paap & Greenberg, 2013; Paap, Johnson, & Sawi, 2015). Given this (sometimes heated) backdrop, it is important to be mindful that the most critical feature of the bilingual experience is still the ability to speak and comprehend in more than one language—a communication advantage that is undeniable.

In educational settings, however, bilingual children typically learn arithmetic facts in only one language—the language that is spoken in their schooling environment. When challenged to solve problems with number words in their other language, bilinguals are able to do so (i.e., they can produce or recognize the correct solution). Nevertheless, they do show differences in accessing math facts across their languages (e.g., Bernardo, 2001). Critically, retrieval fluency for simple arithmetic predicts later higher math competence, which in turn can lead to lifelong consequences on earning capacity and quality of life (e.g., Geary, Hoard, Nugent, & Bailey, 2013; Price, Mazzocco, & Ansari, 2013; Siegler et al., 2012). We do not know whether bilingual children make this critical transition to retrieval fluency as successfully as monolinguals do. Further, if they do transition to using fluent retrieval strategies, we do not know whether this skill is limited to the language in which they were instructed. Given this context, 20% of U.S. children who are bilingual are at a potential disadvantage in early math learning. Characterizing the nature of how this population processes math facts is therefore an important facet of minimizing their potential educational challenges.

How Bilinguals Process Arithmetic

Other chapters in this book indicate that bilingualism may impact symbolic number knowledge in preschool children (Chapter 9 by Sarnecka et al.) and nonsymbolic processing in bilingual adults (Chapter 8 by Salillas and Martínez). In this chapter, we focus on the behavioral and brain evidence for how young adult and child bilinguals process arithmetic, specifically looking at multiplication as an example of exact arithmetic processing. Our lab approaches this research from a bilingualism perspective, informing our questions and paradigms with our studies of language in the bilingual brain. We review bilingual arithmetic processing research, focusing on studies with adult bilinguals that report electrophysiological measures of brain activity during the processing of arithmetic problems. We present evidence from our own work using scalp-recorded event-related brain potentials showing that the language of learning math facts and the frequency of using math facts in a given language can influence the efficiency and the manner in which bilinguals access this information. We conclude with preliminary findings from our investigations of the developmental trajectory for multiplication fact learning in elementary school children. This research has laid the groundwork for understanding bilingual arithmetic, a building block for achieving higher math fluency in 11.2 million (and growing) U.S. children (Aud et al., 2011).

Relevant Models of Mathematical Cognition

Before considering math in the bilingual brain, it is appropriate to discuss the relevant models of math cognition that have set the framework for thinking about arithmetic processing. As mentioned above, simple multiplication problems are solved primarily through direct retrieval from long-term memory, although there is evidence that people use other strategies to solve these problems as well (e.g., repeated addition), especially earlier in development (LeFevre et al., 1996). The details of the primary models of simple arithmetic processing differ somewhat in how they account for different phenomena but largely agree on core principles (see Ashcraft (1992) for a review of differences and similarities in the network-retrieval model (Ashcraft, 1987), the network-interference model (Campbell & Graham, 1985), and the distribution of association model (Siegler, 1988)).

In particular, consistent with modern theories of how semantic information is thought to be retrieved from linguistic representations, simple arithmetic information is accessed through spreading activation across a network of interrelated facts. Differences in response time (RT) to specific problems are then accounted for by the strength of the association between the memory representation for the given problem and its answer, which is strongly influenced by the problem solver's frequency of exposure to these problems and experience with retrieving the answers (see Ashcraft, 1995, for review; c.f., Barrouillet, Fayol, & Lathulière, 1997; Kaufmann, Lochy, Drexler, & Semenza, 2004; Lemaire, Abdi, & Fayol, 1996). Importantly, then, not all information in the network is retrieved with the same level of ease; the problem-size effect, for instance, where problems with smaller solutions are retrieved faster and are less error-prone than larger problems, suggests that some parts of the network are more accessible than others, perhaps due to less interference from other stored facts (Campbell & Graham, 1985).

This spreading of information across the arithmetic network leads to variable RTs in verification tasks, wherein subjects must respond with a confirmation or rejection of the provided answer (e.g., $4 \times 3 = 12$?), based on the strength of the connection between the provided answer and the correct solution. In initial work using this type of verification task (instead of production tasks that had dominated the literature, wherein the answer is not provided and people produce their own answer), Campbell (1987) observed delays in RT when participants needed to reject an answer as incorrect if that answer shared properties with the (not presented) correct solution. For example, if not only the provided answer is incorrect (e.g., $3 \times 4 = 16$) but also it happens to be the correct solution for a table-related problem ($4 \times 4 = 16$), then RTs are longer than if the provided answer were completely unrelated to the problem (e.g., $3 \times 4 = 17$). Typically, then, the fastest RTs are observed for verification of correct answers (due to familiarity and/or retrieval success), slower RTs are observed for incorrect answers with low association to the correct solution, and the slowest RTs are observed for incorrect answers that are associated with the problem itself or with properties of the correct solution (e.g., table-related numbers, Ashcraft & Battaglia, 1978; Lemaire et al., 1996). This pattern of results is broadly inconsistent with early views in which answer retrieval and verification for a given problem unfolded without influence from other known arithmetic facts (for more on this history, see Brysbaert, 2005; Fayol & Seron, 2005).

The exact structure of the representations in this network has been a matter of some debate, especially given that the same facts can be presented in different forms (e.g., Arabic digits and number words). In addition, the extent to which language itself affects access to or retrieval of these facts is debated (Campbell & Xue, 2001; Dehaene & Cohen, 1995). In particular, within the context of multilingualism, there are potentially multiple avenues to retrieve arithmetic-related information from long-term semantic memory. Although the theories of fact representation most relevant to this discussion propose a representation of arithmetic facts in *some* form of verbal memory, they disagree about the nature of the relationship between learned arithmetic facts and verbal memory in each language. Specifically, one class of theories presupposes that arithmetic facts are encoded in a verbal memory store restricted to the language in which the facts were learned (e.g., Dehaene & Cohen, 1995), whereas the alternative views suggest that these same facts exist in separate verbal memory stores for each language (as well as digit formats), even though the facts were primarily learned only in one of the languages (e.g., Campbell & Xue, 2001).

The dominant example of the former theory is the triple-code model (TCM), which suggests that multiplication tables are "syntactically organized sequence(s) of words" and that the language processing system is the basis of arithmetic fact retrieval (Dehaene & Cohen, 1995, p. 85; Dehaene, Piazza, Pinel, & Cohen, 2003). According to this model, the ability to retain and retrieve arithmetic facts is highly dependent on the language system, as indicated by brain injury case studies (Delazer & Benke, 1997). Evidence in support of the model (specifically, the proposal that simple arithmetic facts can simply be retrieved rather than calculated through manipulations of quantity) comes from the finding that lesions in the inferior intraparietal region impair knowledge of number quantities but do not affect the retrieval of multiplication facts (for review, see Dehaene, Dehaene-Lambertz, & Cohen, 1998; Dehaene et al., 2003). On this view, all well-rehearsed simple math facts are verbally encoded, and all exact arithmetic problems, including those presented visually in the Arabic digit format, rely on transformation into this verbal code before retrieval is possible. Most importantly for bilingualism research, the verbal code is specific to the language in which the information was learned (Dehaene, Spelke, Pinel, Stanescu, & Tsivkin, 1999).

In contrast, the encoding-complex model (ECM) proposes that simple multiplication facts are encoded in format-specific representations, including one for the digit form (or forms, as is the case for Mandarin speakers) and one for each written number-word form—that is, each language has its own representation of arithmetic facts (Campbell & Clark, 1988). The facility with which facts are successfully retrieved is modulated by the strength of access to each format-specific representation, through a mechanism similar to how individual problems within the network of arithmetic might vary in retrieval strength as a function of experience (Campbell & Clark, 1988, 1992).

Since the digit format answers can be accessed independently of their verbal counterparts and since people have more experience using the digit than word format, the ECM makes the prediction that answers to arithmetic problems presented as digits should be more easily accessed and less susceptible to retrieval interference than answers to the same problems presented as number words (Campbell & Clark, 1992). This would seem to be in direct contrast to the TCM, in which processing of digits requires a kind of translation into the verbal code for successful answer retrieval.

Indeed, the TCM proposes that the disadvantages for the less frequently used language are the result of additional translation processes from the inexperienced language to the experienced language, from which the rehearsed verbal code is then accessed (Dehaene et al., 1999, p. 971). In contrast, the ECM predicts that retrieval can happen in each language without the need for translation *across* languages, although access to each language should vary based on the strength of the connections within and between modules (Campbell & Epp, 2004). That is, if a problem is presented using number words in the language with which the solver has more experience, retrieval for that language will be superior (at least on some measures) to retrieval for problems presented in the alternate language, due to frequency of exposure.

Owing to the similarities in predicted behavioral outcomes (e.g., both theories suggest slower responses for the inexperienced language), adjudicating between these views can be challenging with behavioral evidence alone. Next, we review behavioral results and note the level of support provided for each theory when possible.

Behavioral Evidence From Bilinguals as Tests of the TMC and ECM Models

In general, bilinguals are more efficient at processing arithmetic in the language of learning or acquisition (herein LA+), which is often, but not always, their first language (L1) (c.f., Bernardo, 2001).[1] Perhaps, the most influential study in this area was conducted by Spelke and Tsivkin (2001). Bilingual college students were taught a set of new math facts (and other information) in one of their languages (English or Russian) and then later attempted to recall the facts in the same language or the untrained language. No difference was observed across languages for math that required calculation, implying that this new knowledge was not encoded in a language-specific manner. However, math that would be more likely to rely on verbal retrieval strategies (i.e., exact arithmetic) was solved faster and more accurately in the language of learning than the other language (herein LA−). This advantage was interpreted as evidence that the math facts themselves were stored and processed in the language of learning, in line with the TCM. However, the emergence of this kind of processing advantage does not necessarily have to be the result of selective processing in LA+ and an absence of processing in LA−. The delays for LA− relative to LA+ could instead have been caused by comparatively weaker access to these math fact representations in LA−, consistent with the ECM. Thus, simple comparisons of behavioral performance under LA+ and LA− formats can lead to interpretations in support of either theory.

Importantly, a third condition, wherein researchers additionally examine behavioral performance when problems are presented in the Arabic digit format, has produced results that are predominantly consistent with the predictions made by only the ECM theory. For example, Bernardo (2002) observed that Filipino high school students, who received their education in English but spoke Tagalog as a first language, performed better in arithmetic when it was presented in English (LA+) than when it was presented in Tagalog (LA−). As before, this result taken in isolation could still be interpreted as support for either view of bilingual arithmetic processing. However, the Arabic digit format *also* resulted in superior performance with number words (regardless of language), which is more consistent with the ECM, given that

1. Students with L2 as their LA+ can enter secondary school ill-prepared for higher math when their mastery of L2 is low (i.e., it is hard for students to advance when they are not able to communicate their understanding of lower material to their teachers). In this chapter, we are mostly addressing the scenario wherein bilingual students are competent in both languages—which is more common for the age at which math facts begin to be taught (Moschkovich, 2002).

the TCM predicts that LA+ should be the format with most direct access for fact retrieval.

In other research with bilinguals, Frenck-Mestre and Vaid (1993) similarly observed the fastest RTs for the verification of problems presented as Arabic digits, the slower RTs for problems presented as words in LA+, and the slowest RTs for problems presented as words in LA−. Our lab has also shown that the brain responses for multiplication verification occur earlier for problems presented as Arabic digits than number words (Salillas & Wicha, 2012); this is true for both bilinguals and monolinguals (Martinez-Lincoln, Giattino, Chapman, & Wicha, 2014). In brief, although it is true that the LA+ word format is processed more efficiently than the LA− word format, the consistent advantages for the digit format over the number words present a challenge to the view that arithmetic problems are always solved by verbal translation into some kind of LA+ code. On that view, it might be expected that number words themselves would be most efficient, given that they would provide access to the memorized arithmetic facts without the need for transcoding prior to retrieval (c.f., Siegler, 1988). This digit advantage over number words was a critical motivating factor for the development of the original ECM as a model of monolingual arithmetic processing (Campbell & Clark, 1992).

Thus, the ECM appears to do a more complete job than the TCM of explaining format-related effects in bilingualism and arithmetic in general. However, it is still an imperfect guide to explaining the mechanisms that bilinguals engage for exact arithmetic. There is accumulating evidence that the bilingual lexica are interconnected and that bilinguals are unable to completely suppress one language when processing the other (e.g., Dijkstra & Van Heuven, 2002; Hernandez, Woods, & Bradley, 2015; Hoshino & Kroll, 2008; Kroll, Sumutka, & Schwartz, 2005; Marian & Spivey, 2003; Perani & Abutalebi, 2005; Sunderman & Kroll, 2006). It is problematic, then, that the ECM proposes separate verbal memory stores for each language that are only connected to each other indirectly through the digit format or through the magnitude code (Campbell & Epp, 2004). The model's assumption that math facts are accessible only through one independently operating language at a time, with no direct associations across languages, is thus broadly inconsistent with what is otherwise known about the dynamics of bilingual verbal memory.

In general, then, no existing theory of bilingual arithmetic processing is perfect or complete. What seems likely to lead to improvements is to incorporate more knowledge from existing fields (i.e., the bilingualism literature more broadly) and to acquire more evidence from other methods for studying cognition (i.e., brain imaging). Neuroimaging research is important, not only because it provides unique insights into the brain mechanisms underlying behavioral performances (see also Berch, Geary, & Mann Koepke, 2016), but also because it can provide measures of distinct cognitive processes indiscernible by behavior methods alone. Namely, the multidimensional nature of the results can help determine if differences in performance across languages are caused by quantitative differences in the same underlying cognitive and neural processing mechanism, such as more efficient use of the same mechanisms, or qualitative differences in processing, such as different neural subsystems for each language.

Although both the TCM and the ECM predict a quantitative processing disadvantage for the LA−, the proposed *source* of the disadvantage is different. The TCM predicts qualitative differences in processing, with LA− relying on translation rather than direct retrieval, which would imply substantial differences in the cortical substrates as well. In contrast, the ECM predicts similar verbal memory retrieval mechanisms for both languages, perhaps using neighboring regions of cortex. We will now review brain evidence in bilingual processing of arithmetic, some of which can be taken as support of a single, language-specific system of arithmetic processing and some of which is more consistent with a system with multiple formats whose access is weighted by experience.

Review of Brain Evidence in Bilingual Arithmetic Processing

Neuroimaging Techniques

For the most part, neuroimaging techniques, such as functional magnetic resonance imaging (fMRI) (and other spatial localization techniques), have focused on locating activations in different regions of the brain during arithmetic processing by monolinguals. In general, verification tasks that are thought to rely on retrieval strategies (i.e., judging simple arithmetic problems for correctness) have been associated with a large frontoparietal network (e.g., Chochon, Cohen, Van De Moortele, & Dehaene, 1999; Fehr, Code, & Herrmann, 2007). Interestingly, particular subregions of the parietal lobe seem critical for successful arithmetic processing, especially when relatively more calculation and estimation is involved, as is the case when problems are less rehearsed (e.g., division problems) or when answer retrieval is otherwise not yet fully mastered (Delazer et al., 2003; Grabner, Saalbach, & Eckstein, 2012; Prado, Mutreja, & Booth, 2014; Qin et al., 2014; Rosenberg-Lee, Chang, Young, Wu, & Menon, 2011; Tschentscher & Hauk, 2014). The role of frontal regions during arithmetic processing has not been as well characterized but is often attributed to general cognitive demands on attention and executive function (Fehr et al., 2007; Prado et al., 2011; see Arsalidou & Taylor, 2011, for review).

As children learn arithmetic facts such as multiplication tables, they become increasingly reliant on direct retrieval strategies. As retrieval success increases, there is in turn a decrease in the engagement of parietal regions (Prado et al., 2014). At the same time, a compelling finding has been that increased use of verbal retrieval strategies can be associated with increased engagement of left temporal areas that are also associated with language processing (Berteletti, Prado, & Booth, 2014; Prado et al., 2014, 2011; also see Price, 2012, for a review of the rich and complicated neuroimaging findings in the language literature). These results largely confirm what the behavioral literature already suggested: some subset of cognitive resources associated with verbal language are used when retrieving arithmetic facts from memory, and these language resources are increasingly tapped as arithmetic facts are rehearsed and successfully memorized.

Not surprisingly, then, several studies have also provided evidence for a language of learning advantage when bilinguals are retrieving these facts, in the form of increased activity in frontal and parietal areas when processing in LA –. This pattern of activity is often attributed to an increase in difficulty for LA – fact retrieval, and it is typically interpreted as a sign that participants might be engaging in more complex strategies to solve these problems (i.e., calculation rather than fact retrieval) (e.g., Grabner et al., 2012; Lin, Imada, & Kuhl, 2012; Mondt et al., 2011; Venkatraman, Siong, Chee, & Ansari, 2006). Importantly, however, the network involved in processing math facts in either language is similar—what appears to change is how much relative demand there is on different parts of the *same* anatomical network. It seems fair to take these results as an indication that each language's retrieval mechanism is not completely autonomous and might share important commonalities. In general, the results from this initial work do not provide support for substantially different networks when processing arithmetic facts across languages. It is still possible, though, that additional research could reveal distinctions.

Temporal Precision in Neuroimaging Techniques

A limitation of localization techniques such as fMRI is that they trade off higher precision in space for lower precision in time, providing at best a 1 s resolution (precision in time) of real-time brain activity. However, changes in neural activity occur on the order of milliseconds (ms), with several processing stages taking place within the first second of experiencing a stimulus. In other words, the timing of these rapid changes in brain activity is lost to most neuroimaging techniques that prioritize anatomical localization. Time-sensitive brain imaging measures, such as neuronal electrophysiology, can then provide the complementary temporal precision that is needed to separately measure how processing unfolds prior to, during, and after a stimulus is perceived or a response is executed.

Electrophysiology and Event-Related Potentials

For this reason, electroencephalography (EEG) has been successfully used to study multiple stages of arithmetic processing in both monolinguals and bilinguals. EEG is a direct measure of neural activity, typically recorded from the scalp, which reflects the electric changes over time of large populations of neurons with millisecond precision. From intracranial recordings, we know that EEG generally captures postsynaptic potentials from cortical pyramidal neurons (for review, see Luck & Kappenman, 2011). The ongoing EEG can be time-locked to particular events of interest and then averaged across trials and subjects to generate the average brain response or event-related potential (ERP) to an experimental condition of interest. These derived ERPs are a time-sensitive multidimensional measure of brain electric activity, with functionally specific effects (i.e., changes in the recorded activity compared to a baseline). The amplitude of the change in voltage, polarity (whether the difference in voltage between conditions is negative or positive), latency (the timing of the effect), and scalp distribution (which electrodes show the effect) of the waveform are all independently informative about the nature of ongoing cognitive activities. Importantly, ERPs can provide information about processing without dependence on explicit responses or self-reported of strategies, which have been a particular issue of some contention in the domain of arithmetic problem solving (e.g., Fayol & Thevenot, 2012).

In the first 300 ms after the onset of a stimulus, ERP components (the typically observed modulations in amplitude) generally reflect modality-specific sensory processing. For example, with visual stimuli, early components are modulated by contrast, size, brightness, and attention (Pratt, 2011). After about 300 ms, more modality-independent effects emerge, including well-characterized cognitive components that have been the most relevant to studies of arithmetic processing—the N400 (see Fig. 1) and a subsequent positive shift or late positive component.

ERP studies of arithmetic have typically measured two effects: the congruency effect, primarily observed in verification tasks during recordings of the brain response to a provided answer (e.g., Niedeggen, Rosler, & Jost, 1999), and the problem-size effect, observed during recordings of the brain response either to provided answers or after presentation of the operands (e.g., 7×6) but prior to the answer itself (Jost, Hennighausen, & Rösler, 2004; Zhou et al., 2006). We focus here on the congruency effect, found in comparisons of brain responses for congruent solutions (correct answers) and incongruent solutions (incorrect answers), as this has been the primary effect measured in bilingual populations. Importantly, typical studies of the congruency effect report the time-locked brain response to the answer presented in isolation, after retrieval/calculation processes have already been initiated by prior presentation of the operands.

The congruency effect is characterized by a more negative response to incorrect solutions (e.g., $7 \times 6 = 36$) compared with correct solutions (e.g., $7 \times 6 = 42$), with a maximum peak amplitude difference occurring about 350–400 ms after the presentation of the answer (Niedeggen et al., 1999). This arithmetic congruency effect has drawn comparisons to similar effects in other domains that study the processing of potentially meaningful items (e.g., language and object processing). Notably, words that are unexpected/incongruent with a prior sentential context (e.g., the brain response to "dog" given the

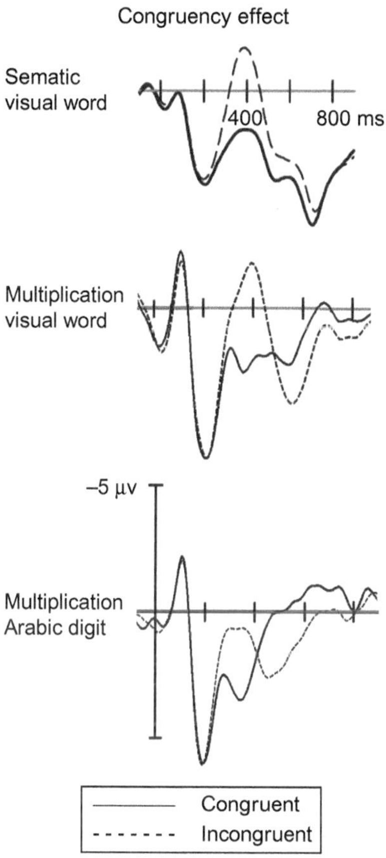

FIG. 1 Grand average ERPs from monolingual adults recorded in response to expected and unexpected sentence-final words (*top*, from Wicha, Moreno, & Kutas, 2004) and in response to correct and incorrect multiplication solutions presented as number words (middle) or Arabic digits (bottom) (unpublished data from Wicha lab). The data come from a representative electrode, with 1 s of time in milliseconds along the *X*-axis and voltage (in microvolts, with negative plotted up) on the *Y*-axis.

prior context, "he takes his coffee with cream and ___") also elicit a more negative brain response than words that are expected/congruent with a prior context (e.g., the brain response to "sugar" given the same context), around 400 ms after the congruent/incongruent word is presented. Thus, the effect of answer congruency has been traditionally labeled an N400 effect (a negative-going wave sensitive to semantics/meaningfulness, peaking at 400 ms; see Kutas & Federmeier, 2011, for a review of N400 response properties).

Both the timing and the size of this congruency effect (i.e., the average difference in voltage across a time window surrounding the ERP effect) can be taken as independent measures of the brain's readiness to categorize an arithmetic problem as correct or not. If the maximal difference occurs later in time (i.e., at 400 ms instead of 350 ms), then it is reasonable to infer that the brain had to do additional processing before the correctness judgment could be fully rendered (much like an RT delay). Moreover, if the average voltage difference is a smaller or larger size in response to different contrasts of answer subtypes (e.g., table-related or table-unrelated incorrect solutions) or to the same answer subtypes under different contexts (e.g., digits/words and fast/slow presentation latencies), then the size of the difference itself can be taken as an indication of the brain's ability to distinguish between the correct and incorrect solutions. This initial and short-lived (~200 ms) effect is generally followed by a slow positive ERP selective to unambiguously incorrect answers. Notably, this later effect is not as consistent or as well characterized and also seems to reflect sensitivity to more subtle differences in answer types (as in traditional language studies, Van Petten & Luka, 2012, for review).

Using ERPs to Investigate Bilingual Processing of Multiplication Problems

We were the first to measure these ERP effects in bilinguals (Salillas & Wicha, 2012). Our samples of adult bilinguals were taught multiplication during childhood in only one of their languages. At the time of testing, they were fluent in both languages and used both languages frequently. Simple multiplication problems were visually presented to them, with one operand presented at a time, followed by the critical stimulus that was either a correct or incorrect solution (e.g., "3...2...6" and "3...2...7," respectively). The multiplication problems were presented in three different formats: Arabic digit format (above) and written number-word formats in English (e.g., "three...two...six") and in Spanish (e.g., "tres...dos...seis").

As in other typical verification paradigms, the participants' task was to determine if the last number was the correct product of the first two numbers or not. We then measured the effect of congruency from the onset of the solution (the third and last number in the sequence) and compared this effect across three critical format conditions: digits, language of learning (LA+), and other languages (LA−). Notably, then, for some participants, LA+ was Spanish, whereas, for other participants, the LA+ was English. This was therefore not a study comparing the response to specific languages per se, but was instead a study of the effect of arithmetic-specific language experience on arithmetic performance under different contexts.

With respect to the primary effect of interest, the N400-like congruency effect, all three formats succeeded in eliciting robust modulations.[2] Critically,

2. Salillas and Wicha (2012) also reported an effect of relatedness for related compared with unrelated incorrect solutions. However, this effect has not been replicated in subsequent studies.

this effect was more similar than different across formats, indicating that the neural responses giving rise to the effects were at least partially overlapping. However, the ERP responses did differ across formats in two key ways.

First, the timescale for processing digits was different than the timescale for processing number words. Problems presented in the digit format elicited a congruency effect that started and peaked earlier (approximately 55 ms) than the effect for number words. This indicates that problems presented in the digit format were more efficiently processed than problems presented as number words were and is consistent with RT findings. Moreover, these ERP findings are able to pinpoint when congruency is processed in each format—prior to when any response is executed. Recall that ERP congruency effects reflect at least two stages of processing in adults, the N400 and the late positivity. The effect of format was observable on this initial congruency effect. Thus, any theory that attempts to account for format-related processing differences needs to incorporate the impact of the effects at the earliest stages of sensitivity to congruency (see also Megías & Macizo, 2016; Megías, Macizo, & Herrera, 2015).

When initially interpreting this latency difference, there was concern that the delayed processing of number words relative to digits might be related to the generally slower processing of words observed in bilinguals (as compared with monolinguals, e.g., Martin et al., 2012). However, we have since observed an equivalent delay in the latency of the effect for number words compared with Arabic digits in monolinguals (Martinez-Lincoln et al., 2014), suggesting that, indeed, the faster processing of digits was not simply an artifact of slower bilingual word reading. The effect for digits was also distributed more focally over the medial posterior sites on the scalp, whereas the effect for number words was more broadly distributed (spreading more to frontal electrode sites than did the effect in response to digits), suggesting at least partially different neural resources were recruited for arithmetic processing in digit and word formats.

The second key finding was that the number words in LA+ elicited a larger-sized effect than did number words in LA−. This suggested that there were weaker activations in the memory network for multiplication facts in LA− although, mechanistically, the lower sensitivity to answer correctness in LA− could result from many possible sources (e.g., less stable representations and weaker access to those representations). Whatever the precise reason, it was apparent that when readers encountered problems in LA−, the assessment of the correctness of the answer was not as robust.

Critically, the effect was no different between LA+ and LA− with respect to its distribution across the scalp or its timing. That the scalp distributions were similar suggests that accessing answers to arithmetic problems in LA− relied on similar processing mechanisms (or at least similar neural sources) as accessing answers in LA+ did. Although this finding is in line with the ECM in the sense that problems were processed in a similar manner in each language, it is also possibly inconsistent with the notion that separate cortical regions are involved when accessing arithmetic representations in each language (but see Luck & Kappenman, 2011, about the ill-posed nature of source localization of ERPs). Moreover, the similar time course across languages does not support the suggestion that translation from LA− into LA+ is necessary for fact retrieval, as suggested by the TCM, which would call for additional processing time when using LA− to accommodate that additional processing stage. In brief, the language of learning advantage in processing simple multiplications appeared to be quantitative in nature (i.e., reflected in the amplitude of the effect), with differences in processing efficiency, as suggested by the ECM, rather than qualitative, such as selective encoding of the facts into only one language, as suggested by the TCM.

An additional observation from Salillas and Wicha (2012) was that the correctness effect in LA−, but not LA+, was positively correlated with the frequency of the daily use of that language. That is, the more a bilingual adult used the LA− in daily life, the better the access to these facts (or stronger representations of these facts) in verbal memory and, consequently, the larger the magnitude of the N400-like congruency effect. This suggested that the processing advantage in LA+ (or disadvantage in LA−) might be mitigated by experience using multiplication facts in the other language.

In a subsequent study, Martinez-Lincoln, Cortinas, and Wicha (2015) tested this possibility using the same multiplication verification task, this time with a population of elementary school teachers who were fluent Spanish-English bilinguals. The key element was that half of the instructors taught arithmetic in their LA− (e.g., having learned arithmetic in Spanish, they now taught in English), while the other half of the instructors still taught in their language of learning (LA+). The hypothesis, then, was that teaching in LA− could strengthen access to (and/or representations of) arithmetic facts in that language, perhaps overriding the advantage usually observed for LA+.

As expected, we observed the LA+ advantage in teachers whose LA+ was also their teaching language (i.e., they retained a larger magnitude congruency effect for LA+ compared with LA−) (as per Salillas & Wicha, 2012). In contrast, there was no difference in magnitude for the congruency effect between LA+ and LA− for the instructors teaching in their LA− (see Fig. 2). That is, either their LA− congruency effect became larger to match the LA+ congruency effect (as a result of their *experience* with LA−) or the advantage they otherwise would have for LA+ was lost from disuse (due to *inexperience* with LA+ over time). Given that noninstructors possess the LA+ advantage despite similar amounts of the lack of use, it seems more probable that the LA− congruency effect grew more robust in this population.[3]

Another finding from this work was that the timing of the effect itself differed across languages based on experience, such that the effect peaked earlier

3. Direct comparisons of these effects across populations of teachers while tempting are difficult to interpret due to the potential influence of individual differences.

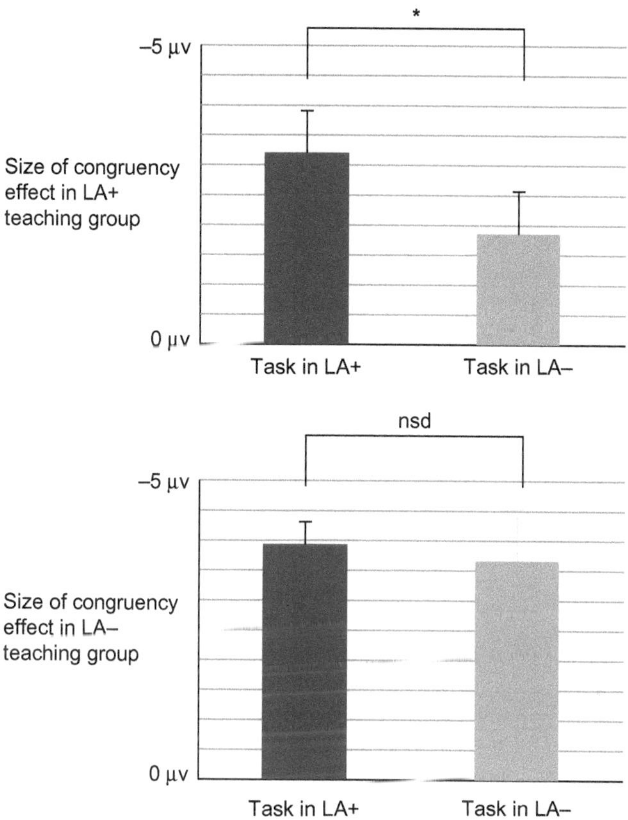

FIG. 2 The effect of congruency is plotted in two groups of bilingual teachers performing a simple multiplication verification task in their LA+ (*left*) and LA− (*right*). Although a robust congruency effect is present in all cases, a difference in the size of the congruency effect for performing the task in LA+ compared with LA− was only observed in the teaching group who continued to use their LA+ in the classroom (top chart). In the LA− teaching group (bottom chart), there was no difference in the size of the congruency effect as a function of language used in the task (i.e., no LA+ advantage).

in the teaching language than the nonteaching language regardless of what their language of learning had been. Thus, long-term experience instructing in a language not only benefited the strength of arithmetic fact representations in memory for that language (leading to a larger congruency effect) but also sped (facilitated) access to math facts in that language. That is, when participants had experience teaching in their LA−, the LA− became at least as efficient as the LA+ for processing simple multiplication problems. This experience-driven explanation of arithmetic processing is more consistent with the ECM than the TCM, which (in its various iterations) has tended to be a more static model of arithmetic fact representations in the mind and brain (i.e., language of learning and experience seem not to be treated as independent properties of the system).

As alluded to in the introduction of ERP work with arithmetic, sometimes there are additional effects following this initial effect of congruency. Indeed, this was the case in the teacher study: there was a later positivity for incorrect solutions, selective to people who were instructing arithmetic in their LA −. Later positivities of this kind are often thought to reflect more explicit and controlled analyses of the presented item—which might be expected for a population who has needed to rely on self-monitoring strategies for success in teaching in a less dominant language. We include this result here, in part, as a demonstration of the value and power of ERPs for exploring the multidimensional nature of assessing the solution to a problem. Whereas behavioral results can tell us critical end-state outcomes, this dynamic view into the cognitive system allows us further insight into the complex processes unfolding as the answers are evaluated prior to a response being executed.

In general, our study of bilingual teachers provided the first evidence that language experience can affect arithmetic fact processing in the adult bilingual, suggesting that an adult bilingual's ability to perform arithmetic not only is influenced by early language experience (i.e., effects of LA+ are long-lasting) but also remains sensitive to language use through adulthood (i.e., the effect of congruency was earlier in the teaching language regardless of the language in which arithmetic had originally been learned). These results support a view wherein bilingual arithmetic facts can be reorganized and strengthened in either language depending on how individuals use their languages at any point in time.

Using ERPs to Investigate Monolingual and Bilingual Children's Processing of Multiplication Problems

These findings in young adult bilinguals, when the encoding of multiplication facts is well established, have shifted our attention to the early learning of math facts in young children. Even with monolingual children, there has been relatively little brain imaging research conducted on this topic in the most relevant age group—namely, from third- to fifth-grade elementary school children. The few extant studies are not longitudinal and tend to be in older samples (Moore, Drollette, Scudder, Bharij, & Hillman, 2014; Prieto-Corona et al., 2010), leaving basic questions—even in monolingual samples—open to future study. Fortunately, we know that N400 effects emerge and stabilize even in quite young children (although Friedrich & Friederici, 2005, 2006 directly compare adults and children and suggest there are subtle but important underlying differences; for review, see Kutas & Federmeier, 2011). Thus, to the extent that arithmetic congruency effects are indexed by the N400 component, it should be possible to conduct these types of studies in children and obtain reasonably interpretable effects.

To address the many gaps in this literature, our lab has recently started an ERP study of multiplication verification in third- through fifth-grade children. Our goal is to establish the electrophysiological signatures of the developmental transition as children grow from novice to expert multipliers and, in turn, move from calculation to verbal memory retrieval. Our study uses a modified version of the multiplication verification task from Salillas and Wicha (2012).

Prior to ERP recording, the children begin by performing a series of cognitive tasks that measure their language fluency, working memory, math fluency, and other relevant skills. The children then perform an Arabic digit version of the verification task (e.g., 3…2…6 vs 3…2…7) while we record their ongoing brain waves. The off-line measures can then be used to estimate the impact of various factors on task performance and, most critically for us, brain activity. Fig. 3 shows preliminary ERP data ($n=49$, collapsed across age and language groups;[4] see below) measured from the onset of the Arabic digit solution. A robust congruency effect is clearly visible, with a more negative-going response for incorrect compared with correct solutions, which is consistent with the typical pattern for an N400 effect. However, although

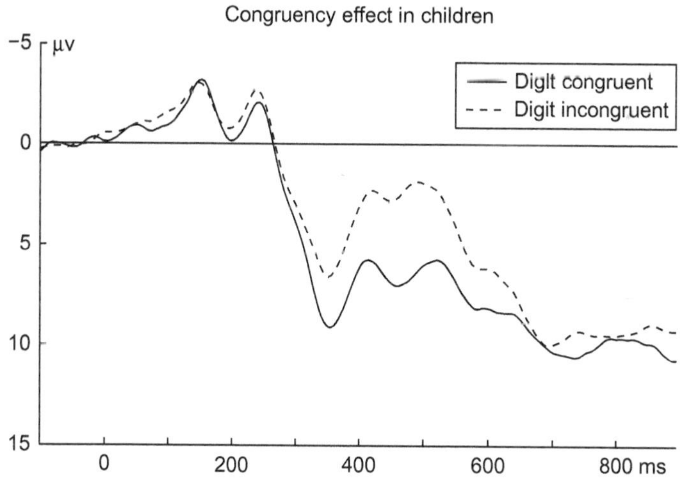

FIG. 3 Grand average ERPs for 49 children (ranging from third to fifth grades) time-locked to the onset of correct (solid black) and incorrect (dashed black) solutions during the multiplication verification task. The ERPs measured at the vertex electrode are shown. A robust N400 congruency effect is observed between 300 and 600 ms—this effect occurs later than the same effect in young adults (e.g., Salillas & Wicha, 2012). In addition, there is no noticeable target P300 response for the correct trials, as is observed in young adults (see Fig. 1).

4. We include in this sample a mix of monolingual English speakers and Spanish-English bilingual children. We present them collapsed across both groups because the samples are not large enough to compare across groups at this time.

the direction of this effect is the same across adults and children (i.e., incorrect solutions elicit a more negative-going ERP than correct solutions), the precise morphology differs between populations (as in Prieto-Corona et al., 2010).

There are two noticeable differences in the brain response in children (Fig. 3) compared with young adults (Fig. 1), both of which seem likely to reflect key developmental changes. First, the N400-like effect occurs later in time for the children than for adults, starting about 100 ms later (Fig. 3) and lasting longer than the typical 200 ms effect in adults (Fig. 1). It is known that age and language proficiency are two factors that can affect the timing of the N400 (Kutas & Federmeier, 2011; Moreno & Kutas, 2005), so this developmental shift is consistent with normal and expected age-related processing differences. Moreover, this result can be taken as an indication that access to this particular subset of semantic memory also becomes more efficient with age and experience, as indexed by the speed with which the brain attempts to extract meaning for the solution to the problems.

A second difference between children and young adults seems to be selective to the processing and recognition of the congruent trials (i.e., the brain response to correct solutions). In the adult data (Fig. 1), the effect of congruency appears to be driven not only by a larger negativity for the incorrect trials but also by another effect overlapping in the same time window (see Jasinski & Coch, 2012, for an expanded discussion of component identification issues in this literature). Most notably, in addition to the co-occurring N400-like effect, there appears to be a classic target P300 response to the correct solutions, specifically when they are presented in the Arabic digit format (bottom plot, Fig. 1).

The target P300 is a well-known ERP component that is most often reported after detection of a target stimulus (i.e., the word "apple" in a word list if the goal is to press a button in response to food items). In this case, the P300 is largest to the correct solution to the provided multiplication problem, which would be a target item based on the task (i.e., finding the correct answers to math problems). This large P300 is clearly visible in the ERP for correct solutions presented in the Arabic digit format (the visually/perceptually expected item). In contrast, the response to number words (middle plot, Fig. 1) is more unambiguously a traditional N400 effect with little evidence for a target P300 response for correct number words. These effects reflect different cognitive processes—whereas the N400 is linked to the obligatory processing of modality-independent meaning (and is largest/most negative for contextually unexpected items), the P300 is linked to stimulus classification and updating of context (and is largest/most positive for task-relevant target items) (Donchin & Coles, 1988; see Polich, 2011, for review).

Intriguingly, in our preliminary child data displayed in Fig. 3, there is no noticeable P300 in response to the correct solutions. Instead, children's responses to congruent digits solutions look more like the adult response to written number words (i.e., the congruency effect itself is primarily an N400 effect with no additional contribution from a P300; see middle plot of Fig. 1). We also see morphological differences in the children across grades (i.e., children with different levels of experience with exact arithmetic)—although, interestingly, even our fifth-grade sample, which we consider successful learners of multiplication, does not show this adult-like response. This suggests that the P300 pattern only emerges after many years of experience with arithmetic facts.

It is possible that children, who have less experience with detecting correct solutions, read multiplication problems for comprehension much like adults read sentences. That is, they process the correct digit like it was an expected word or conceptually congruent picture rather than an item to be detected and categorized. Thus, when encountering either correct or incorrect solutions, they might be attempting to extract the meaning in both cases, resulting in a solitary N400 effect. In contrast, by the time a person reaches adulthood, the solution to a simple multiplication problem is so easily retrievable that digit answers are effectively processed as target items, eliciting a robust P300 response once the end goal (the correct solution) is recognized and categorized. Incorrect solutions do not elicit this target detection response in adults and are instead processed more like unexpected words in a sentence or nontarget items in a word list (which also only produce an N400 component, c.f., Jasinski & Coch, 2012). In other words, adults may come to see arithmetic problems as a prompt to search for the correct answer at the end, whereas children are just trying to build a coherent meaningful representation as the problem unfolds.

Tentatively, then, children's ERPs to the solution appear to differ from adult's both in efficiency (slower latency of the congruency effect in children) and mechanistically (the lack of a P300 in children likely reflects an inability to process correct answers as target items). In addition, like other researchers, in our preliminary overall group average, we do not find the later positivity for incorrect answers that is often present in adults (see Fig. 1, bottom panel, and Moore et al., 2014; Prieto-Corona et al., 2010). This later positivity tends to be smaller than the earlier congruency effect and has more sensitivity to the particulars of task demands, so the possible reasons for its absence in children are likely to be even more complex to account for than P300/N400 differences between children and adults.

What we believe that can be largely taken from the ERP research to date is that the processes that children undergo when evaluating the solutions for multiplication problems are substantially different from the processes adults undergo. Notably, the nuances of these differences are not clearly predicted by any of the existing models of arithmetic, which were based primarily on behavioral responses that reflect the summation of multiple underlying cognitive processes. In contrast, ERPs are a temporally sensitive measure able to expose differences in these underlying cognitive processes with more specificity. Future models of arithmetic will need to take into account these findings to better explain not only the adult state of the system but also how the system arrives at that end state.

CONCLUSIONS AND FUTURE DIRECTIONS

In addition to better characterizing the developmental trajectory that leads to adult-like responses, we also hope to construct a better understanding of the critical link between language experience, arithmetic ability itself, and the congruency effect(s). Thus, in addition to elucidating the developmental trajectory for multiplication fact retrieval when using the Arabic digit format, we are also simultaneously investigating the effect of language use and experience on the morphology of the congruency effect (i.e., its size, timing, distributional properties, and shape). To this end, the population we study includes children who are early Spanish-English bilinguals in addition to monolingual English-speaking children. Given the evidence for effects of bilingualism on other aspects of cognition, discussed above, our ultimate goal is to determine if being bilingual can impact multiplication fact retrieval, even in the digit format. Critical for this comparison, both linguistic populations will have learned—or are still learning—multiplication in English (i.e., all have English as their LA+).[5]

We have already described the ERP responses to problems presented in the Arabic digit format for our entire child sample to date (i.e., Fig. 3). What we have not revealed is that the children's brain responses are also being recorded when they perform the same verification task with the operands presented as number words instead of digits. For this task to better accommodate children, we adapted the number-word version of the multiplication verification task from Salillas and Wicha (2012) in important ways. Critically, when number words are used in the paradigm, they are presented as spoken (auditory) words, rather than written number words as was done with the adult bilinguals. Spoken number words are more naturalistic, given that multiplication problems are rarely presented as fully spelled out written words in an educational setting. And, relating back to the models of arithmetic, the verbal code in the TCM is discussed as a spoken-language memory trace, making spoken number words a more appropriate test for this model.

Additionally, using spoken number words is critical for this child population given that children can vary widely in their reading skills, especially at

5. US schools have various bilingual education programs, some aimed at transitioning non-native speakers into English proficiency and others at fostering bilingualism. Our understanding of how these specific learning environments influence cognitive development, specifically math cognition, is limited. We are also recruiting participants from dual-language learning and other bilingual school environments, but for simplicity, we will not discuss these subgroups, or the predictions related to their experience, here.

the younger end of our sample. Moreover, bilinguals tend not to have equivalent reading experience or ability in their two languages. Therefore, using spoken number words for the operands in each problem allows us to force the child to process the math problem in that language and at the same time eliminates a potential confound of processing differences based on reading ability (or at least reduces its influence), both across individual children and across languages within a bilingual child. We note that this is a more direct approach to studying the effect of language on processing arithmetic than is often used in the literature, where the language manipulation is done by providing the instructions in a particular language or by teaching a novel arithmetic skill in only one language (e.g., Spelke & Tsivkin, 2001; Venkatraman et al., 2006; c.f., Frenck-Mestre & Vaid, 1993, whose work does use number words but still has the potential confound of reading ability differences). It seems to us that our manipulation is closer to what is truly intended when we think about participants "performing math" in a specific language—each problem is deliberately presented in a particular language of interest, hopefully reducing and/or eliminating the potential for use of the alternative language.

To our knowledge, there have been no prior attempts to use auditory number words as operand stimuli in such a paradigm with ERPs, but auditory presentation of sentence stimuli tends to result in similar (if not identical) findings in the broader language literature (e.g., Federmeier, McLennan, Ochoa, & Kutas, 2002; Kutas, Neville, & Holcomb, 1987), so it seemed plausible that the same would be true for our arithmetic "sentences."

The second modification from the adult study is that regardless of the format of the operands, which appear in the Arabic digit, LA+ or LA− formats across alternate blocks, the solution is always presented as an Arabic digit. This is critical because the brain response to the solution will reflect processing specific to that format, as previously described and as depicted above in Fig. 1. Using Arabic digits for the solution across all conditions thus allows us to compare the brain response with physically identical stimuli (i.e., the solution) while still measuring the effect of retrieving (or otherwise processing) the problems in the specific format and/or language of the preceding operands. Thus far, we have observed a robust effect of congruency in both bilinguals and monolinguals in this novel cross modal arithmetic verification task that will lend itself to investigations of the influences of both math and linguistic experience on the developmental trajectory discussed above.

Due to the number of conditions, the number of potential subgroups of populations, and the developmental approach we are taking, this research opens numerous questions, the first of which is when and how being bilingual might affect how children process the provided solutions. To address this matter, our design allows us to compare effects of congruency in the tasks that both bilinguals and monolinguals perform (i.e., using data from both Arabic numeral and English, LA+, and number-word tasks). Given its emphasis on the role of language in math fact processing, the TCM would predict that there should be no differences between monolinguals and bilinguals when performing in English, the LA+ for both groups. In contrast, a bilingual model that takes into account that the amount of exposure in a given format will determine the strength of access to that format, as in the ECM, might predict differences between monolinguals and bilinguals because bilinguals may have less exposure to arithmetic in any one language compared with monolinguals. That is, of their total time spent gaining experience with arithmetic, monolinguals typically spend all of it in English-only contexts, whereas there is a possibility that bilinguals could spend some of that time using arithmetic in their other language, by default having relatively less English exposure than monolinguals (see Gollan, Montoya, Cera, & Sandoval, 2008, for a similar argument regarding bilingual lexical access). Or, perhaps, given the evidence for differences between monolinguals and bilinguals in other domains (e.g., vocabulary acquisition, discussed at the beginning of this chapter), it is possible that bilinguals might outperform or underperform monolinguals in different ways and at different points in development that are not predicted by either numerical processing model.

We are also interested in delineating the developmental trajectory for arithmetic processing in both languages used by bilinguals. We thus plan to compare the effect of congruency when the operands are presented in the LA+ (English) and the LA− (Spanish) as bilingual children gain more expertise in multiplication (from third to fifth grades). The primary evidence of interest will come from tracking the relative size of the congruency effect in the LA− as they gain experience multiplying and retrieving answers in the context of LA+. According to the TCM, the LA+ might establish the sole representation for arithmetic facts during learning, leaving the LA− representations to emerge only via translation, perhaps later in development (thus predicting a larger congruency effect in LA+ than LA−, especially in the least experienced cohort). However, if bilingual children process math facts the same way other information is processed in verbal memory, then it is possible that math fact representations are established in both languages during learning, even when only the LA+ is explicitly used for learning the facts (predicting fairly similar congruency effects in both languages). Given that this work is in its nascent state, there are many other questions that we expect will emerge as we begin to understand the basic mechanisms of early learning and use of multiplication facts in the bilingual brain.

It seems to us that most theories of numerical cognition more carefully define how number-related information is structured in memory than the temporal structure through which this information is accessed (e.g., Campbell & Clark, 1992; Dehaene et al., 2003). By using ERP measures in combination with RT data, we are able to gain insight into the dynamics that lead up to (and follow) the response itself, which we think generates many new constraints to be considered within any model of math cognition. How do RTs and other behavioral outcomes relate to the latency and magnitude of the congruency effect, and does this differ for adults and children? Do children with superior behavioral performance show a target P300 response to correct answers, similar to adults, or does the P300 come later as part of some other developmental change? Why do adults have a secondary stage of processing after the congruency effect, and under what conditions does it emerge in children? And how do these various effects interact, or not, with language experience and bilingualism? At present, there seem to be more questions than answers in this domain, especially compared with the traditional language literature with ERPs. Although this state of affairs can make it challenging for interpreting the various effects, on the whole, we prefer to think that this gap provides current and future researchers a unique opportunity to make substantial contributions to this area in a comparatively brief amount of time.

REFERENCES

Arsalidou, M., & Taylor, M. J. (2011). Is 2+ 2= 4? Meta-analyses of brain areas needed for numbers and calculations. *Neuroimage*, *54*(3), 2382–2393.

Ashcraft, M. H. (1987). Children's knowledge of simple arithmetic: A developmental model and simulation. In J. Bisanz, C. J. Brainerd, & R. Kail (Eds.), *Formal Methods in Developmental Psychology* (pp. 302–338). New York, NY: Springer Series in Cognitive Development.

Ashcraft, M. H. (1992). Cognitive arithmetic: A review of data and theory. *Cognition*, *44*, 75–106.

Ashcraft, M. H. (1995). Cognitive psychology and simple arithmetic: A review and summary of new directions. *Mathematical Cognition*, *1*(1), 3–34.

Ashcraft, M. H., & Battaglia, J. (1978). Cognitive arithmetic: Evidence for retrieval and decision processes in mental addition. *Journal of Experimental Psychology: Human Learning and Memory*, *4*(5), 527–538.

Aud, S., Hussar, W., Kena, G., Bianco, K., Frohlich, L., Kemp, J., et al. (2011). *The Condition of Education 2011 (NCES 2011-033)*. U.S. Department of Education, National Center for Education Statistics Washington, DC: U.S. Government Printing Office [p. 30].

Barrouillet, P., Fayol, M., & Lathulière, E. (1997). Selecting between competitors in multiplication tasks: An explanation of the errors produced by adolescents with learning difficulties. *International Journal of Behavioral Development*, *21*(2), 253–275.

Berch, D. B., Geary, D. C., & Mann Koepke, K. (Eds.), (2016). *Development of mathematical cognition: Neural substrates and genetic influences*. San Diego, CA: Elsevier Academic Press.

Bernardo, A. B. I. (2001). Asymmetric activation of number codes in bilinguals: Further evidence for the encoding complex model of number processing. *Memory & Cognition*, *29*(7), 968–976.

Bernardo, A. B. I. (2002). Language and mathematical problem solving among bilinguals. *The Journal of Psychology*, *136*(3), 283–297.

Berteletti, I., Prado, J., & Booth, J. R. (2014). Children with mathematical learning disability fail in recruiting verbal and numerical brain regions when solving simple multiplication problems. *Cortex*, *57*, 143–155.

Bialystok, E., Craik, F., & Luk, G. (2008). Cognitive control and lexical access in younger and older bilinguals. *Journal of Experimental Psychology: Learning, Memory, and Cognition*, *34*(4), 859–873.

Brysbaert, M. (2005). Number recognition in different formats. In J. I. D. Campbell (Ed.), *Handbook of mathematical cognition* (pp. 23–42). New York, NY: Psychology Press.

Campbell, J. I. D. (1987). Production, verification, and priming of multiplication facts. *Memory & Cognition, 15*(4), 349–364.

Campbell, J. I. D., & Clark, J. M. (1988). An encoding-complex view of cognitive number processing: Comment on McCloskey, Sokol, and Goodman (1986). *Journal of Experimental Psychology: General, 117*(2), 204–214.

Campbell, J. I. D., & Clark, J. M. (1992). Cognitive number processing: An encoding-complex perspective. In J. I. D. Campbell (Ed.), *The nature and origins of mathematical processing* (pp. 457–491). Amsterdam: Elsevier Science Publishers.

Campbell, J. I. D., & Epp, L. J. (2004). An encoding-complex approach to numerical cognition in Chinese-English bilinguals. *Canadian Journal of Experimental Psychology, 58*(4), 229–244.

Campbell, J. I. D., & Graham, D. J. (1985). Mental multiplication skill: Structure, process, and acquisition. *Canadian Journal of Experimental Psychology, 39*(2), 338–366.

Campbell, J. I. D., & Xue, Q. (2001). Cognitive arithmetic across cultures. *Journal of Experimental Psychology: General, 130*(2), 299–315.

Chochon, F., Cohen, L., Van De Moortele, P. F., & Dehaene, S. (1999). Differential contributions of the left and right inferior parietal lobules to number processing. *Journal of Cognitive Neuroscience, 11*(6), 617–630.

De Smedt, B., & Boets, B. (2010). Phonological processing and arithmetic fact retrieval: Evidence from developmental dyslexia. *Neuropsychologia, 48*(14), 3973–3981.

De Smedt, B., Noël, M., Gilmore, C., & Ansari, D. (2013). How do symbolic and non-symbolic numerical magnitude processing skills relate to individual differences in children's mathematical skills? A review of evidence from brain and behavior. *Trends in Neuroscience and Education, 2*(2), 48–55.

Dehaene, S., & Cohen, L. (1995). Toward an anatomical and functional model of number processing. *Mathematical Cognition, 1*(1), 83–120.

Dehaene, S., Dehaene-Lambertz, G., & Cohen, L. (1998). Abstract representations of numbers in the animal and human brain. *Trends in Neurosciences, 21*(8), 355–361.

Dehaene, S., Piazza, M., Pinel, P., & Cohen, L. (2003). Three parietal circuits for number processing. *Cognitive Neuropsychology, 20*(3-6), 487–506.

Dehaene, S., Spelke, E. S., Pinel, P., Stanescu, R., & Tsivkin, S. (1999). Sources of mathematical thinking: Behavioral and brain-imaging evidence. *Science, 284*(5416), 970–974.

Delazer, M., & Benke, T. (1997). Arithmetic facts without meaning. *Cortex, 33*, 697–710.

Delazer, M., Domahs, F., Bartha, L., Brenneis, C., Lochy, A., Trieb, T., et al. (2003). Learning complex arithmetic: An fMRI study. *Cognitive Brain Research, 18*(1), 76–88.

Dijkstra, T., & Van Heuven, W. J. B. (2002). The architecture of the bilingual word recognition system: From identification to decision. *Bilingualism: Language and Cognition, 5*(3), 175–197.

Donchin, E., & Coles, M. G. H. (1988). Is the P300 component a manifestation of context updating. *Behavioral and Brain Sciences, 11*(3), 357–427.

El Yagoubi, R., Lemaire, P., & Besson, M. (2005). Effects of aging on arithmetic problem-solving: An event-related brain potential study. *Journal of Cognitive Neuroscience, 17*(1), 37–50.

Fayol, M., & Seron, X. (2005). About numerical representations: Insights from neuropsychological, experimental, and developmental studies. In J. I. D. Campbell (Ed.), *Handbook of mathematical cognition* (pp. 3–22). New York, NY: Psychology Press.

Fayol, M., & Thevenot, C. (2012). The use of procedural knowledge in simple addition and subtraction problems. *Cognition, 123*(3), 392–403.

Federmeier, K. D., McLennan, D. B., Ochoa, E., & Kutas, M. (2002). The impact of semantic memory organization and sentence context information on spoken language processing by younger and older adults: An ERP study. *Psychophysiology, 39*(2), 133–146.

Fehr, T., Code, C., & Herrmann, M. (2007). Common brain regions underlying different arithmetic operations as revealed by conjunct fMRI–BOLD activation. *Brain Research, 1172*, 93–102.

Frenck-Mestre, C., & Vaid, J. (1993). Activation of number facts in bilinguals. *Memory & Cognition, 21*(6), 809–818.

Friedrich, M., & Friederici, A. D. (2005). Semantic sentence processing reflected in the event-related potentials of one-and two year-old children. *Neuroreport, 16*(16), 1801–1804.

Friedrich, M., & Friederici, A. D. (2006). Early N400 development and later language acquisition. *Psychophysiology, 43*(1), 1–12.

Fuchs, L. S., Fuchs, D., Compton, D. L., Powell, S. R., Seethaler, P. M., Capizzi, A. M., et al. (2006). The cognitive correlates of third-grade skill in arithmetic, algorithmic computation, and arithmetic word problems. *Journal of Educational Psychology, 98*(1), 29–43.

Geary, D. C., Hoard, M. K., Nugent, L., & Bailey, D. H. (2013). Adolescents' functional numeracy is predicted by their school entry number system knowledge. *PLoS ONE, 8*(1), e54651.

Gelman, R., & Butterworth, B. (2005). Number and language: How are they related? *TRENDS in Cognitive Sciences, 9*(1), 6–10.

Gollan, T. H., Montoya, R. I., Cera, C., & Sandoval, T. C. (2008). More use almost always means a smaller frequency effect: Aging, bilingualism, and the weaker links hypothesis. *Journal of Memory and Language, 58*(3), 787–814.

Grabner, R. H., Saalbach, H., & Eckstein, D. (2012). Language-switching costs in bilingual mathematics learning. *Mind, Brain, and Education, 6*(3), 147–155.

Hecht, S. A., Torgesen, J. K., Wagner, R. K., & Rashotte, C. A. (2001). The relations between phonological processing abilities and emerging individual differences in mathematical computation skills: A longitudinal study from second to fifth grades. *Journal of Experimental Child Psychology, 79*(2), 192–227.

Hernandez, A. E., Woods, E. A., & Bradley, K. A. L. (2015). Neural correlates of single word reading in bilingual children and adults. *Brain and Language, 143*, 11–19.

Hilchey, M. D., & Klein, R. M. (2011). Are there bilingual advantages on nonlinguistic interference tasks? Implications for the plasticity of executive control processes. *Psychonomic Bulletin & Review, 18*(4), 625–658.

Hoff, E., Core, C., Place, S., Rumiche, R., Señor, M., & Parra, M. (2012). Dual language exposure and early bilingual development. *Journal of Child Language, 39*(1), 1–27.

Hoshino, N., & Kroll, J. F. (2008). Cognate effects in picture naming: Does cross-language activation survive a change of script? *Cognition, 106*(1), 501–511.

Jasinski, E. C., & Coch, D. (2012). ERPs across arithmetic operations in a delayed answer verification task. *Psychophysiology, 49*(7), 943–958.

Jost, K., Hennighausen, E., & Rösler, F. (2004). Comparing arithmetic and semantic fact retrieval: Effects of problem size and sentence constraint on event-related brain potentials. *Psychophysiology, 41*, 46–59.

Kaufmann, L., Lochy, A., Drexler, A., & Semenza, C. (2004). Deficient arithmetic fact retrieval—storage or access problem?: A case study. *Neuropsychologia, 42*(4), 482–496.

Kleemans, T., Segers, E., & Verhoeven, L. (2011). Precursors to numeracy in kindergartners with specific language impairment. *Research in Developmental Disabilities*, *32*(6), 2901–2908.

Kroll, J. F., Sumutka, B. M., & Schwartz, A. I. (2005). A cognitive view of the bilingual lexicon: Reading and speaking words in two languages. *International Journal of Bilingualism*, *9*(1), 27–48.

Kutas, M., & Federmeier, K. D. (2011). Thirty years and counting: Finding meaning in the N400 component of the event related brain potential (ERP). *Annual Review of Psychology*, *62*, 621–647.

Kutas, M., Neville, H. J., & Holcomb, P. J. (1987). A preliminary comparison of the N400 response to semantic anomalies during reading, listening and signing. *Electroencephalography and Clinical Neurophysiology Supplement*, *39*, 325–330.

LeFevre, J., Bisanz, J., Daley, K. E., Buffone, L., Greenham, S. L., & Sadesky, G. S. (1996). Multiple routes to solution of single-digit multiplication problems. *Journal of Experimental Psychology: General*, *125*(3), 284–306.

LeFevre, J., Fast, L., Skwarchuk, S., Smith-Chant, B. L., Bisanz, J., Kamawar, D., et al. (2010). Pathways to mathematics: Longitudinal predictors of performance. *Child Development*, *81*(6), 1753–1767.

LeFevre, J., Sadesky, G. S., & Bisanz, J. (1996). Selection of procedures in mental addition: Reassessing the problem size effect in adults. *Journal of Experimental Psychology: Learning, Memory, and Cognition*, *22*(1), 216–230.

Lemaire, L., Abdi, H., & Fayol, M. (1996). Working memory and cognitive arithmetic: Evidence from the disruption of the associative confusion effect. *European Journal of Cognitive Psychology*, *8*, 73–103.

Lin, J. L., Imada, T., & Kuhl, P. K. (2012). Mental addition in bilinguals: An fMRI study of task-related and performance-related activation. *Cerebral Cortex*, *22*(8), 1851–1861.

Luck, S. J., & Kappenman, E. S. (2011). ERP components: The ups and downs of brainwave recordings. In S. J. Luck & E. S. Kappenman (Eds.), *The Oxford handbook of event-related potential components* (pp. 3–30). New York, NY: Oxford University Press.

Marian, V., & Spivey, M. (2003). Competing activation in bilingual language processing: Within- and between-language competition. *Bilingualism: Language and Cognition*, *6*(02), 97–115.

Martin, C. D., Costa, A., Dering, B., Hoshino, N., Wu, Y. J., & Thierry, G. (2012). Effects of speed of word processing on semantic access: The case of bilingualism. *Brain and Language*, *120*(1), 61–65.

Martinez-Lincoln, A., Cortinas, C., & Wicha, N. Y. Y. (2015). Arithmetic memory networks established in childhood are changed by experience in adulthood. *Neuroscience Letters*, *584*, 325–330.

Martinez-Lincoln, A., Giattino, C., Chapman, C., & Wicha, N. Y. Y. (2014). 2 X 3 = six: An ERP study of arithmetic processing across formats. In *Paper presented at the development of mathematical cognition: Neural substrates and genetic influences conference, Arlington, VA*.

Megías, P., & Macizo, P. (2016). Activation and selection of arithmetic facts: The role of numerical format. *Memory & Cognition*, *44*(2), 350–364.

Megías, P., Macizo, P., & Herrera, A. (2015). Simple arithmetic: Evidence of an inhibitory mechanism to select arithmetic facts. *Psychological Research*, *79*(5), 773–784.

Mondt, K., Struys, E., Van den Noort, M., Balériaux, D., Metens, T., Paquier, P., et al. (2011). Neural differences in bilingual children's arithmetic processing depending on language of instruction. *Mind, Brain, and Education*, *5*(2), 79–88.

Moore, R. D., Drollette, E. S., Scudder, M. R., Bharij, A., & Hillman, C. H. (2014). The influence of cardiorespiratory fitness on strategic, behavioral, and electrophysiological indices of arithmetic cognition in preadolescent children. *Frontiers in Human Neuroscience*, *8*, 258.

Moreno, E. M., & Kutas, M. (2005). Processing semantic anomalies in two languages: An electrophysiological exploration in both languages of Spanish-English bilinguals. *Cognitive Brain Research*, *22*, 205–220.

Moschkovich, J. (2002). A situated and sociocultural perspective on bilingual mathematics learners. *Mathematical Thinking and Learning*, *4*(2-3), 189–212.

Niedeggen, M., Rosler, F., & Jost, K. (1999). Processing of incongruous mental calculation problems: Evidence for an arithmetic N400 effect. *Psychophysiology*, *36*, 307–324.

Paap, K. R., & Greenberg, Z. I. (2013). There is no coherent evidence for a bilingual advantage in executive processing. *Cognitive Psychology*, *66*(2), 232–258.

Paap, K. R., Johnson, H. A., & Sawi, O. (2015). Bilingual advantages in executive functioning either do not exist or are restricted to very specific and undetermined circumstances. *Cortex*, *69*, 265–278.

Perani, D., & Abutalebi, J. (2005). The neural basis of first and second language processing. *Current Opinion in Neurobiology*, *15*(2), 202–206.

Polich, J. (2011). Neuropsychology of P300. In S. J. Luck & E. S. Kappenman (Eds.), *The Oxford handbook of event-related potential components* (pp. 159–188). New York, NY: Oxford University Press.

Prado, J., Lu, J., Liu, L., Dong, Q., Zhou, X., & Booth, J. R. (2013). The neural bases of the multiplication problem-size effect across countries. *Frontiers in Human Neuroscience*, *7*, 189.

Prado, J., Mutreja, R., & Booth, J. R. (2014). Developmental dissociation in the neural responses to simple multiplication and subtraction problems. *Developmental Science*, *17*(4), 537–552.

Prado, J., Mutreja, R., Zhang, H., Mehta, R., Desroches, A. S., Minas, J. E., et al. (2011). Distinct representations of subtraction and multiplication in the neural systems for numerosity and language. *Human Brain Mapping*, *32*(11), 1932–1947.

Pratt, H. (2011). Sensory ERP components. In S. J. Luck & E. S. Kappenman (Eds.), *The Oxford handbook of event-related potential components* (pp. 89–114). New York, NY: Oxford University Press.

Price, C. J. (2012). A review and synthesis of the first 20 years of PET and fMRI studies of heard speech, spoken language and reading. *Neuroimage*, *62*(2), 816–847.

Price, G. R., Mazzocco, M. M. M., & Ansari, D. (2013). Why mental arithmetic counts: Brain activation during single digit arithmetic predicts high school math scores. *Journal of Neuroscience*, *33*(1), 156–163.

Prieto-Corona, B., Rodríguez-Camacho, M., Silva-Pereyra, J., Marosi, E., Fernández, T., & Guerrero, V. (2010). Event-related potentials findings differ between children and adults during arithmetic-fact retrieval. *Neuroscience Letters*, *468*(3), 220–224.

Qin, S., Cho, S., Chen, T., Rosenberg-Lee, M., Geary, D. C., & Menon, V. (2014). Hippocampal-neocortical functional reorganization underlies children's cognitive development. *Nature Neuroscience*, *17*(9), 1263–1269.

Rasmussen, C., & Bisanz, J. (2005). Representation and working memory in early arithmetic. *Journal of Experimental Child Psychology*, *91*(2), 137–157.

Rosenberg-Lee, M., Chang, T. T., Young, C. B., Wu, S., & Menon, V. (2011). Functional dissociations between four basic arithmetic operations in the human posterior parietal cortex: A cytoarchitectonic mapping study. *Neuropsychologia*, *49*(9), 2592–2608.

Salillas, E., & Wicha, N. Y. Y. (2012). Early learning shapes the memory networks for arithmetic: Evidence from brain potentials in bilinguals. *Psychological Science*, *23*(7), 745–755.

Siegler, R. S. (1988). Strategy choice procedures and the development of multiplication skill. *Journal of Experimental Psychology: General*, *117*(3), 258–275.

Siegler, R. S., Duncan, G. J., Davis-Kean, P. E., Duckworth, K., Claessens, A., Engel, M., et al. (2012). Early predictors of high school mathematics achievement. *Psychological Science, 23*(7), 691–697.

Simmons, F. R., & Singleton, C. (2007). Do weak phonological representations impact on arithmetic development? A review of research into arithmetic and dyslexia. *Dyslexia, 14*(2), 77–94.

Simmons, F., Singleton, C., & Horne, J. (2008). Brief report—Phonological awareness and visual-spatial sketchpad functioning predict early arithmetic attainment: Evidence from a longitudinal study. *European Journal of Cognitive Psychology, 20*(4), 711–722.

Spelke, E. S., & Tsivkin, S. (2001). Language and number: A bilingual training study. *Cognition, 78*, 45–88.

Sunderman, G., & Kroll, J. F. (2006). First language activation during second language lexical processing: An investigation of lexical form, meaning, and grammatical class. *Studies in Second Language Acquisition, 28*(3), 387–422.

Thordardottir, E., Rothenberg, A., Rivard, M., & Naves, R. (2006). Bilingual assessment: Can overall proficiency be estimated from separate measurement of two languages? *Journal of Multilingual Communication Disorders, 4*(1), 1–21.

Tschentscher, N., & Hauk, O. (2014). How are things adding up? Neural differences between arithmetic operations are due to general problem solving strategies. *Neuroimage, 92*, 369–380.

Umbel, V. M., Pearson, B. Z., Fernández, M. C., & Oller, D. K. (1992). Measuring bilingual children's receptive vocabularies. *Child Development, 63*(4), 1012–1020.

Van Petten, C., & Luka, B. J. (2012). Prediction during language comprehension: Benefits, costs, and ERP components. *International Journal of Psychophysiology, 83*, 176–190.

Venkatraman, V., Siong, S. C., Chee, M. W. L., & Ansari, D. (2006). Effect of language switching on arithmetic: A bilingual fMRI study. *Journal of Cognitive Neuroscience, 18*(1), 64–74.

Vukovic, R. K., & Lesaux, N. K. (2013). The relationship between linguistic skills and arithmetic knowledge. *Learning and Individual Differences, 23*, 87–91.

Wicha, N. Y. Y., Moreno, E. M., & Kutas, M. (2004). Anticipating words and their gender: An event-related brain potential study of semantic integration, gender expectancy, and gender agreement in Spanish sentence reading. *Journal of Cognitive Neuroscience, 16*(7), 1272–1288.

Zhou, X., Chen, C., Dong, Q., Zhang, H., Zhou, R., Zhao, H., et al. (2006). Event-related potentials of single-digit addition, subtraction, and multiplication. *Neuropsychologia, 44*(12), 2500–2507.

Chapter 8

Linguistic Traces in Core Numerical Knowledge: An Approach From Bilingualism

Elena Salillas and Alejandro Martínez
Basque Center on Cognition, Brain and Language, San Sebastián, Spain

INTRODUCTION

To what extent does language play a role in mathematics? A considerable number of studies favor a view of numbers without language. Perhaps, the most evident is the research with animals and preverbal infants (e.g., Agrillo & Beran, 2013; Izard, Sann, Spelke, & Streri, 2009; Nieder & Dehaene, 2009). These works show how a sense of quantity exists in humans and animals, suggesting that core numerical knowledge is based on mechanisms with a long phylogenetic history. This essence of numbers has a concrete brain locus that is repeatedly engaged whenever fundamental numerical markers are targeted with neuroimaging techniques, namely, the intraparietal sulcus (IPS). Comparable areas exist in the primate and human brain, and analogous brain regions are found across a wide range of species (Geary, Berch, & Koepke, 2015). However, the quantitatively equipped infant has to eventually cope with a world that includes increasingly complex mathematical concepts whose neurofunctional bases are not yet well understood. The initial core magnitude system is not enough. Thus, language might be a fundamental tool that expands and could modify this central evolutionarily ancient numerical knowledge (Carey, 2011; Gelman & Butterworth, 2005).

There are multiple ways in which linguistic components influence the cognitive basis of math functioning; nonverbal and verbal symbols are necessary to grasp greater mathematical complexity. The first potential merging of language abilities and quantitative knowledge occurs with children's learning of number words and the quantities they represent. With time, a comparison between number symbols, such as numerals or number words, can be faster than a comparison between the actual numerosities they represent (Holloway & Ansari, 2009). In turn, the acquisition of numerical

Mathematical Cognition and Learning, Vol. 4. https://doi.org/10.1016/B978-0-12-812574-8.00008-0

symbols allows a fast computation of small and large numerosities, reducing cognitive load. They allow one to go beyond the concrete objects they represent. Such integration between numerical symbols and numerosities might turn specific in bilingualism.

Bilingualism acquires considerable importance not only as a window to the study of associations between mathematics and language but also regarding possible specificities that the management of more than one verbal code might entail in math functioning. Multilinguals and bilinguals attach two or more verbal codes to the same magnitude, and those linkages often do not have the same strength. Verbal arithmetic storage and exact calculation could differ between these two verbal codes, and the majority of research on bilingual math initially has focused on those differences (e.g., Bernardo, 2001; Frenck-Mestre & Vaid, 1993; Marsh & Maki, 1976; McClain & Huang, 1982). Further, as we will show, the two codes could differentially modify the essential quantity representation. This, in turn, not only would provide strong evidence of language traces in numerical magnitude but also would have obvious consequences for bilingual math functioning. Furthermore, as one of the languages would be predominant for math, the question is also which language is the predominant one.

Finally, integrative processes appear to have a role in developmental dyscalculia (DD), a specific learning disability affecting the development of arithmetic skills. Available empirical evidence and theoretical proposals highlight a likely role of symbolism, involving the symbolic number words or nonverbal Arabic numerals, in what could be taken as a dynamic and core deficit in DD. We will explore and characterize what we will call bilingual DD (bDD) as a special case of the deficit, in which the symbol-to-quantity integration likely passes through increasing complexity. The reported data with bilingual adults and new data we will present here suggest that in bDD, a very superficial processing of number words might occur in the nonpreferred language.

NUMBERS WITHOUT LANGUAGE

The starting point when addressing the extension of math and language links should be the nature of number knowledge prior to language acquisition. There is enough evidence suggesting that nonhuman animals and preverbal infants already possess a domain-specific representation of number and elementary arithmetic operations. One can certainly say that no animal can balance a checkbook; nevertheless, there is evidence that animals can perform simple arithmetic with and without training (Flombaum, Junge, & Hauser, 2005; Rumbaugh, Savage-Rumbaugh, & Hegel, 1987; Woodruff & Premack, 1981).

How would animal and human abilities in number processing be related? Some authors (Feigenson, Dehaene, & Spelke, 2004; Geary, 1995) suggest that these shared number processing mechanisms imply a biological start-up tool that is fundamental for the unique human ability to develop higher-level arithmetic. Comparative research has shown that animals can discriminate numerosities (Brannon & Terrace, 1998; Davis & Pérusse, 1988), attend to the numerical attributes of their world spontaneously and automatically (Agrillo & Beran, 2013), and can use numerical information to make informed decisions (Hauser, Carey, & Hauser, 2000; Wilson, Hauser, & Wrangham, 2001). Shoaling species of fish, for example, prefer larger to smaller shoals that in turn reduce predation risk. Ratio-dependent number discrimination can be found in tamaris monkeys. Performance in these types of conditions is ratio-dependent, that is, it is influenced by the ratio of the two quantities not the absolute difference between them (Hauser, Tsao, Garcia, & Spelke, 2003). This is found without training in numerical tasks and inexperienced rhesus macaques (Flombaum et al., 2005). Hauser et al. (2003) further suggested that monkeys detect violations in operations of small sets of items, but only when the ratio between the observed and the expected outcome is favorable. Furthermore, single-cell recording studies have shown number-selective neurons in monkeys' prefrontal and parietal areas, comparable with the number-sensitive areas in the human brain (Nieder & Dehaene, 2009). These number neurons respond selectively to certain numerosities and, importantly, follow the characteristic logarithmic organization of the number domain, that is, differences between small numbers is large and becomes increasingly compressed as quantities increase.

These findings speak to the evolutionary importance of numerical competence; in order to guarantee survival, it is important to process quantitative information. Additionally, these studies point to an abstract, nonsymbolic representation of numbers that is available to animals. When animals as diverse as monkeys and rats are tested in the same nonverbal number tasks as adult humans, striking parallels are revealed, suggesting that humans and nonhuman animals share a common basic number representation. Quantitative processes have also been found in human infants (Feigenson & Halberda, 2004; Starkey & Cooper, 1980). Sensitivity of preverbal infants toward quantity occurs at a very early age (Libertus & Brannon, 2009). They can discriminate between groups of objects when the involved quantities are small (one, two, or three items in Antell & Keating, 1983; Starkey & Cooper, 1980). Lipton and Spelke (2003) and Xu and Spelke (2000) showed that 6-month-old infants can also discriminate between large sets of objects such as an array of 8 dots from an array of 16 but fail to discriminate 8 from 12. At 9 months, infants are able to discriminate 8 from 12 but fail to discriminate 8 from 10 (Lipton & Spelke, 2003). They can also engage with rudimentary arithmetic (Wynn, 1992a).

Similarly, individuals living in cultures with languages containing few number words also have sensitivity to magnitude. For instance, the Piranha culture in Brazil only has words for "one" and "two," and they have a single term for larger quantities, "many." Gordon (2004) asked Piranha participants to construct sets of nuts or batteries that matched an example set in number, and they were perfectly able to construct sets of one, two, and sometimes three objects; however, when the sets of numbers were larger than three, precision dropped. Members of another Brazilian culture, the Munduruku, whose language only has words for numbers from 1 to 5, were able to approximately compare the relative magnitude of large numbers of dots, performing similarly to the French-speaking controls (Pica, Lemer, Izard, & Dehaene, 2004). Both groups' accuracy increased as the ratio between the two values increased. Thus, adults without language for numbers also possess a ratio-dependent approximate magnitude representation.

These abilities in turn appear to be dependent on two basic core systems: the "approximate number system" (ANS), which is approximate and becomes more imprecise as numerosity increases (following Weber's law), and another system, perhaps more debatable, restricted to the precise and automatic processing of sets of 1–3 objects, often referred to as the "object tracking system" (Feigenson et al., 2004). These systems will be the bases for symbolic mathematics, a cultural achievement that will allow humans to reach mathematical complexity.

LANGUAGE AND SYMBOL ACQUISITION

The studies mentioned above robustly indicate that there is a core numerical system that is evolutionarily ancient and situated in a well-studied area of the brain that precedes language acquisition. However, we do acquire language, and we manipulate numerosities through language symbols and within language syntax (Sarnecka, 2014). Thus, an impact of linguistic factors should be considered in an evolved numerical system that interacts with other cognitive domains.

Classical numerical cognition models have restricted any linguistic component to simple arithmetic (Dehaene & Cohen, 1995) and clearly differentiate an approximate number system, independent from language, from an exact system developed from counting and the verbal acquisition of exact arithmetic. Specifically, with the triple code model Dehaene and Cohen (1995) proposed that an auditory verbal frame supports the processing of arithmetic facts and exact calculation. The abstract code model (McCloskey, 1992; McCloskey, Caramazza, & Basili, 1985) proposed that once the abstract semantic representation is accessed, it operates independently from the input code. Thus, both proposals postulated additive effects between format and numerical semantics when manipulating verbal numerical input. In contrast, exact arithmetic, learned through verbal rote mechanisms, has been proposed to be verbally stored and subsequently verbally retrieved, hence, linguistic in essence (e.g., Campbell & Clark, 1992; Dehaene, 1997) and format-dependent (Campbell, 1994; Campbell & Epp, 2004).

At the neural level, very explicit proposals delimit the core numerical system from any verbal component. This distinction is supported first by neuropsychological studies showing double dissociations between patients with perisylvian lesions and patients with intraparietal lesions (Cipolotti, Butterworth, & Denes, 1991; Lee, 2000; Lemer, Dehaene, Spelke, & Cohen, 2003). While the latter showed disruption of the number sense (addition, subtraction, comparison, or numerosity estimation), the former showed an alteration in rote operations such as multiplication. Early neuroimaging studies converged in showing consistent activations in the intraparietal sulcus for any numerical operation and even in the mere detection of digits (see Dehaene, Piazza, Pinel, & Cohen, 2003, for a metaanalysis). In turn, these parietal regions would be involved in the processing of number semantics and would work in coordination with ventral temporooccipital areas for the decoding of symbolic input (e.g., Dehaene, 2009; Piazza & Eger, 2016). As for exact arithmetic, left perisylvian areas, specifically the left angular gyrus, were proposed as the neural substrate for the linguistic processing of arithmetic fact retrieval (Dehaene et al., 2003). Training studies even showed that initial intraparietal and frontal activations to unlearned material transferred to the angular gyrus when facts were verbally stored (Delazer et al., 2003). However, and in addition to this verbal component, we have recently documented the interactive nature of arithmetic facts processing (e.g., Semenza, Salillas, Di Pellegrin, & Della Puppa, 2016), also suggesting approximation in the right hemisphere.

Despite these initial proposals of apparent parcellation (i.e., division of the brain into nonoverlapping regions) between pure numerical and linguistic components, an impact of symbolic acquisition on basic numerical knowledge has also been directly addressed. This perspective implies possible modifications of our basic numerical system by the acquisition of numeric symbols, number words among them. This possibility was first noticed by Verguts and Fias (2004) through a neural network model simulating the numerosity labeling process. In their model, and without the consideration of numerical labeling, the network developed numerosity detectors following a logarithmic scale. This logarithmic scale is typical in the core numerical system shared with preverbal infants and nonhuman species (e.g., Nieder & Dehaene, 2009). However, after the pairing of numerosity input with symbols, the network became sensitive to symbols as well. Furthermore, the pre- and postsymbolic behavior of the detectors changed. After the pairing with symbols, the simulated neurons evolved from a logarithmic behavior to a more linear response in which the discriminability between quantities was sharper. In turn, it was proposed that symbolic learning profoundly changes our analogue core numerical knowledge in its most characteristic trace, that is, in its organization according to Weber's law. The basic idea is that a very automatic link between the core representation of numbers and their corresponding symbols is established and that the system should gain precision when it is paired with symbols.

This linearization has been subsequently observed in psychological studies. Halberda and Feigenson (2008) showed that math achievement correlated with the acuity of nonsymbolic comparison and that ANS acuity increased with age. Albeit in an indirect way, Piazza, Pica, Izard, Spelke, and Dehaene (2013) suggested that this ANS increases in accuracy with education. These authors contrasted groups of unschooled vs educated Amazonian Munduruku and observed higher accuracy in the estimation of sets of concrete objects for the educated group. Halberda, Mazzocco, and Feigenson (2008) further showed that poorer ANS acuity at age 14 was associated with lower mathematics achievement in kindergarten. Recently, Shusterman, Slusser, Halberda, and Odic (2016) showed how the acquisition of the cardinality principle in preschoolers correlates with an increase in acuity in the ANS. These studies highlight the possibility of a key role of culture and language in developmental changes in our numerical knowledge. In fact, different authors have observed deficits in the processing of symbolic material by dyscalculic children (Furman & Rubinsten, 2012), and a deficit in the access to the quantities associated number symbols has been proposed as one essential component in dyscalculia (Noël & Rousselle, 2011; Rousselle & Noël, 2007). It turns out, then, that the correct labeling of numerosities might be an essential part of the cultural evolution of our mathematical system, allowing for accuracy and efficient math abilities.

From neuroimaging studies, there is evidence on the overlap of intraparietal activations for symbolic and nonsymbolic numerical processing (e.g., Lyons, Ansari, & Beilock, 2015; Piazza, Pinel, Le Bihan, & Dehaene, 2007). But the same studies show differences between notations in the left parietal lobe (Piazza et al., 2007). Hence, a tendency for left parietal lateralization has been proposed; left parietal activations increase during arithmetic development (Rivera, Reiss, Eckert, & Menon, 2005 or Ansari & Dhital, 2006), while higher stability of the right parietal responses appears throughout education (Emerson & Cantlon, 2015). Similarly, lateralization of language functions correlates with the asymmetry between the two parietal lobes for calculation (Pinel & Dehaene, 2010). An interaction between the lateralization for the processing of symbols in a ventral pathway and the same hemisphere parietal areas has been proposed as an explanation for these hemispheric differences that arise during education (Dehaene, 2009). However, current behavioral and neuroimaging studies contain only incomplete and indirect evidence of the causal role of education and/or symbolic acquisition in the evolution of the brain basis for math. As we will see, all these issues acquire relevance and can be addressed from the study of bilingual math.

BILINGUALISM AS A WINDOW TO NUMERICAL COGNITION

Bilingualism in numerical cognition is concerned with one or more symbolic systems that refer to the same numerosity. This happens usually for number words but sometimes for digits as well (such as in Eastern Arabic numerals). These different codes may be acquired at different moments during education and over different length time periods. They can differ in frequencies of use, and they might even differ in the syntactic structures. For example, Dutch or German imply inversions in the unit-decade order where 35 would be spoken as "five and thirty" differing from English or Spanish where 35 is spoken as "thirty-five." These differences might imply variations to the default base-10 system. Taken together, these disparities give rise to questions concerning the extent to which language might modulate a basic organization of our numerical system, such as place value.

This management of more than one language has called attention to numerical cognition research in several ways. First, studies have used bilingualism as a way of testing the distinction between approximate and exact calculation (Dehaene, Spelke, Pinel, Stanescu, & Tsivkin, 1999; Spelke & Tsivkin, 2001; Venkatraman, Siong, Chee, & Ansari, 2006). The effect that the two available bilingual linguistic contexts have on calculation has been approached as well (Van Rinsveld, Schiltz, Landerl, Brunner, & Ugen, 2016). Finally, another research line, including our own studies, has considered the relative dominance of the two languages for math (Bernardo, 2001; Grabner, Saalbach, & Eckstein, 2012; Martinez-Lincoln, Cortinas, & Wicha, 2015; Salillas, Barraza, & Carreiras, 2015; Salillas & Carreiras, 2014; Salillas & Wicha, 2012; see also Chapter 7 by Wicha et al.). Among these studies, we have also focused on the impact of the number wording system on the nonverbal numerical system (Salillas et al., 2015; Salillas & Carreiras, 2014).

More specifically, it has been shown (Dehaene et al., 1999; Spelke & Tsivkin, 2001) that training exact arithmetic in one language, but not training approximate calculation, leads to a greater difficulty of subsequent retrieval in the other language. Using fMRI, Venkatraman et al. (2006) showed how switching between languages during exact retrieval of arithmetic facts in the nontrained language involved a left lateralized network comprising the left angular gyrus and the left inferior frontal gyrus (see also Grabner et al., 2012). However, a comparable situation of switching for approximate calculation in the nontrained language implied bilateral activations in the intraparietal sulci and left inferior frontal regions. Note, however, that Venkatraman et al. found inferior frontal activations for both types of calculation, though they were not overlapping, and these activations might indicate the translation (switching) to the right code in both cases (Grabner et al., 2012; Luk, Green, Abutalebi, & Grady, 2012). In any event, both studies suggest some dissociation in the neurocognitive systems for approximate and exact calculation.

Bilingualism is also a window for the exploration of differential attachment strengths or integration to quantity that each number word system has, that is, a test for a possible numerical system that runs independently from surface code vs an integration of the verbal and nonverbal numerical representations. This hypothesis is testable by using core magnitude markers such as the distance effect (Moyer & Landauer, 1967) or ratio indexes (Halberda & Feigenson, 2008) after a numerical input in each of the languages for very proficient bilinguals. If only additive effects appear, that is, if there is a cost for transcoding from each bilingual code, orthogonal and added to a certain numerical nonverbal effect, then the math verbal and core nonverbal components are independent. However, if distance effects do not arise with a nonpreferred language or the underlying neural bases vary between codes, then independence cannot be sustained. In that case, differential integrations between the two language codes and the quantity code can be proposed (see Fig. 1). In addition, differential attachments to the quantity code from each of the languages mean that the factor that rules that attachment (i.e., frequency of use, early practice with that language) is indeed important for said attachment. In turn, by uncovering which factor determines a preferred language, one can infer *why* and *when* the verbal codes attach to the preverbal numerical representation. This factor would then also apply to monolinguals.

Finally, the syntax of numerical words differs sometimes between languages. Languages that reverse tens and units have been studied in monolinguals and, to some extent, also in bilinguals. For example, in German, Dutch, and Arabic, the order of tens and units is inverted in their number words ("one-and-twenty" for 21). Some researchers have studied transcoding effects from these inverted systems as compared with the transcoding of transparent wording (e.g., two-ten one, for 21) for the other language in bilinguals. It appears that transcoding is more efficient in the transparent system (Moeller, Zuber, Olsen, Nuerk, & Willmes, 2015; Pixner, Moeller, Hermanova, Nuerk, & Kaufmann, 2011; Pixner, Zuber, et al., 2011). Despite the fact that these difficulties might be accounted for by a superficial route between the different codes, some studies have also shown small behavioral implications of inverted systems in number semantics (number comparison

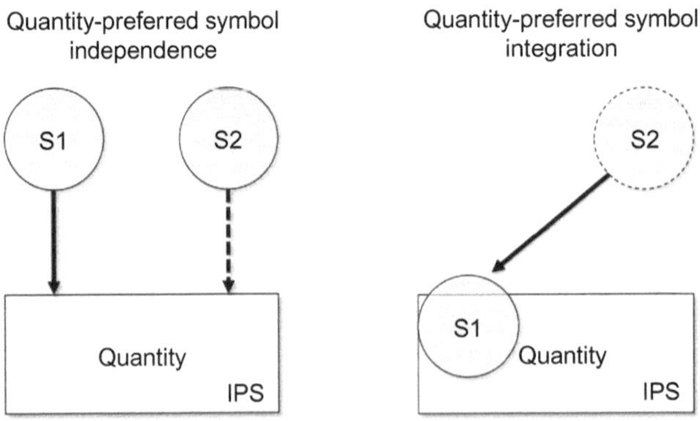

FIG. 1 Two possibilities on quantity—numeric symbols relationship. *Left*, one of the symbolic systems in bilinguals has a preferred access to core numerical knowledge. Only additive effects will be obtained in this case, derived from independence between surface formats and quantity. *Right*, the preferred symbolic system for a bilingual, integrates with the quantity code. *S1*, preferred symbolic system; *S2*, nonpreferred symbolic system; *IPS*, intraparietal sulcus.

in Nuerk, Weger, & Willmes, 2005) or in number line estimations (Helmreich et al., 2011) in crosslinguistic comparisons. In addition, number words' structure has also been shown to impact complex calculation (Van Rinsveld et al., 2016). Thus, both from the less culturally affected bilingual studies and from crosslinguistic research, some possible interactions between number wording systems, asemantic transcoding, or numerical semantic effects have been reported.

The Case of Basque

In line with these studies, we have been focusing on the case of Spanish-Basque bilingualism (Salillas et al., 2015; Salillas & Carreiras, 2014) as entailing a peculiar combination of number word systems. The reason is that number words for the two languages follow different base systems. In Spanish, as in English, number words follow the base-10 system (e.g., the number word for 58 is 50 and 8, "cincuenta y ocho"). In Basque, number words follow the base-20 system (e.g., the number word for 58 is 2×20 and 10 and 8, i.e, 40 and 10 and 8, "berrogeita hamasei"). This makes the study of Spanish-Basque bilingualism an optimal opportunity for uncovering possible language-math interactions; the base-10 system does not need to be linguistic in origin. In fact, the base-10 system is implied in the widespread use of fingers for counting. Base-10 is used for place coding in multidigit numbers. In turn, base-10 structure is a feature of a particular representational system rather than a fundamental mathematical fact. Nevertheless, it is a feature that is incorporated into many of the algorithms that children learn for arithmetic performance (Miller, Kelly, & Zhou, 2005). *Base-20, however, is restricted to number words in Basque, so any impact on any number effect under study that is related to this vigesimal system should be interpreted as coming from the number word and therefore linguistic.* Consequently, Spanish-Basque bilingualism appeared as an interesting case to study traces of these two systems on number representations in the same individuals.

THE LANGUAGE OF LEARNING MATHEMATICS

Counting and cardinality are acquired before digit systems. The acquisition of the counting series and the understanding of numeric labeling is a difficult process (Gelman & Butterworth, 2005; Wynn, 1992b) even for monolingual children. A recent study shows that even this initial process differs between languages in bilinguals (Wagner, Kimura, Cheung, & Barner, 2015). Despite the fact that a cardinality insight seems to appear at the same time in both languages, there is an advantage for one of the languages in the previous steps, which is ultimately dependent on the exposure to each of the two languages' number words. However, counting is only the beginning of math education. There are many mathematical concepts that will be taught in one of the languages thereafter (i.e., place value, simple and complex arithmetic, fractions, geometry, and ratios). Hence, the question that we have been empirically addressing is, *Will the language of learning math (hereafter* LL^{math}*) entail a preferred code for math? And if so, to what depth will that code get into the approximate number system?*

The clearest case was the already mentioned arithmetic facts, with an acknowledged verbal component. Thus, our first study explored the nature of arithmetic representations in each of the languages (Salillas & Wicha, 2012). Other authors had already shown, behaviorally, that the structure of arithmetic fact memory networks and the speed and accuracy in retrieving arithmetic facts differ between languages (Bernardo, 2001; Frenck-Mestre & Vaid, 1993; but see Rusconi, Kwan, Giordano, Umiltà, & Butterworth, 2006). However, importantly, both Bernardo's work and our event-related potential (ERP) study stressed the idea that LL^{math} may not necessarily match the more proficient (i.e., fluent) language of a bilingual, an idea initially put forward by Dehaene et al. (1999). (See Chapter 7 by Wicha et al., for a basic introduction to the ERP electrophysiological methodology and its various waveform components.) Instead, a pattern of dominance between the two codes for math could have been established depending on the language used for the learning of those arithmetic facts (LA+).

We measured both the strength of the arithmetic networks and their quality. We used a verification task for one-digit multiplication facts, and we contrasted (1) the ERPs to correct vs incorrect solutions and (2) the ERPs to incorrect solutions that were related to the correct solution (related, $3 \times 2 = 9$) versus incorrect solutions that were unrelated to the correct solution (unrelated, $3 \times 2 = 7$). The first contrast referred to the strength of those networks (i.e., strength of activation of arithmetic facts); the second contrast referred to the complexity of those networks (i.e., how related and correct solutions were connected in those networks). ERP results showed that the negative wave peaking around 400 ms (known as N400 component) was larger for incorrect solutions than for correct solutions for LL^{math} and that, crucially, the amplitude of the N400 was modulated by relatedness only in the LA+. During arithmetic fact retrieval, the N400 amplitude is thought to reflect the automatic spread of activation among representations of arithmetic facts (Niedeggen & Rösler, 1999). In our data, this spread of activation differed between languages. More crucially, the N400 amplitude difference between related and unrelated solutions was absent for LA−; hence, the quality and actual form of those networks was different, less precise, for the language in which arithmetic facts were firstly acquired. Importantly, these effects were independent on the relative overall proficiency between languages as measured by picture naming and fluency tasks. We concluded that arithmetic facts in bilinguals were stronger and their network more complex in the LA+.

In a subsequent study, Martinez-Lincoln et al. (2015) tested, however, if explicit experience using LA− (i.e., bilingual teachers repeating arithmetic facts and teaching them in the other language to the students) could equate the strength of the networks between the two languages (see also Chapter 7 by Wicha et al.). Thus, the same paradigm as in Salillas and Wicha (2012) was applied to the special case of elementary school teachers, adding a new variable: the language in which the participants had been teaching arithmetic for an average of 9 years. This language could be either LA+ or LA−. Martinez-Lincoln and collaborators showed that extensive teaching of arithmetic facts in LA− equated the strength of arithmetic facts activations to the LA+. This was demonstrated by similar N400 amplitude differences between correct and incorrect solutions in both languages. The latency of the congruency effect was even shorter for LA+. Despite the fact that these authors did not report the N400 relatedness effects (complexity of the two memory networks), these results interestingly show that it was exposure to arithmetic facts in one language that determined the imbalance between languages in Salillas and Wicha (2012).

In a personal communication, the authors stated that the relatedness effect was not found in this study. This fact is interesting because it points to the possibility that extensive production of and exposure to arithmetic facts would neutralize interference from irrelevant solutions (e.g., Frenck-Mestre & Vaid, 1993; Geary, 1996; Graham & Campbell, 1992; Niedeggen & Rösler, 1999, using ERPs). Crucially, once more in this study, the pattern of dominance between languages for arithmetic facts was independent from overall proficiency in each language.

The two studies described above agreed with the view that the surface code plays an essential role in the storage and subsequent retrieval of arithmetic facts (Campbell, 1994; Campbell & Epp, 2004). However, this language preference could go beyond arithmetic facts. The cognitive manipulation of math concepts during math learning in a specific language could have brought that language into more essential number representations. That is, exchanging math knowledge through one of the verbal symbolic systems could have linked one of those linguistic systems to core math meaning. With that working hypothesis, we explored those links using electrophysiology in Salillas and Carreiras (2014) and Salillas et al. (2015).

The two number wording systems in Basque-Spanish bilingualism allowed us to explore (1) whether number word structure attaches to our representation of quantity and (2) whether number word structure from LL^{math} has a preferential impact on numerical knowledge. We explored potential interactions between number word forms and the distance effect (a marker of core numerical activation). In Salillas and Carreiras (2014), we measured ERPs while bilingual participants performed number comparisons on pairs of Arabic digits. During the comparison, the ERP distance effect was modulated by the verbal relationships that reside in Basque number words. The most important finding was that an earlier N1-P2 distance effect appeared for pairs of digits related through the base-20 system and only for those participants whose LL^{math} was Basque. This component involves the transition wave from a negative peak at 100 ms to a positive peak at 200 ms and usually indexes numerical semantic processes (access to quantity), together with a subsequent P2p. These effects were again unrelated to the participants' relative proficiency in Basque and Spanish. Thus, these effects converged in the independence of the LL^{math} in numerical computations and general linguistic dominance in bilinguals. Importantly, given that the stimuli were not verbal, the data demonstrated that the base-20 structure, only present in Basque number words, modulated core numerical effects such as the distance effect. Moreover, the N1-P2 distance effect correlated with verbal retrieval efficiency, again suggesting that verbal factors could be linked to the initial computation of distance (see Fig. 2).

In Salillas et al. (2015), we used a measure of phase synchronization as a dependent variable. Synchronization means adjustment of frequencies of periodic self-sustained oscillators due to interaction. Thus, when certain sites show synchronized phases, it means that they are interacting and therefore reflecting the activity of an underlying brain network. In our study, oscillatory EEG activity converged to show an activation of the base-20 system during this basic numerical comparison of Arabic digits (Salillas et al., 2015). Again, this activation only occurred in bilinguals whose LL^{math} (which was not necessarily their L1) contained this base-20 system in the number wording system—that is, those whose LL^{math} is Basque (see Fig. 2). Specifically, we observed that the brain waves whose frequency was close to the band of 40 Hz (specifically 43–46 Hz, within the gamma band) were synchronized (gamma-band synchronization) to contrasted numbers that were related through the vigesimal system, *only* for the group whose LL^{math} was Basque. Synchronized electrodes were localized to the left hemisphere in frontoparietal sites and then bilaterally extended. As numerical comparison of Arabic digits does not imply a verbal code, all these effects would originate in number words. In addition, a more posterior synchronization in a different frequency band, the beta band (22–26 Hz) at occipitoparietal sites, appeared when bilinguals were computing base-10 relations, which are overspread in our math functioning and math learning systems.

That is to say that base-10 and base-20 relations contained in number words appeared to activate different brain networks at different frequency bands. However, in our digit comparison task, the vigesimal system relating the two digits to be compared triggered synchronization in gamma only when the vigesimal relationship was in fact contained in the number word from the bilingual's LL^{math}. Contrarily, a Basque-Spanish bilingual, whose LL^{math} was Spanish, was "blind" to the vigesimal relation contained in pair of digits such as $58-40$. For them, 58 seemed to not activate the vigesimal relation

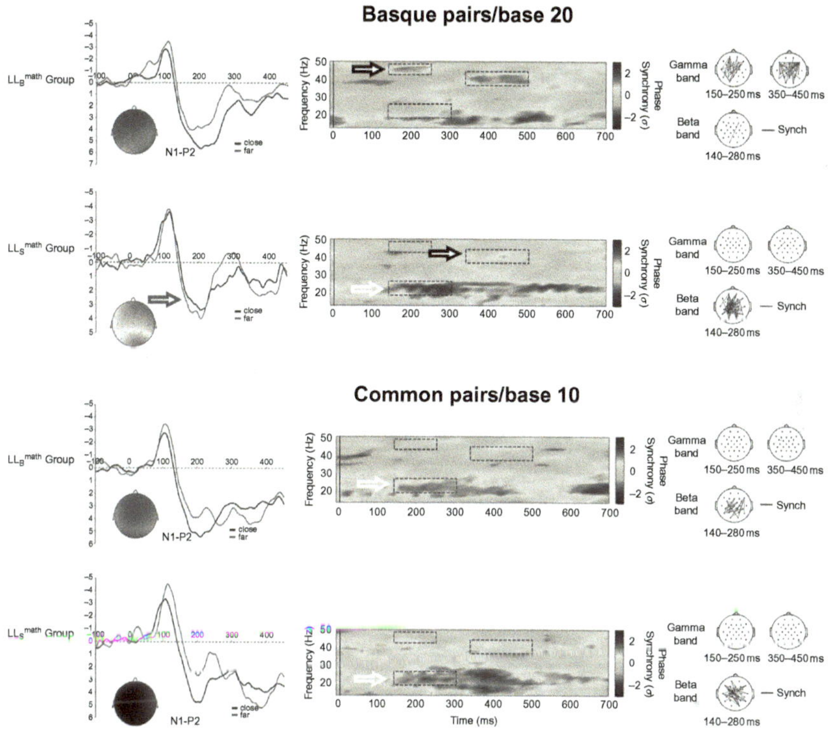

FIG. 2 Impact of the number wording system in the processing of quantity in Basque-Spanish bilinguals. Participants performed a number of comparison task in which the Arabic digits to compare could be related through the base-20 system or through the base-10 system, according to the corresponding number words. While some Basque number words follow the base-20 wording system and some Basque words follow the base-10 wording system, all Spanish number words follow the base-10 wording system. Thus, Arabic pairs related through the base-20 were called "Basque pairs," while pairs related through the base-10 were called "common pairs." Participants were divided in two groups depending on their LL^{math} (B—Basque or S—Spanish). N1-P2 distance effect appeared only for the LL_B^{math} group for the Basque pairs and was absent for the LL_S^{math} group (gray arrow). For common pairs, the distance effect appeared in both groups of participants. Moreover, we observed gamma-band synchronization when the LL_B^{math} group processed the base-20-related Basque pairs (black arrows). This synchronization linked left hemisphere electrodes first, becoming bilateral later on. For all the rest of conditions, this gamma band did not appear. Instead and likely reflecting the activation of the less language-dependent base-10 system, beta-band synchronization was found (white arrow) between bilateral centroparietal sites. Thus, despite being very proficient bilinguals, the LL_S^{math} group appeared to be "blind" to the vigesimal relations contained in Basque number words. Furthermore, LL^{math} and input language interacted in the ultimate generation of distance effects even with nonverbal material, suggesting linguistic traces in the quantity code. *Adapted from Salillas, E., & Carreiras, M. (2014). Core number representations are shaped by language.* Cortex, 52*(1), 1–11; Salillas, E., Barraza, P., & Carreiras, M. (2015). Oscillatory brain activity reveals linguistic prints in the quantity code.* PLoS ONE, 10*(4), e0121434.*

"2 × 20 and 10 and 8" from Basque, but the common base-10 relationship contained in Spanish number words. Importantly, and similar to Salillas and Carreiras (2014), very proficient bilinguals whose LL^{math} is Spanish do not show this synchronization in gamma. This in turn, provides information about *when* number words attach to quantity: in our participants, during early math learning.

Ultimately, these studies show that verbal input during simple numerical tasks is not necessary for obtaining verbal effects, which influence our quantity manipulation. Remarkably, bilinguals possess one verbal symbolic system that impacts the quantity code. In most circumstances, this would be the language of early math learning. However, an evolving and malleable link is possible in more specific training circumstances (Campbell & Epp, 2004). Importantly, bilinguals' preference for LL^{math} might have consequences for cases in which basic numerical knowledge is altered, such as in developmental dyscalculia. In what follows, we describe an empirical test of this interaction between LL^{math} and basic number knowledge in children with dyscalculia.

NEW EVIDENCE: BILINGUAL DYSCALCULIA

Developmental dyscalculia (DD), which is as prevalent as dyslexia (6%), is a disorder of numerical development and mathematical learning. There are different explanations of its causes that are based on a possible core numerical deficit (Butterworth, 1999, 2005, 2010; Geary, 2004; Piazza et al., 2010; Wilson & Dehaene, 2010). There is also a high degree of comorbidity between DD and dyslexia (a developmental disorder that affects the ability to read in otherwise normally developing individuals), and it is characterized by slow and error-prone learning and retrieval of arithmetic facts from memory (Jordan & Montani, 1997). Additionally, there is use of immature calculation, problem-solving and counting strategies, and delay in the transition from finger counting to verbal counting and fact retrieval (Geary, 2004). People with DD present heterogeneous symptomatology, including spatial working memory deficits or symptoms due to a failure in broad executive functions such as attention (Henik, Rubinsten, & Ashkenazi, 2011; Rubinsten & Henik, 2009).

The prevalent view for some years has been that DD is ultimately caused by an anomaly in the core magnitude representation (Butterworth, 1999, 2005, 2010; Geary, 2004; Wilson & Dehaene, 2010). People with DD potentially lack essential numerosity representation due to functional (Price, Holloway, Räsänen, Vesterinen, & Ansari, 2007) and morphological alterations in the numerical neural networks (Molko et al., 2003; Rotzer et al., 2008).

There are alternative explanations that emphasize the role of numerosity-to-symbol matching and the subsequent modification of the ANS (Noël & Rousselle, 2011). Thus, a core deficit could arise later on during symbol acquisition. In normal circumstances, the acquisition of number words for counting precedes nonverbal Arabic number symbols, which might lead to qualitative modifications in ANS (Piazza et al., 2013) that also allow exact calculation. For DD children, a process of linearization is impaired (Piazza et al., 2010). In fact, there are researchers who have observed specific alteration of the processing of quantity when using symbols in contrast to an apparent spared quantity processing when nonsymbolic numerical stimuli (i.e., dot patterns) are used (Noël & Rousselle, 2011; Rousselle & Noël, 2007).

At the neural level, IPS differences during the processing of nonsymbolic numerosities have been reported (Price et al., 2007). Kucian et al. (2006) compared DD children and controls and found similar activation patterns for both groups for approximate and exact calculations, but DD children presented weaker and more diffuse activation than the controls. Additionally, when symbolic stimuli are used, notation modulates the activity of frontoparietal activations, as metaanalysis shows (Kaufmann, Wood, Rubinsten, & Henik, 2011). In turn, DD children differ from controls in not only parietal function but also frontal function during number processing (see also Rotzer et al., 2008). In fact, some residual numerical processing in adults with dyscalculia has been found in frontal areas, possibly reflecting more use of working memory or inhibition processes than controls (Cappelletti & Price, 2014). Further, there is evidence of disconnection between the areas linked by the superior longitudinal fasciculus in DD. That is, deficient fiber projection between parietal, temporal, and frontal areas has recently been reported in DD (Kucian et al., 2014). In addition, transcranial electric stimulation over left posterior parietal areas when learning symbol-quantity associations ameliorates those associations in DD by facilitating neuroplasticity (Iuculano & Cohen Kadosh, 2014).

In addition, an ERP study that investigated numerical magnitude processing in DD children showed distance effects that did not differ between controls and DD at an early time window during a symbolic number comparison task (Soltész, Szucs, Dékány, Márkus, & Csépe, 2007). However, at a later time window, the controls showed a significant distance effect over the right parietal areas, while the DD group showed no such effect. After further analysis, the authors showed DD participants also presented an effect on the frontocentral electrodes that the controls did not. Therefore, the authors suggested executive control differences between groups, in keeping with working memory deficits contributing to DD (Geary, 2004). All these studies point to the fact that DD is characterized at the neuronal level by atypical properties in parietal and frontal regions and that DD children show atypical behavior and atypical functional activation in the parietal areas during basic numerical processing.

However, the management of two codes in bDD is a circumstance no study has yet contemplated behaviorally or at the neuronal level. Given the aforementioned linguistic imbalances for math in bilinguals, it is predicted that the input language will matter even more when a core number processing deficit occurs in bDD. Thus, we will here introduce a new ERP study that tested basic numerical operations with an input in either LL^{math} or the other language (OL) in bDD individuals and matched control participants to determine if (1) distance effects occur in each of the languages for each group and (2) what are the brain sources of the exhibited ERP distance effects.

Ultimately, the differences between the two languages in bilinguals described above should be based in differences in the brain networks sustaining those effects: either the computation of quantity implies a wider network of interacting systems or core number parietal areas show a better integration with the preferred language in a sort of neural recycling for that specific symbolic system (Dehaene, 2009). This can be tested in a control group, while the distance effect is passively computed from LL^{math} or OL inputs. Finally, how these two languages facilitate the management of basic numerical knowledge in bDD were also assessed.

We recently conducted a study on a sample of 14 Basque-Spanish bilingual children aged between 8 and 13 years. Seven were children who had been previously diagnosed with DD and tested using the Dyscalculia Screener (DS—Butterworth, 2003), and seven were age- and gender-matched controls (four males and three females aged between 8 and 13 years). Disregarding other cognitive impairments and unlike a matched control group, all DD children showed serious difficulties in math at school. All participants were more proficient in Spanish, however, and crucially, both groups had learned/were being taught math at school in Basque. Thus, there was a mismatch between the most fluent language and the language used for math. In what follows, we will use LL^{math} for Basque and OL for Spanish.

The experimental paradigm was an adaptation paradigm inspired by the version proposed by Hsu and Szucs (2012). Number words were presented instead of Arabic digits. Stimuli included six number words in Spanish and six number words in Basque that corresponded to the digits 1, 2, 3, 7, 8, and 9 ("uno, dos, tres, siete, ocho, and nueve" in Spanish and "bat, bi, hiru, zazpi, zortzi, and bederatzi" in Basque). These stimuli were grouped to create stimulus lists that were repeated using the number words corresponding to 1 and 2 or 8 and 9 in each language as adaptation stimuli and number words 3 and 7 as deviants. Three could serve as a close-distance deviant when the adaptation items were 1 and 2 and far when the adaptation items were 8 and 9. The opposite happened with 7; it could serve as a close-distance deviant when the adaptation items were 8 and 9 and far when the adaptation items were 1 and 2. Every list was repeated 8 times for a total of 32 close-distance trials and 32 far-distance trials in each language. Thus, the sequences could be as follows: 1-2-1-2-1-2-1-2-1-3 (sequence with a close-distance deviant) and 9-8-9-8-9-8-9-8-9-3 (sequence with a far-distance deviant).

Each stimulus was presented for 200 ms with an interstimulus interval of 1000 ms. Stimuli were presented in different locations around the center of the screen at random sizes. During each trial, a fixation cross was present in the center of the screen. Participants were required to fixate on the cross throughout the trials. Their (irrelevant) active task to assure attention was to press a button when the color of the stimulus was orange, which randomly happened 90 times during the experiment. Each trial was followed by an inter-trial interval of 1000 ms. Then, 26 more lists of fillers with random series were included in the experiment to distract participants from the real objective.

Results showed ERP distance effects in both groups when using the LL^{math}. However, no distance effect was found when using the OL as an input for the DD group. Thus, both groups have a preference for the LL^{math} when manipulating quantity. In addition, there were some differences between groups in the brain source localization of those distance effect with LL^{math} input. Although both groups showed activation in parietal areas, the DD group showed activation in the left hemisphere, and the control group showed activation in the right hemisphere. Moreover, the DD group relied on a left hemisphere network that involved the frontal, perisylvian, and inferoparietal areas, which were not shown by the control group for LL^{math}, for whom the distance effect was more focal in right parietal sites. As for OL in the control group, the distance effect involved a right-lateralized network comprising frontal and inferoparietal areas, suggesting a less automatic processing of magnitude in the OL. Importantly, this outcome must be put in the context of an LL^{math} mismatching L1, hence showing again, the independence between a dominant language for math and general language use.

The above introduced data suggest interactions between the verbal codes and a core numerical marker, the distance effect. Furthermore, input language and DD interacted, as shown by the absence of a distance effect in OL for this group. Right-lateralized distance effects in superior parietal areas (including the occipital part of the IPS and the supramarginal gyrus) for the control group in LL^{math} suggested an integration of this language with classical quantity and calculation areas (Dehaene et al., 2003; Menon, 2014). The control group processed distance differently when the input was in OL, which again involved a right-lateralized network comprising frontal and inferofrontal areas. This suggests that executive processes are involved when input is in OL. In general, this also points to a somehow different management of distance with OL input for the control group. Moreover, the DD group seems to also use executive processing and more explicit linguistic processes when computing distance in LL^{math}. In turn, our data suggest a worse integration between the preferred code and core quantity knowledge in bDD. All things considered, distance effects that are detectable in the ERPs imply very different neurofunctional bases (and networks) for the two groups when functioning in LL^{math} and for the two verbal codes in the control group.

CONCLUSIONS

Several facts point to the management of two language codes as a relevant factor in educational contexts, including the learning of mathematics. The PISA report (2012) with a focus on mathematics compared immigrant (9% in 2003 to 12% in 2012) with nonimmigrant students, showing an 11-point math disadvantage for immigrant students (after controlling for socioeconomic status). Moreover, the TIMSS International Mathematics Report (Mullis, Martin, Gonzalez, & Chrostowski, 2003) reveals that bilinguals with mismatched language of instruction and language at home outperform their peers in mathematics. It is noteworthy that the relevance of the verbal medium during math learning was evidenced in the PISA report: exposure to word problems was somewhat defined as "formal math situated in a word problem when it is obvious to students what mathematical knowledge and skills are needed." Given that math content is immersed in a linguistic medium, we wondered whether one of the languages of bilinguals is a preferred vehicle for math and, importantly, whether this linguistic context always remains on the surface or at some point merges with numerical representation.

The available empirical evidence from animal and infant studies to date suggests that we depart from abstract and amodal numerical knowledge, and that this knowledge precedes the use of symbols and language acquisition. However, the acquisition of number words during early education and, likely, repeated exposure to math content in one of the languages may facilitate an integration of that language in the storage of arithmetic facts and also, in our core numerical knowledge. Even in the absence of explicit verbal content, our electrophysiological studies uncovered traces of LL^{math} number words in very basic number representations. Vigesimal syntax in number words appears to get activated when manipulating Arabic input and accessing the quantity code. These base-20 relations activate a complex network during the processing of Arabic digits. However, this network is only activated when number words containing such base-20 structures were acquired early in education. These vigesimal relations are also active when we compute the distance between two digits and thus uncover a trace on core magnitude processing. Again, they are only evident in the distance effect when vigesimal number words are contained in bilinguals' LL^{math}. Consequently, our bilingual studies indicated one window to uncover *when* number words are preferentially coded together with basic numerical knowledge. However, our data do not disregard that in concrete circumstances, this complex quantity-word mapping can be reorganized due to extensive recoding. Yet, in usual circumstances, early learning will set the network.

Our studies suggest a window for the study of the math-language relationship. Moreover, they emphasize the relevance of studying bilingual math. Though this peculiar imbalance for math in bilinguals may lead to similar observed behavior for both language inputs, the underlying brain networks appear to differ between the preferred and nonpreferred language for math. This makes the study of bilingual math relevant, as more complex math functioning is based on the differential neural bases that we have outlined. As we have shown, this relevance is more evident in bDD, in which the differences between languages in the passive computation of basic numerical processes are exacerbated. In other words, when everything goes well, very proficient bilinguals should show efficient behavior using both codes, albeit relying on different brain networks. However, the preference should be evidenced when math complexity increases or when core math functions are essentially altered, as in bDD.

ACKNOWLEDGMENTS

Both authors contributed equally to this paper. This research was supported by the Marie Curie grant FP7-PEOPLE-2010-IEF and the Ministry of Science and Innovation grants PSI2011-23995 and PSI2014-53351 to ES and by the Severo Ochoa program grant SEV-2015-0490. AM was supported by a Basque government grant PRE_992 from the Department of Education, Policy, and Culture. The authors declare no competing financial interests. Correspondence concerning this chapter should be addressed to Alejandro Martínez or Elena Salillas, Basque Center on Cognition, Brain and Language, Paseo Mikeletegi 69, 20009, Donostia, Spain (e-mail: a.martinez@bcbl.eu or e.salillas@bcbl.eu/esalillas@gmail.com).

REFERENCES

Agrillo, C., & Beran, M. J. (2013). Number without language: Comparative psychology and the evolution of numerical cognition. *Frontiers in Psychology*, *4*, 295. https://doi.org/10.3389/fpsyg.2013.00295.

Ansari, D., & Dhital, B. (2006). Age-related changes in the activation of the intraparietal sulcus during nonsymbolic magnitude processing: An event-related functional magnetic resonance imaging study. *Journal of Cognitive Neuroscience*, *18*(11), 1820–1828.

Antell, S. E., & Keating, D. P. (1983). Perception of numerical invariance in neonates. *Child Development*, *54*(3), 695–701.

Bernardo, A. B. (2001). Asymmetric activation of number codes in bilinguals: Further evidence for the encoding complex model of number processing. *Memory Cognition*, *29*, 968–976.

Brannon, E. M., & Terrace, H. S. (1998). Ordering of the numerosities 1 to 9 by monkeys. *Science (New York, N.Y.)*, *282*(5389), 746–749.

Butterworth, B. (1999). *The mathematical brain*. London: Macmillan.

Butterworth, B. (2003). *Dyscalculia screener*. London: GL Assessment.

Butterworth, B. (2005). Developmental dyscalculia. In J. I. D. Campbell (Ed.), *Handbook of mathematical cognition* (pp. 298–318). New York: Psychology Press.

Butterworth, B. (2010). Foundational numerical capacities and the origins of dyscalculia. *Trends in Cognitive neuroscience*, *14*, 534–541.

Campbell, J. I. D. (1994). Architectures for numerical cognition. *Cognition*, *53*(1), 1–44.

Campbell, J. I. D., & Clark, J. M. (1992). Cognitive number processing: An encoding-complex perspective. In J. I. D. Campbell (Ed.), *The nature and origins of mathematical skills* (pp. 457–491). Amsterdam: Elsevier.

Campbell, J. I. D., & Epp, L. J. (2004). An encoding-complex approach to numerical cognition in chinese-english bilinguals. *Canadian Journal of Experimental Psychology/Revue Canadienne de Psychologie Expérimentale*, *58*(4), 229–244.

Cappelletti, M., & Price, C. J. (2014). Residual number processing in dyscalculia. *NeuroImage: Clinical*, *4*, 18–28.

Carey, S. (2011). Concept innateness, concept continuity, and bootstrapping. *Behavioral and Brain Sciences*, *34*(3), 152–162.

Cipolotti, L., Butterworth, B., & Denes, G. (1991). A specific deficit for numbers in a case of dense acalculia. *Brain*, *114*(6), 2619–2637.

Davis, H., & Pérusse, R. (1988). Numerical competence in animals: Definitional issues, current evidence, and a new research agenda. *Behavioral and Brain Sciences*, *11*(4), 561.

Dehaene, S. (1997). *The number sense*. New York, NY: Oxford University Press.

Dehaene, S. (2009). Origins of mathematical intuitions: The case of arithmetic. *Annals of the New York Academy of Sciences*, *1156*, 232–259.

Dehaene, S., & Cohen, L. (1995). Towards an anatomical and functional model of number processing. *Mathematical Cognition*, *1*, 83–120.

Dehaene, S., Piazza, M., Pinel, P., & Cohen, L. (2003). Three parietal circuits for number processing. *Cognitive Neuropsychology*, *20*(3–6), 487–506. https://doi.org/10.1080/02643290244000239.

Dehaene, S., Spelke, E. S., Pinel, P., Stanescu, R., & Tsivkin, S. (1999). Sources of mathematical thinking: Behavioral and brain imaging evidence. *Science*, *284*(5416), 970–974.

Delazer, M., Domahs, F., Bartha, L., Brenneis, C., Lochy, A., Trieb, T., et al. (2003). Learning complex arithmetic—An fMRI study. *Cognitive Brain Research*, *18*(1), 76–88.

Emerson, R. W., & Cantlon, J. F. (2015). Continuity and change in children's longitudinal neural responses to numbers. *Developmental Science*, *18*(2), 314–326.

Feigenson, L., Dehaene, S., & Spelke, E. (2004). Core systems of number. *Trends in Cognitive Sciences*, *8*, 307–314.

Feigenson, L., & Halberda, J. (2004). Infants chunk object arrays into sets of individuals. *Cognition*, *91*(2), 173–190.

Flombaum, J. I., Junge, J. A., & Hauser, M. D. (2005). Rhesus monkeys (Macaca mulatta) spontaneously compute addition operations over large numbers. *Cognition*, *97*(3), 315–325.

Frenck-Mestre, C., & Vaid, J. (1993). Activation of number facts in bilinguals. *Memory & Cognition*, *21*(6), 809–818.

Furman, T., & Rubinsten, O. (2012). Symbolic and non symbolic numerical representation in adults with and without developmental dyscalculia. *Behavioral and Brain Functions: BBF*, *8*(1), 55.

Geary, D. C. (1995). Reflections of evolution and culture in children's cognition. Implications for mathematical development and instruction. *The American Psychologist*, *50*(1), 24–37.

Geary, D. C. (1996). The problem-size effect in mental addition: Developmental and cross-national trends. *Mathematical Cognition*, *2*(1), 63–94.

Geary, D. C. (2004). Mathematics and learning disabilities. *Journal of Learning Disabilities*, *37*(1), 4–15.

Geary, D. C., Berch, D., & Koepke, K. (Eds.), (2015). *Mathematical cognition and learning. Evolutionary origins and early development of number processing*. San Diego, CA: Elsevier/Academic Press.

Gelman, R., & Butterworth, B. (2005). Number and language: How are they related? *Trends in Cognitive Sciences*, *9*, 6–10.

Gordon, P. (2004). Numerical cognition without words: Evidence from Amazonia. *Science, 306*(5695), 496–499.

Grabner, R. H., Saalbach, H., & Eckstein, D. (2012). Language-switching costs in bilingual mathematics learning. *Mind, Brain, and Education, 6*(3), 147–155.

Graham, D. J., & Campbell, J. I. D. (1992). Network interference and number-fact retrieval: Evidence from children's alphaplication. *Canadian Journal of Psychology, 46*(1), 65–91.

Halberda, J., & Feigenson, L. (2008). Developmental change in the acuity of the "Number Sense": The approximate number System in 3-, 4-, 5-, and 6-year-olds and adults. *Developmental Psychology, 44*(5), 1457–1465.

Halberda, J., Mazzocco, M. M. M., & Feigenson, L. (2008). Individual differences in non-verbal number acuity correlate with maths achievement. *Nature, 455*(October), 665–668.

Hauser, M. D., Carey, S., & Hauser, L. B. (2000). Spontaneous number representation in semi-free-ranging rhesus monkeys. *Proceedings. Biological Sciences/The Royal Society, 267*(1445), 829–833.

Hauser, M. D., Tsao, F., Garcia, P., & Spelke, E. S. (2003). Evolutionary foundations of number: spontaneous representation of numerical magnitudes by cotton-top tamarins. *Proceedings. Biological Sciences/The Royal Society, 270*(1523), 1441–1446.

Helmreich, I., Zuber, J., Pixner, S., Kaufmann, L., Nuerk, H.-C., & Moeller, K. (2011). Language effects on children's nonverbal number line estimations. *Journal of Cross-Cultural Psychology, 42*(4), 598–613.

Henik, A., Rubinsten, O., & Ashkenazi, S. (2011). The "where" and "what" in developmental dyscalculia. *The Clinical Neuropsychologist, 25*(6), 989–1008.

Holloway, I. D., & Ansari, D. (2009). Mapping numerical magnitudes onto symbols. The numerical distance effect and individual differences in children's mathematics achievement. *Journal of Experimental Child Psychology, 103*(1), 17–29.

Hsu, Y. F., & Szucs, D. (2012). The time course of symbolic number adaptation: Oscillatory EEG activity and event-related potential analysis. *NeuroImage, 59*(4), 3103–3109.

Iuculano, T., & Cohen Kadosh, R. (2014). Preliminary evidence for performance enhancement following parietal lobe stimulation in Developmental Dyscalculia. *Frontiers in Human Neuroscience, 8*, 38.

Izard, V., Sann, C., Spelke, E. S., & Streri, A. (2009). Newborn infants perceive abstract numbers. *Proceedings of the National Academy of Sciences of the United States of America, 106*(25), 10382–10385.

Jordan, N. C., & Montani, T. O. (1997). Cognitive arithmetic and problem solving: A comparison of children with specific and general mathematics difficulties. *Journal of Learning Disabilities, 30*, 6224–6634.

Kaufmann, L., Wood, G., Rubinstein, O., & Henik, A. (2011). Meta-analyses of developmental fMRI studies investigating typical and atypical trajectories of number processing and calculation. *Developmental Neuropsychology, 36*(6), 763–787.

Kucian, K., Kucian, K., Ashkenazi, S. S., Hänggi, J., Rotzer, S., Jäncke, L., et al. (2014). Developmental dyscalculia: A dysconnection syndrome? *Brain Structure & Function, 219*(5), 1721–1733.

Kucian, K., Loenneker, T., Dietrich, T., Dosch, M., Martin, E., & von Aster, M. (2006). Impaired neural networks for approximate calculation in dyscalculic children: A functional MRI study. *Behavioral and Brain Functions : BBF, 2*(1), 31.

Lee, K. M. (2000). Cortical areas differentially involved in multiplication and subtraction: A functional magnetic resonance imaging study and correlation with a case of selective acalculia. *Annals of Neurology, 48*(4), 657–661.

Lemer, C., Dehaene, S., Spelke, E., & Cohen, L. (2003). Approximate quantities and exact number words: Dissociable systems. *Neuropsychologia, 14*, 1942–1958.

Libertus, M. E., & Brannon, E. M. (2009). Behavioral and neural basis of number sense in infancy. *Current Directions in Psychological Science, 18*(6), 346–351.

Lipton, J., & Spelke, E. S. (2003). Origins of number sense: Large number discrimination in human infants. *Psychological Science, 14*(5), 396–401.

Luk, G., Green, D. W., Abutalebi, J., & Grady, C. (2012). Cognitive control for language switching in bilinguals: A quantitative meta-analysis of functional neuroimaging studies. *Language and Cognitive Processes, 27*(10), 1479–1488.

Lyons, I. M., Ansari, D., & Beilock, S. L. (2015). Qualitatively different coding of symbolic and nonsymbolic numbers in the human brain. *Human Brain Mapping, 36*(2), 475–488.

Marsh, L. G., & Maki, R. H. (1976). Efficiency of arithmetic operations in bilinguals as a function of language. *Memory & Cognition, 4*(4), 459–464.

Martinez-Lincoln, A., Cortinas, C., & Wicha, N. Y. Y. (2015). Arithmetic memory networks established in childhood are changed by experience in adulthood. *Neuroscience Letters, 584*, 325–330.

McClain, L., & Huang, J. Y. (1982). Speed of simple arithmetic in bilinguals. *Memory & Cognition, 10*(6), 591–596.

McCloskey, M. (1992). Cognitive mechanisms in numerical processing: Evidence from acquired dyscalculia. *Cognition, 44*(1–2), 107–157.

McCloskey, M., Caramazza, A., & Basili, A. (1985). Cognitive mechanisms in number processing and calculation: Evidence from dyscalculia. *Brain and Cognition, 4*(2), 171–196.

Menon, V. (2014). Arithmetic in the child and adult brain. In R. Cohen Cadosh & A. Dowker (Eds.), *The oxford handbook of numerical cognition* (pp. 502–530). Oxford: Oxford University Press.

Miller, K. F., Kelly, M., & Zhou, X. (2005). Learning mathematics in China and the United States: Cross-cultural insights into the nature and course of mathematical development. In J. I. D. Campbell (Ed.), *Handbook of mathematical cognition* (pp. 163–178). New York, NY: Psychology Press.

Moeller, K., Zuber, J., Olsen, N., Nuerk, H. C., & Willmes, K. (2015). Intransparent German number words complicate transcoding—A translingual comparison with Japanese. *Frontiers in Psychology, 6*, 740. https://doi.org/10.3389/fpsyg.2015.00740.

Molko, N., Cachia, A., Rivière, D., Mangin, J. F., Bruandet, M., Le Bihan, D., et al. (2003). Functional and structural alterations of the intraparietal sulcus in a developmental dyscalculia of genetic origin. *Neuron, 40*(4), 847–858.

Moyer, R. S., & Landauer, T. K. (1967). Time required for judgements of numerical inequality. *Nature, 215*(5109), 1519–1520.

Mullis, I. V. S., Martin, M. O., Gonzalez, E. J., & Chrostowski, S. J. (2003). *TIMSS 2003 International Mathematics Report. Findings From IEA's Trends in International Mathematics and Science Study at the Fourth and Eighth Grades*, (pp. 127, 133–134). (http://files.eric.ed.gov/fulltext/ED494650.pdf).

Niedeggen, M., & Rösler, F. (1999). N400 effects reflect activation spread during retrieval of arithmetic facts. *Psychological Science, 10*(3), 271–276.

Nieder, A., & Dehaene, S. (2009). Representation of number in the brain. *Annual Review of Neuroscience, 32*(1), 185–208.

Noël, M. P., & Rousselle, L. (2011). Developmental changes in the profiles of dyscalculia: An explanation based on a double exact-and-approximate number representation model. *Frontiers in Human Neuroscience, 5*, 1–4.

Nuerk, H. C., Weger, U., & Willmes, K. (2005). Language effects in magnitude comparison: Small, but not irrelevant. *Brain and Language, 92*(3), 262–277.

Piazza, M., & Eger, E. (2016). Neural foundations and functional specificity of number representations. *Neuropsychologia*, *83*, 257–273. https://doi.org/10.1016/j.neuropsychologia.2015.09.025.

Piazza, M., Facoetti, A., Trussardi, A. N., Berteletti, I., Conte, S., Lucangeli, D., et al. (2010). Developmental trajectory of number acuity reveals a severe impairment in developmental dyscalculia. *Cognition*, *116*(1), 33–41.

Piazza, M., Pica, P., Izard, V., Spelke, E. S., & Dehaene, S. (2013). Education enhances the acuity of the nonverbal approximate number system. *Psychological Science*, *24*(6), 1037–1043.

Piazza, M., Pinel, P., Le Bihan, D., & Dehaene, S. (2007). A magnitude code common to numerosities and number symbols in human intraparietal cortex. *Neuron*, *53*(2), 293–305.

Pica, P., Lemer, C., Izard, V., & Dehaene, S. (2004). Exact and approximate arithmetic in an Amazonian indigene group. *Science*, *306*(5695), 1–16.

Pinel, P., & Dehaene, S. (2010). Beyond hemispheric dominance: Brain regions underlying the joint lateralization of language and arithmetic to the left hemisphere. *Journal of Cognitive Neuroscience*, *22*(1), 48–66.

Pixner, S., Moeller, K., Hermanova, V., Nuerk, H. C., & Kaufmann, L. (2011). Whorf reloaded: Language effects on nonverbal number processing in first grade-A trilingual study. *Journal of Experimental Child Psychology*, *108*(2), 371–382.

Pixner, S., Zuber, J., Heřmanová, V., Kaufmann, L., Nuerk, H. C., & Moeller, K. (2011). One language, two number-word systems and many problems: Numerical cognition in the Czech language. *Research in Developmental Disabilities*, *32*(6), 2683–2689.

Price, G. R., Holloway, I., Räsänen, P., Vesterinen, M., & Ansari, D. (2007). Impaired parietal magnitude processing in developmental dyscalculia. *Current Biology*, *17*, 1043.

Rivera, S. M., Reiss, A. L., Eckert, M. A., & Menon, V. (2005). Developmental changes in mental arithmetic: Evidence for increased functional specialization in the left inferior parietal cortex. *Cerebral Cortex*, *15*(11), 1779–1790.

Rotzer, S., Kucian, K., Martin, E., von Aster, M., Klaver, P., & Loenneker, T. (2008). Optimized voxel-based morphometry in children with developmental dyscalculia. *NeuroImage*, *39*(1), 417–422.

Rousselle, L., & Noël, M. P. (2007). Basic numerical skills in children with mathematics learning disabilities: A comparison of symbolic vs non-symbolic number magnitude processing. *Cognition*, *102*(3), 361–395.

Rubinsten, O., & Henik, A. (2009). Developmental Dyscalculia: Heterogeneity might not mean different mechanisms. *Trends in Cognitive Sciences*, *13*, 92–99.

Rumbaugh, D. M., Savage-Rumbaugh, S., & Hegel, M. T. (1987). Summation in the chimpanzee (Pan troglodytes). *Journal of Experimental Psychology: Animal Behavior Processes*, *13*(2), 107–115.

Rusconi, E., Kwan, B., Giordano, B. L., Umiltà, C., & Butterworth, B. (2006). Spatial representation of pitch height: The SMARC effect. *Cognition*, *99*(2), 113–129.

Salillas, E., Barraza, P., & Carreiras, M. (2015). Oscillatory brain activity reveals linguistic prints in the quantity code. *PLoS ONE*, *10*(4), e0121434.

Salillas, E., & Carreiras, M. (2014). Core number representations are shaped by language. *Cortex*, *52*(1), 1–11.

Salillas, E., & Wicha, N. Y. Y. (2012). Early learning shapes the memory networks for arithmetic: Evidence from brain potentials in bilinguals. *Psychological Science*, *23*(7), 745–755.

Sarnecka, B. W. (2014). On the relation between grammatical number and cardinal numbers in development. *Frontiers in Psychology*. https://doi.org/10.3389/fpsyg.2014.01132.

Semenza, C., Salillas, E., Di Pellegrin, S., & Della Puppa, A. (2016). Balancing the 2 Hemispheres in simple calculation: Evidence from direct cortical electrostimulation. *Cerebral Cortex*, 1–9. https://doi.org/10.1093/cercor/bhw277.

Shusterman, A., Slusser, E., Halberda, J., & Odic, D. (2016). Acquisition of the cardinal principle coincides with improvement in approximate number system acuity in preschoolers. *PLoS ONE*, *11*(4), e0153072.

Soltész, F., Szucs, D., Dékány, J., Márkus, A., & Csépe, V. (2007). A combined event-related potential and neuropsychological investigation of developmental dyscalculia. *Neuroscience Letters*, *417*(2), 181–186.

Spelke, E. S., & Tsivkin, S. (2001). Language and number: A bilingual training study. *Cognition*, *78*(1), 45–88.

Starkey, P., & Cooper, R. (1980). Perception of numbers by human infants. *Science*, *210*, 1033–1035.

Van Rinsveld, A., Schiltz, C., Landerl, K., Brunner, M., & Ugen, S. (2016). Speaking two languages with different number naming systems: What implications for magnitude judgments in bilinguals at different stages of language acquisition? *Cognitive Processing*, *17*(3), 225–241.

Venkatraman, V., Siong, S. C., Chee, M. W. L., & Ansari, D. (2006). Effect of language switching on arithmetic: A bilingual FMRI study. *Journal of Cognitive Neuroscience*, *18*(1), 64–74.

Verguts, T., & Fias, W. (2004). Representation of number in animals and humans: A neural model. *Journal of Cognitive Neuroscience*, *16*(9), 1493–1504.

Wagner, K., Kimura, K., Cheung, P., & Barner, D. (2015). Why is number word learning hard? Evidence from bilingual learners. *Cognitive Psychology*, *83*, 1–21.

Wilson, A. J., & Dehaene, S. (2010). Number sense and developmental Dyscalculia. In D. Coch, G. Dawson, & K. W. Fischer (Eds.), *Human behavior, learning, and the developing brain: A typical development* (pp. 212–238). New York, NY: Guilford Press.

Wilson, M. L., Hauser, M. D., & Wrangham, R. W. (2001). Does participation in intergroup conflict depend on numerical assessment, range location, or rank for wild chimpanzees? *Animal Behaviour*, *61*(6), 1203–1216.

Woodruff, G., & Premack, D. (1981). Primitive mathematical concepts in the chimpanzee: Proportionality and numerosity. *Nature*, *293*(5833), 568–570.

Wynn, K. (1992a). Addition and subtraction by human infants. *Nature*, *358*(6389), 749–750.

Wynn, K. (1992b). Children's acquisition of number words and counting system. *Cognitive Psychology*, *24*, 220–251.

Xu, F., & Spelke, E. S. (2000). Large number discrimination in 6-month-old infants. *Cognition*, *74*, B1–B11.

FURTHER READING

Kaplan, E., Goodglass, H., & Weintraub, S. (1983). *Boston naming test*. Philadelphia: Lea & Febiger.

Kaufmann, L., Wood, G., Rubinsten, O., Henik, A., Rubinstein, O., & Henik, A. (2011). Meta-analyses of developmental fMRI studies investigating typical and atypical trajectories of number processing and calculation. *Developmental Neuropsychology*, *36*(6), 763–787.

Noël, M. P., Fias, W., & Brysbaert, M. (1998). About the influence of the presentation format on arithmetical-fact retrieval processes. *Cognition*, *63*(3), 335–374.

Wang, Y., Xue, G., Chen, C., Xue, F., & Dong, Q. (2007). Neural bases of asymmetric language switching in second-language learners: An ER-fMRI study. *NeuroImage*, *35*(2), 862–870. https://doi.org/10.1016/j.neuroimage.2006.09.054.

Whalen, J., McCloskey, M., Lindemann, M., & Bouton, G. (2002). Representing arithmetic table facts in memory: Evidence from acquired impairments. *Cognitive Neuropsychology*, *19*(6), 505–522. https://doi.org/10.1080/02643290244000086.

Chapter 9

Early Number Knowledge in Dual-Language Learners From Low-SES Households

Barbara W. Sarnecka*, James Negen†, and Meghan C. Goldman*
**University of California, Irvine, Irvine, CA, United States*
†*Durham University, Durham, United Kingdom*

INTRODUCTION

As of 2015, one in every four American children is Hispanic, and approximately two-thirds of Hispanic children live in or near poverty (defined as less than two times the federal poverty level; Gennetian, Rodrigues, Hill, & Morris, 2015). In fact, Hispanic children now constitute the largest population of U.S. children living in poverty (Lopez & Velasco, 2011; Stepler & Brown, 2015). These children face serious educational challenges. Currently, Hispanic students lag at least a half a standard deviation behind their white and Asian-American peers in mathematics at the beginning of kindergarten and throughout K-12 schooling, with no narrowing of the gap in sight (Garcia & Jensen, 2009; Wiley, 2009). Given that students' educational performance during the early years of school is strongly correlated with their later achievement (e.g., Duncan et al., 2007), improved early education for Hispanic children from low-SES backgrounds may be critical to their long-term educational success.

The situation is complicated by the fact that many Hispanic children are dual-language learners (DLLs), meaning that they are exposed to and learning through two distinct languages before age 5. As of 2013, 63% of Hispanic children were DLLs (childstats.gov., 2015). The numbers of these children in Head Start and Early Head Start have also increased dramatically in recent years. As of 2015, children whose home language was Spanish accounted for 48% of all Head Start and Early Head Start participants in California and about 25% of children in those programs nationwide (CA Head Start Assoc, 2017; U.S. DHHS, 2017).

As educators and policymakers consider the needs of these children, one important area to think about is early mathematics and specifically early

Mathematical Cognition and Learning, Vol. 4. https://doi.org/10.1016/B978-0-12-812574-8.00009-2

numeracy—the understanding of counting and whole numbers that children acquire during the preschool years. A gap in number knowledge between young children from high- and low-SES households, independent of ethnic background, is already measurable by age 3, and it only widens as children get older (Council, 2009; Jordan & Levine, 2009). Children from low-SES households come to kindergarten with significantly less knowledge of counting, numbers, and number operations than middle-SES children and are four times more likely to show flat growth in these areas throughout kindergarten and early first grade (Jordan, Kaplan, Locuniak, & Ramineni, 2007; Jordan, Kaplan, Nabors Oláh, & Locuniak, 2006; see also Aunola, Leskinen, Lerkkanen, & Nurmi, 2004). Other foundational math tasks on which higher-SES children outperform their low-SES peers include reciting the counting list, counting objects, knowing the meanings of small numbers, understanding the cardinal principle (also called the cardinality principle), identifying written numerals, adding and subtracting small numbers, comparing numerical magnitudes, and estimating the positions of numbers on a number line (Griffin, Case, & Siegler, 1994; Jordan, Huttenlocher, & Levine, 1992, 1994; Jordan et al., 2007; Jordan et al., 2006; Jordan & Levine, 2009; Kirk, Hunt, & Volkmar, 1975; Ramani & Siegler, 2008; Starkey, Klein, & Wakeley, 2004).

DOES BILINGUALISM MATTER FOR EARLY NUMERACY?

Plenty of research has shown that growing up in a low-SES household is bad for early math learning, but much less is known about the effect of being a dual-language learner. Broadly speaking, early number development does interact with language development. Spoken and written numbers are linguistic symbols, and the number knowledge of 2- to 4-year-old monolingual English speakers is highly correlated with their expressive and receptive vocabulary, even when controlling for age and other variables (Negen & Sarnecka, 2012). It also seems that the grammar of a child's native language has some influence on how long it takes the child to learn the meanings of the first few number words: children learning English or Russian—languages that make a grammatical distinction between singular and plural referents (e.g., my *child is* waiting in the car/my *children are* waiting in the car) learn the meaning of "one" slightly earlier than children who speak Japanese or Mandarin—languages that rarely mark plurality (Le Corre, Li, Huang, Jia, & Carey, 2016; Sarnecka, Kamenskaya, Yamana, Ogura, & Yudovina, 2007). Similarly, children learning Saudi Arabic or Slovenian—languages that grammatically distinguish between sets of one, two, and more than two—appear to learn the word for "two" slightly earlier than children learning other kinds of languages (Almoammer et al., 2013).

The practical question for Spanish/English dual-language learners who are learning pre-K math is not about language-specific effects (Spanish and English do not differ much in the ways that might matter for early numeracy) but rather about the possible effects of learning two languages, period. For example, if a child's home language is different from the language of instruction in his or her preschool classroom, how does this mismatch affect early math instruction? Does it hinder the child's early development of numeracy?

There is some evidence that bilingualism might confer cognitive benefits on children. Under controlled conditions, bilingual children may perform better than monolinguals on tasks involving inhibition, attentional control, and perspective-taking (Bialystok, 2001, 2009; Fan, Liberman, Keysar, & Kinzler, 2015; Hilchey & Klein, 2011). However, these effects are disputed (e.g., Goldman, Negen, & Sarnecka, 2014; Morton & Harper, 2007; Paap & Greenberg, 2013), and assuming they do exist, the effects are too small to make much of a difference in educational or life outcomes. They are mainly interesting because of the insight they offer into how development in the domains of executive functioning and social cognition is shaped by experience. For our purposes, the relevant question about bilingualism (or, more precisely, about being a dual-language learner) is whether it causes children to have a different profile of strengths and needs when it comes to prekindergarten math.

Not many studies have focused specifically on preschool math development in dual-language learners, but the available data suggest that numeracy development in this population of children is slow relative to national norms. For example, Xue and colleagues (Xue, Atkins-Burnett, and Moiduddin, 2012) used the Applied Problems test from the Woodcock-Johnson III Tests of Academic Achievement (WJ III ACH) to assess 675 Spanish-English bilingual preschoolers from Los Angeles Universal Preschool as part of the Universal Preschool Child Outcomes Study. They found that the DLLs scored below the national average for their age. Similarly, Iglesias (2012) studied 132 Spanish-English DLLs attending Head Start programs in Florida and found that their performance on the Applied Problems and Quantitative Concepts subtests of the Woodcock-Johnson III and its equivalent in Spanish ranged from average to low average when compared with national normative samples of monolingual peers.

Somewhat more encouraging findings came from a mathematics teaching intervention study with somewhat older children, which was conducted by Fuson and colleagues (Fuson, Smith, & Lo Cicero, 1997) in two predominantly Latino, low-SES, urban first-grade classrooms. One classroom was English-speaking; the other was Spanish-speaking. The intervention sought to support first graders' thinking of two-digit quantities as tens and ones, and it was successful. By the end of the year, children in both classrooms accurately added and subtracted two-digit numbers and used units of tens and ones on various tasks. This performance was substantially above the norms reported in other studies for U.S. first graders of higher SES and for older U.S. children, suggesting that well-designed interventions in either English or Spanish may be able to narrow or close the early math achievement gap, at least during the time that the interventions are being provided. (See some relevant training intervention studies in Chapter 3 by LeFevre et al.).

Taking this rather sparse literature as a starting point, we (the authors) wanted to examine the early numeracy development of DLL children from low-SES households in greater detail. Our specific questions were shaped by research on early number development during the last several decades; in the next section, we briefly review evidence about how young children represent numbers, including their representations of approximate number, spoken numbers, and written numbers.

WAYS FOR YOUNG CHILDREN TO REPRESENT NUMBERS

The Innate, Approximate Number System

Broadly speaking, mental representations of number are either approximate or exact. Approximate number representations are the outputs of a built-in perceptual system that humans share with other animals (Dehaene, 2011; Feigenson, Dehaene, & Spelke, 2004; Gallistel, 1989). In recent years, people have taken to calling this "approximate number system" or ANS for short. (It has also been variously called the *analog-magnitude number system*, *nonverbal counting system*, and simply *number sense*.) This system represents approximate numbers without using language. For example, humans and other animals can easily see the difference between 40 and 80 objects without having to count them. But we cannot see the difference between 40 and 41 objects, because the ANS is not accurate enough to perceive such fine distinctions. The system's ability to tell the difference between two magnitudes is based on the ratio of one magnitude to the other. In other words, the difficulty of distinguishing between 40 and 80 items is the same as distinguishing between 20 and 40 or between 400 and 800 or any other pair where the ratio is 1:2.

Human infants and children have this approximate number system, but it is not as accurate as in adults. On average, 6-month-old infants can discriminate ratios of 1:2 or easier, whereas most 9-month olds can discriminate the more difficult ratio of 2:3 (Brannon, Abbott, & Lutz, 2004; Lipton & Spelke, 2003, 2004; Xu, 2003; Xu & Arriaga, 2007; Xu & Spelke, 2000; Xu, Spelke, & Goddard, 2005). Accuracy improves as children get older. For example, Halberda and Feigenson (2008) reported that on average, 3-year olds could discriminate sets with ratios of 2:3 or easier; 4-year olds could handle ratios of 3:4; 5-year olds succeeded with ratios of 4:5; for 6-year olds, the threshold was 6:7; and for adults, it was 9:10.

But these are only averages. In practice, estimation accuracy varies from person to person. Some studies have suggested an association between an individual's ANS acuity and his or her math achievement (Halberda, Mazzocco, & Feigenson, 2008; Mazzocco, Feigenson, & Halberda, 2011b). This relation, however, is not a strong one and can probably be explained by a small subset of children who have both very poor ANS acuity and a math learning disability (Chen & Li, 2014; Fazio, Bailey, Thompson, & Siegler, 2014; Mazzocco, Feigenson, & Halberda, 2011a; Rousselle & Noël, 2007; van Marle, Chu, Li, & Geary, 2014).

Do Not Trust the Standard Way of Measuring ANS Acuity in Preschoolers

When we try to measure young children's ANS acuity, we run into a problem: children who do not understand how counting represents number (specifically, children who have not learned the cardinal principle of counting) often do not understand what experimenters are asking them to do. The standard way of measuring ANS acuity is to show participants two clouds of dots of varying sizes and to ask which cloud has more (e.g., children might see a bunch of blue dots and yellow dots and be asked "Are there more blue dots, or more yellow dots?").

The problem is that children who do not have a clear notion of cardinal number do not necessarily interpret the word "more" to mean "a greater number of." How they interpret "more" is unclear. They might think it means "bigger dots," in which case they might look for the biggest individual dot or try to decide which color dots are, in general, bigger. Or they might think it means "more blue stuff," in which case they might choose the cloud covering a greater area. In any case, this spells trouble for researchers who want to use the task to measure children's ANS acuity. Assuming the stimuli are controlled for area and average or largest dot size, children will perform at chance. A number of studies have reported that children who know more number-word meanings seem to have better ANS acuity (e.g., Mussolin, Nys, Leybaert, & Content, 2012; Shusterman, Slusser, Halberda, & Odic, 2016; Wagner & Johnson, 2011; see also Abreu-Mendoza, Soto-Alba, & Arias-Trejo, 2013). However, when the data from these studies are closely examined, the pattern is usually one of chance performance by children who do not understand cardinality (see Section "Spoken Numbers" below for a discussion of what this means) and above-chance performance by children who do understand cardinality. This is enough to create a significant positive association between knowledge of spoken numbers and ANS acuity even if the actual ANS acuity of children in the two groups does not differ. (For an in-depth exploration of this problem, see Negen & Sarnecka, 2014). The upshot for researchers is that measuring ANS acuity in young children is very tricky, and studies purporting to find links between young children's ANS acuity and any other measure should have their methods closely and skeptically examined.

Spoken Numbers

Although the ANS is the only innate system that humans have for representing magnitudes *per se*, cognitive scientists broadly agree that the ANS cannot be the source (or at least not the sole source) of exact-number concepts. This is because ANS representations are only approximate. If ANS representations are the only representations a person has, there is no way that person can distinguish between, for example, 31 and 32 objects, because a ratio of 31:32 is too fine-grained for anyone to perceive. Such precise distinctions require *exact* numerical representations.

In order to represent numbers exactly, we need more than the ANS. We need some way of keeping track of individual things. We can do this on our own, using just our innate cognitive capacities, but only up to a point. Imagine a puppy playing on a grassy lawn. With a glance, you can say that there is only one puppy. If a second or even a third puppy is added, you still have no trouble saying exactly how many there are. But when a fourth puppy is added, it starts to get a little more challenging. Is that four puppies or five? At six or seven puppies, it is hopeless. Now the only way to tell exactly how many there are is to catch and count them—that is, to use the resources of spoken language.

When we think of children's math learning, we typically think of the exact, symbolic number system. Unlike the innate ANS, the exact-number system is a cultural invention and must be acquired anew by each generation. Children must learn the counting list (*one*, *two*, *three*, *four*, *five*, etc.), rules about how to count correctly (e.g., always say number words in the right order; count each object once and only once), and the cardinality principle (also called the cardinal principle), which says that the final word of a count tells you the number of items in the set that was counted (Gelman & Gallistel, 1978; Schaeffer, Eggleston & Scott, 1974).

Recognizing the central importance of counting and spoken numbers for early math, the California guidelines for preschool mathematics include a strand called *number sense*,[1] which states that by 48 months of age, children should be able to recite the number-word list up to "ten," to match the words "one," "two," and "three" with the appropriate set sizes; count up to five objects without making a mistake; and answer the question *how many* with the last number word of a count (California Dept of Education, 2008).

But acquiring number sense is a lengthy process. In order to meet the guidelines at age 48 months, children must begin learning about numbers and counting much earlier than that—and in high-SES households, at least, they do. One study showed that high-SES children in Australia and Japan begin to recognize the counting list of their language and expect it to be paired with a correct counting procedure sometime between 15 and 18 months of age (Slaughter, Itakura, Kutsuki, & Siegal, 2011).

Many studies of early number development (albeit mostly with high-SES children) have shown the following pattern of development. First, children learn to recite the number list and to recite it in situations where the focus

1. Note that the term "number sense" is used differently by different authors. As mentioned earlier, cognitive neuroscientists (e.g., Dehaene, 2011) have used it to refer to the innate, approximate number system. But in the education world, "number sense" refers to the child's understanding of the counting system and exact, whole-number concepts expressed as spoken and written numerals (e.g., Dyson, Jordan, & Glutting, 2013; Jordan, Glutting, & Ramineni, 2010; Jordan et al., 2006; Locuniak & Jordan, 2008).

is on quantity. For example, one of us (BWS) had a child who, at 18 months old, discovered a tall stack of compact disks. He looked back and forth from his mother to the stack, pointing to it repeatedly and saying, "onetwothree, onetwothree, onetwothree!" He seemed to be expressing something like, "Hey! This is a lot of CD's, isn't it?"

Similarly, that child's brother at age 24 months turned off the tap after filling the bathtub and said the following: "We needed *enoughs* of water. We needed *onetwothreefourfivesixseveneightnineTEN* of water." These anecdotes illustrate that even toddlers can learn the counting list and connect it somehow to quantification. But children this young do not typically know exactly what set sizes the words "one" or "two" or any other number words actually stand for (e.g., Baroody & Price, 1983; Briars & Siegler, 1984; Frye, Braisby, Lowe, Maroudas, & Nicholls, 1989; Fuson, 1988; Wagner & Walters, 1982; Wynn, 1990, 1992).

As children begin to learn the meanings of the number words, they learn the first three or four words one at a time, and in order (Sarnecka & Lee, 2009; Wynn, 1992). Many studies have demonstrated this pattern of learning, using a task that has many names: *Give-N*, *Give-a-number*, *Give-me-X*, the *verbal production task*, and probably others. We generally use the shortest name, Give-N.

In the Give-N task, the experimenter asks the child to produce a set of some number of objects. We use a stuffed animal (e.g., Peter the Anteater, the UC Irvine mascot) and a bowl of 15 identical small toys (e.g., green plastic trucks). We ask the child to "Give *one* truck to Peter," "Give *two* trucks to Peter," "Give *five* trucks to Peter," and so on. When the child gives something to Peter (by placing it on Peter's plate and sliding the plate over to him), we give them a chance to check or change their answer by asking, e.g., "Is that five?" when we asked for five. We also record whether children spontaneously use counting to produce the set. After five or six requests, we bring out a new animal and a new bowl of toys and repeat the process all over again, asking for the same numbers in a different order. We do the whole thing three times.

In thinking about this task, researchers have sometimes worried that children might perform badly on it even if they understand the logic of counting, because it might not occur to the children to use counting in order to produce the sets (Gelman, 1993). In order to avoid this problem, researchers sometimes use additional follow-up questions, telling children to count and check their answers. For example, when a child was asked to "Give FIVE trucks to Peter" but actually put eight trucks on the plate, the experimenter might say "Can you count and make sure it's five?" or "Hmm. I don't think that's five. Can you count and fix it so it's five?" (Le Corre, Van de Walle, Brannon, & Carey, 2006). These extra prompts make little difference to children's performance, which looks quite similar to when children are asked only a single follow-up question (e.g., "Is that five?"). However, the extra prompts do generate some amusing video data, because many children who do not understand the cardinal principle (and thus do not know how to use counting to produce a set of five) *do* have a vague notion that the last word of a count should correspond to the number of items in the set. So faced with a set of say, eight objects, they do things like count the set properly but change the last word to five ("one, two, three, four, five, six, seven, five") or just keep saying five until they get to the end of the set ("one, two, three, four, five, five, five, five").

Studies using the Give-N task have revealed a predictable pattern of development, such that any given child's performance can be described in terms of his or her *number-knower level* (e.g., Condry & Spelke, 2008; Le Corre & Carey, 2007; Le Corre et al., 2006; Lee & Sarnecka, 2010, 2011; Negen & Sarnecka, 2012; Sarnecka & Gelman, 2004; Sarnecka & Lee, 2009; Slusser, Ditta, & Sarnecka, 2013; Slusser & Sarnecka, 2011; Wynn, 1990, 1992). These number-knower levels are found across language environments, not only in children speaking English or Spanish (or both) but also in child speakers of Arabic, Japanese, Mandarin Chinese, Russian, Slovenian, and even Tsimané, the language of a farming and foraging people by the same name who live in rural Bolivia (Almoammer et al., 2013; Barner, Libenson, Cheung, & Takasaki, 2009; Li, Le Corre, Shui, Jia, & Carey, 2003; Piantadosi, Jara-Ettinger, & Gibson, 2014; Sarnecka et al., 2007).

Here is how a child's number-knower level is diagnosed using the Give-N task. At the earliest or "pre-number-knower" level, the child makes no distinctions among the meanings of different number words. On the Give-N task, pre-number-knowers might give one object for every request or always give a handful, but the number given is unrelated to the number requested. At the next or "one-knower" level, the child gives exactly one object when asked for "one" and gives two or more objects when asked for any other number. After this comes, the "two-knower" level. Two-knowers give one for "one" and two objects for "two" but do not reliably produce any larger exact sets. The two-knower level is followed by a "three-knower" and a "four-knower" level, although not all studies find four-knowers. It may be that some children make the leap from "three-knower" directly to the next level—the level where they use the cardinal principle to determine the meanings of number words higher than "four."

Note that the cardinal principle is not just a simple rule saying "the last number of a count tells you how many are in the set." A rule like that is easy enough to teach kids, and it leads to the kind of amusing "fixes" in the Give-N task that we described earlier. The cardinal principle is actually much more profound—it is the understanding (certainly implicit, because no parent talks about it this way) that the cardinal meanings of number words like *five* and *eight* depend on their place in the counting list. The fifth word in a counting list means 5; the eighth word means 8. And so, the way to produce a set of objects that matches a particular number word is to count out one object for each word in the number list, starting at one and stopping when you get to the number requested.

There are many reasons to think that a child's number-knower level tells us something important about his or her conceptual understanding of numbers. For example, one-knowers understand that each number word means something different from all the others, and that the number words have something to do with quantity (Sarnecka & Gelman, 2004). One-knowers also seem to understand that numbers apply to discontinuous quantities such as crayons, but not to continuous quantities such as water (Slusser et al., 2013). Two-knowers from high-SES households in fact do often answer the question "how many" by repeating the last word of a count, but do not seem to know that the last word tells the number of items in the set (Sarnecka & Carey, 2008). Understanding seems to dawn gradually. For example, three- and four-knowers seem to understand that number words appearing later in the count list denote larger quantities, although they do not quite know how to map exact quantities to specific number words (Sarnecka & Carey, 2008).

An important step forward seems to come when children figure out how the cardinal principle connects each number in the list to an exact set size. Children who understand the cardinal principle identify *number* (as opposed to some other property of sets such as total summed area or contour length) as the dimension of experience that number words describe (Slusser & Sarnecka, 2011). Similarly, only cardinal-principle-knowers seem to understand that each "next word" in the list corresponds to the addition of one more item to the named set (Sarnecka & Carey, 2008; see Chapter 2 by Marchand and Barner for an alternative account of how understanding of this so-called successor function is acquired). And there is a near-perfect correlation between understanding the cardinal principle (as tested by the Give-N task) and understanding that the members of two sets of the same number (e.g., five peaches and five apples) can be matched up in one-to-one correspondence with each other (Sarnecka & Wright, 2013). One way we think of it is that when children understand the cardinality principle, they understand what exact numbers are.

Cardinality understanding is a prerequisite for the kindergarten and first-grade number learning that predicts math achievement in later years (Geary & vanMarle, 2016; Jordan et al., 2007; Jordan et al., 2006; Jordan, Kaplan, Ramineni, & Locuniak, 2009; Merkley & Ansari, 2016). So a central concern of the present study was to investigate what dual-language learners from low-SES households know about these key aspects of the spoken-language number system: counting, number meanings (i.e., their number-knower levels), and the cardinality principle.

Written Numbers

The other number representations that may be relevant to early education are written numerals (Geary & vanMarle, 2016; Merkley & Ansari, 2016; Purpura & Lonigan, 2015; Raghubar & Barnes, 2017). Children must learn to recognize and later to write the numerals 0–9, and in the early elementary grades, children will learn how to represent place value in multidigit numerals. But at the preschool level, most of researchers' attention has been focused either on children recognizing single-digit numbers or on children's understanding of how number lines represent the counting list.

OUR STUDY

The study we describe in this chapter examined the early number knowledge of Spanish-English dual-language-learners from very low-SES households who were attending Head Start programs in Orange County, California. For each aspect of early number development that we measured, we were interested in two comparisons. The first was a within-child comparison of performance in Spanish and English. We wanted to know whether children's performance depended on the language of testing, and for that reason, we tested all linguistic/symbolic measures twice: once in Spanish and once in English. We also wanted to know how children's performance was related to the language of math instruction in their Head Start program, which was English.

The second comparison of interest was between different populations of children in Orange County. We wanted to know how the low-SES DLL children's number knowledge compared with that of three other groups of Orange County preschoolers: English monolingual Head Start students, English monolingual students at tuition-based preschools, and dual-language learners (with a variety of home languages, of which Chinese was the most common) at tuition-based preschools. This was not a controlled comparison in the experimental sense, because these groups of children differed in many ways. Our reasoning was simply that these are actual groups of children who live in Orange County and will enter Orange County schools together, and so, this is the range of early numeracy development that educators and policymakers need to understand.

What We Measured and What We Found

All of our assessments were administered by research assistants who were native bilingual speakers of English and Spanish, almost all of whom grew up in the same communities as the participants and their families and still lived in those communities during the time of data collection. Following best practices for the assessment of bilingual students (Espinosa, 2014), we conducted all of our language-dependent assessments (i.e., assessments of vocabulary and of spoken and written number knowledge) twice (once in English and again in Spanish) in an order that was counterbalanced across participants. Demographic data were collected in whichever language the respondent preferred, and the approximate number system assessment was done in the child's preferred language or in a combination of languages—whatever worked best to facilitate communication.

Family Demographics

When they signed their child up for the study, parents were asked to complete a demographic questionnaire that asked about the child's racial and ethnic background, household SES, the language(s) spoken in the home, and education levels of the child's caregivers. Most of the children in the low-SES DLL (95%) group were Hispanic/Latino and spoke Spanish in addition to English. Most of the high-SES DLLs were Asian, and Chinese was their most common other language. High-SES DLL families also used more English at home: Only 33% of low-SES DLL families reported speaking mostly English at home, whereas 55% of high-SES DLL families did.

The differences in household income and education between the four groups were stark. For low-SES DLLs, over one-third of families reported an annual household income of less than $10,000. The rest reported incomes from $10,000 to $30,000. Among the low-SES English monolingual families, the most common answer was $20–30,000. So the low-SES DLL households actually had a lot less money than the low-SES monolingual households. In contrast, 100% of the parents in both high-SES groups—monolingual and DLL—ticked the box for the highest income category on the questionnaire, which was "over $75,000."

Differences in caregiver education were even more dramatic, with 43% of caregivers for low-SES DLLs having less than a high school diploma. In the English monolingual low-SES group, the most common level of education was "high school diploma or GED," and over a quarter of caregivers in this group had some college education. So caregivers in the low-SES DLL households had less education than those in the low-SES monolingual households. Among caregivers of children attending the tuition-based preschools, all but one had a college education and most (65% of DLLs and 75% of English monolinguals) had graduate degrees. Thus, our participants illustrate the wide differences in education and income that divide families in Southern California and increasingly—as income inequality grows and higher education continues to become less affordable—in other parts of the United States as well (Duncan & Murname, 2014; Reich, 2015).

Receptive Vocabulary

The first thing we assessed was children's receptive vocabulary (i.e., the words they understood) in both English and Spanish. We did this using standardized measures—the PPVT for English (Dunn & Dunn, 2007) and the TVIP for Spanish (Dunn, Lugo, Padilla, & Dunn, 1986). In these measures, children are presented with a set of four pictures and are told to, e.g., "Put your finger on *bicycle*," or "Put your finger on *climbing*." We included this vocabulary assessment not because we thought it was particularly diagnostic of numeracy development (although general receptive vocabulary and number knowledge are strongly correlated) (Negen & Sarnecka, 2012), but as a general measure of which language the children were more familiar with.

Results of the vocabulary assessments for low-SES DLLs showed that as a group, they knew more Spanish than English. Although their vocabulary scores in both languages were lower than the national norms for their age, both the PPVT and the TVIP were normed on monolingual speakers of each language, so the norms are not necessarily appropriate for DLLs. Comparing low-SES DLLs to other groups of children showed two effects that surprised no one: High-SES children had much bigger vocabularies than low-SES children, and monolingual speakers had bigger vocabularies than DLLs of the same age.

ANS Acuity

The ANS is a system that children have available from infancy representing number information (at least approximately). Given the math achievement gap that we see between children from different backgrounds and the interest among cognitive scientists in the question of how individuals' ANS acuity may relate to their math achievement, we wanted to measure the ANS acuity of the children in our study.

However, as discussed earlier in this chapter, the standard way of measuring ANS acuity does not ensure that child participants are actually trying to point to the dot cloud with the greater *number* of dots. Our solution was to modify the standard task (see Fig. 1). We did this by having the children complete a set of training trials in which the ratio of the dot clouds was a very easy 1:3. For example, children might see a card with 8 dots in the right-hand cloud and 24 dots in the left-hand cloud. These set sizes are so different that they should be very easy to tell apart, even for young children. So if children did not get these trials right, we could be confident that they were not answering on the basis of numerosity. As is standard for these tasks, we controlled the dot clouds in our stimuli for area, such that half of the trials were area-congruent (meaning that the cloud with a greater number of dots also had more total dot area) and the other half were area-incongruent (meaning that

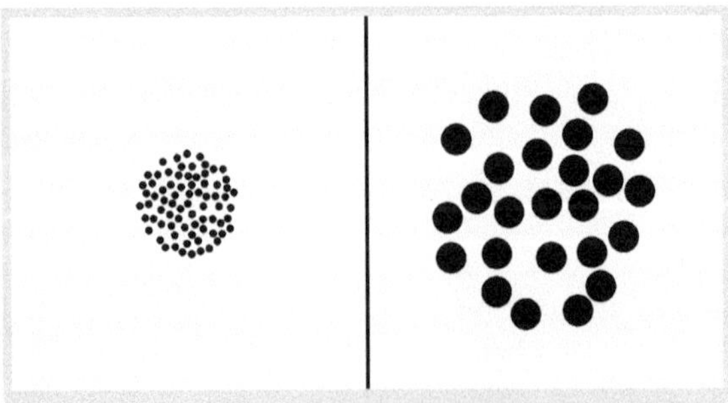

"Which side has more dots?"

FIG. 1 Example of stimuli used in the approximate number system task.

the cloud with the greater number of dots had less dot area). During the training trials, we gave children feedback to help them understand what we wanted them to do. For example, when children incorrectly chose the dot cloud with fewer dots but greater area, we said "Well these dots are *bigger*, but this side has *more* dots. They're *smaller*, but there's *more* of them." We continued with the training trials until the child either got eight trials in a row correct or quit playing. Children who did get eight training trials in a row correct moved on to the test trials.

The ratios tested were 1:2 (=0.50), 7:12 (=0.58), 2:3 (=0.66), 17:24 (=0.71), 3:4 (=0.75), 4:5 (=0.80), 5:6 (=0.83), 7:8 (=0.87), and 9:10 (=0.90). For each trial, the child was shown a large card with two dot clouds, side by side. Each dot cloud contained between 20 and 100 black dots. To keep children motivated, experimenters gave them feedback after every trial. If the child correctly chose the side with more dots, the experimenter said "That's right—this side has more." If the child chose incorrectly the experimenter said "Uh oh, this side has more dots, you see?" Trials were presented too rapidly for children to count the dots, and no children were observed attempting to count.

Results of the ANS Task

The changes that we made to the ANS testing procedure ensured that for those children who did complete the task (i.e., who reached criterion on the training trials and then went on to complete at least one block of test trials), we can be confident that what we measured really was their ANS acuity. However, these changes also meant that a lot of children simply could not complete the task. Over one-third of the children we tested never succeeded at eight training trials in a row, and so never moved on to the test trials. Another handful of children got through the training trials but performed at chance on all the test blocks (even at easy ratios like 1:2), indicating that they still were not drawing on ANS representations to pick the more numerous dot cloud.

Most of the children who failed to get through the training trials were from the low-SES group. In fact, they were mostly children who did not yet understand the principle of cardinality. This replicated the results of an earlier study from our lab (Negen & Sarnecka, 2014), where we found that many children who have not yet grasped the cardinality principle simply do not understand what experimenters want them to do on the ANS task. Understanding the cardinality principle seems to be connected with identifying *numerosity* as the perceptual domain that number words and counting are about. Perhaps another way to say that is that when children come to understand the cardinality principle, they begin to link counting with ANS representations.

In practical terms, this means that we (researchers using this method) cannot say much at all about the relation between preschoolers' ANS acuity and their knowledge of symbolic numbers, because we cannot readily measure the ANS acuity of the children who do not know much about symbolic numbers.

Knowledge of the spoken-number system (specifically, the cardinality principle of counting) seems often to be needed in order for children to understand the ANS measurement task.

Returning to our study results, if we look at the data from children who did get through the training trials and complete the ANS task, we find (not surprisingly) that older children performed better than younger ones. There was no evidence that dual-language learners performed better or worse than English monolingual learners. This finding is consistent with earlier studies from our lab, which found no effect of monolingual or DLL status on ANS performance (Goldman et al., 2014).

Among the children we were able to test, the present study also found that children from high-SES households performed better than those from low-SES households. Does that imply that the math achievement gap between children from high- and low-SES backgrounds arises because of differences in their ANS acuity? It is possible, but we think unlikely. The difference, while statistically significant, is not big enough to matter in practical terms. If we look at the mean coefficient of variance (a standard measure of ANS acuity; lower numbers are better) for the high- and low-SES children in our sample, we find that it was 0.42 for the high-SES group and 0.47 for the low-SES group. What does this mean in real life? Imagine a set of 50 items (e.g., a flock of 50 birds flying through the air). If asked to guess how many birds there are, a person whose ANS coefficient of variance is 0.42 (like the low-SES kids in our study) should estimate somewhere between 26 and 73 birds. A person whose coefficient of variance is 0.47 (like the high-SES kids in our study) should be slightly more accurate, estimating somewhere between 29 and 71 birds. We can measure that kind of difference in the laboratory, but it hardly seems to account for the big differences in math achievement between the groups. This is especially true when we consider the much bigger differences that showed up in children's knowledge of counting and spoken numbers.

Spoken Numbers

To measure children's knowledge of the spoken-language number system, we asked them to count to 10, to count rows of 3 and 6 objects (Fig. 2), and to complete the Give-N task described earlier (Fig. 3). We gave these tasks in English to all participants, and also in Spanish to the children who spoke Spanish (these were mostly the low-income dual-language learners). On the first two counting tasks, children in all groups did pretty well, but the Give-N task was a different story.

First, let us consider the counting tasks. When we asked children to count to 10, virtually all of the children in the high-SES groups performed at ceiling, as did the majority of children (⅔ of monolinguals and ¾ of dual-language learners) in the low-SES groups. Not surprisingly, older children counted better than younger children did, but neither monolinguals nor DLLs did better

"Can you show me how you count these?"

FIG. 2 Example of stimuli used in the counting objects task.

"Can you give him five?"

FIG. 3 Example of stimuli used in the Give-N task.

overall. We saw similar results when we asked children to count rows of objects. Almost all of the children in the high-SES groups performed at ceiling and so did most of the children in the low-SES groups (89% of DLLs and 72% of monolinguals). So on these two counting tasks, although we saw an SES difference, it did not seem insurmountable.

On the Give-N task, however, SES differences in performance were alarming. Recall that the Give-N task measures which number-word meanings (one through six) children know. It also measures whether children understand the cardinality principle, which is an important index of their conceptual understanding of counting and the spoken-number system. When we gave children this task, most children in the high-SES groups performed at ceiling, showing that they understood the cardinality principle. Less than half of the children in the low-SES groups did so.

To us, this difference in Give-N performance between high- and low-SES kids is the most important (and most worrisome) finding of the study. Many studies in our lab and others have linked children's conceptual understanding of numbers and counting to their grasp of the cardinality principle (e.g., Carey, 2009; Condry & Spelke, 2008; Geary & vanMarle, 2016; Izard, Pica, Spelke, & Dehaene, 2008; Sarnecka & Carey, 2008; Sarnecka & Gelman, 2004; Sarnecka & Wright, 2013; Slusser & Sarnecka, 2011; Thomas & Sarnecka, 2015). As we discussed in the context of the ANS task above, children who understand cardinality seem to know *what numbers are* in a way that they did not before. They can interpret the phrase "more dots" to mean *a greater number of dots* (Negen & Sarnecka, 2014). They know that two sets with the same number of items can be placed in one-to-one correspondence with each other (Sarnecka & Wright, 2013). Grasping the cardinality principle truly seems to be an important breakthrough in developing numeracy. So to find out that less than half of low-SES students understand cardinality in the year before they go to kindergarten was truly sobering. If children do not understand how counting represents number and if they do not understand what number words mean, what hope do they have of understanding the kindergarten math curriculum? When one sees the magnitude of the gap in conceptual understanding of numbers among preschoolers, the SES-related math achievement gap in grades K-12 seems almost inevitable.

Written Numbers

The third area that we wanted to measure was children's knowledge of written numbers. In a literate society like ours, children must learn to read written number symbols. We measured children's knowledge using three tasks: written numeral identification, scaffolded number line, and classic number line.

Written numeral identification. In this task, children were asked to identify the written numerals 1–10. The materials used in this task were a foam card in the shape of a house and a set of small laminated cards showing the written numerals 0–10 (see Fig. 4). The experimenter introduced the game by saying "Here are some numbers. I'm going to look for zero. This is zero. (Experimenter shows the laminated card with the number zero on it to the child.) I'm going to put zero in its home, like this. (Experimenter places the number on the sticky square in the house and then puts it back in the pile.) Now it's your turn. Can you find the number *one* and put it in its home?"

"Can you find the number one and put it in its home?"

FIG. 4 Example of stimuli used in the written numeral identification task.

Once the child put a number on the square, the experimenter asked "Is that one?" Children were asked for the numbers 1–10 in numerical order and were given generalized positive feedback (e.g., "thank you!") on every trial, regardless of their responses. Numeral identification was coded as correct or incorrect, based on children's responses.

The results revealed, unsurprisingly, that older children knew more written numerals than younger ones. More worrisome, however, children from higher-SES households knew many more written numerals than children from lower-SES households. A majority of the higher-SES children (including about three-quarters of high-SES dual-language learners and two-thirds of high-SES monolinguals) performed at ceiling, correctly identifying all the numerals 1–10. Among low-income children, few performed at ceiling. Among the low-SES dual-language learners, the most common score in both Spanish and English was 1, meaning that children correctly identified only one of the 10 numerals. This is the score that we would expect to see if children were randomly guessing. It seems that knowledge of written numerals is an area of SES difference already in preschool.

Scaffolded number line. For this task, children were asked to place numbers on a number line. Unlike the classic number line task described below, though, this task was "scaffolded" because the children could not place the numbers just anywhere: There were a series of spaces, and they had to choose which space to put the number in (Fig. 5). The experimenter set up the task by placing a foam board number line with the numbers 0 and 10 already placed on it, with nine empty squares for the other numbers in between. The experimenter also put a set of laminated cards (the same ones used in the numeral identification task) with the numbers one through nine in a pile next to the

"Okay, here's the number one
Can you show me where it goes
on the number line?"

FIG. 5 Example of stimuli used in the scaffolded number line task.

number line. The experimenter introduced the game by saying "This is a number line. It starts with zero and goes to ten. And, this is five. (Experimenter picks the laminated card with the number five out of the pile.) It goes right in the middle, like this." (Experimenter places the five on the fifth square of the line.) Leaving the five where it was, the experimenter said "Now it's your turn. I'll give you a number and you put it in its place on the number line. Okay, here's the number seven. Can you show me where it goes?" Children were asked for the numbers one–nine (excluding five that stayed on the line throughout as an additional scaffold) in a preset, randomized order. They were given generalized positive feedback (e.g., "thank you!") on every trial, regardless of their responses. Children's responses were coded as correct or incorrect, based on whether they placed each number in the correct location or at least in the correct region (above or below five) on the number line.

The results of this task were similar in many ways to those of the written numeral identification task. Older children performed better than younger ones, and children from high-SES households performed better than those from low-SES households. This task seemed to be more difficult than the written numeral identification task: even in the high-SES groups, fewer than half of the children performed at ceiling. In the low-SES groups, very few children performed at ceiling, and the most common score was again 1, the score we would expect if children had no idea at all where to put the numerals.

Interestingly, we noticed a strong relation between children's performance on this task and their knower-level. It is hard to say exactly what chance performance is on this task, but many children tried to place all their numbers in the middle of the board (perhaps trying to imitate the training trial, where the experimenter placed the number 5 in the middle). In practice, they had to choose between the fourth or the sixth spot, since the fifth spot was occupied by the number 5, which stayed on the board the whole time. A child who always puts the number in the fourth spot would get a score of 20. On average, preknowers, 1-knowers, 2-knowers, and 3-knowers (as measured by the Give-N task) scored worse than that mark. But the children who performed at ceiling (correctly placing all of the numbers on the line) were generally either 4-knowers or CP-knowers. This correlation suggests that in order for children to map numerals to a number line, they must first build a counting-list representation of numbers (which is what the Give-N task measures).

Classic number line. This task may be familiar to readers from other studies. Children were given a line with numbers at either end (0 on the left and 10 on the right) and asked to point to the place on the line where another number should go (Fig. 6). For example, if the child was asked where five should go, the correct answer would be right in the middle of the line. Children were asked for the same numbers as in the scaffolded number line task in a preset, randomized order.

As in the other written number tasks, older children performed better than younger ones but more strikingly, high-SES children performed much better than low-SES children. In fact, the average performance for children in the low-SES groups was 1.944. The way this task is scored, a child who always puts his or her mark in the center of the page would get a score of 2.0. In other words, there was no evidence that the children were doing anything other than guessing. There was no effect of being a dual-language learner vs monolingual; what mattered was SES and to a lesser extent, age.

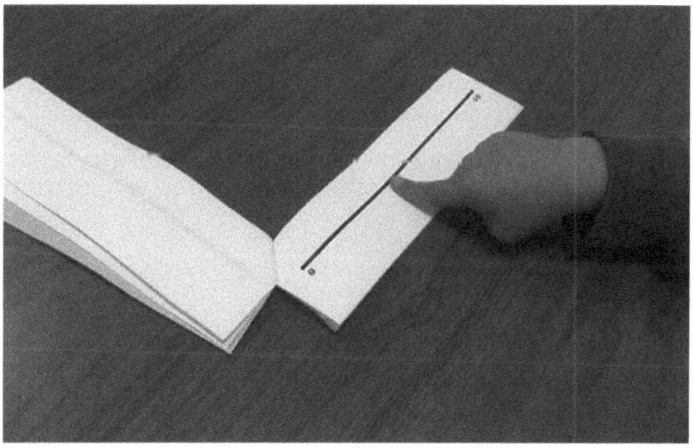

"Now point to where two should go."

FIG. 6 Example of stimuli used in the classic number line task.

Comparing Low-SES DLLs' Performance in English and Spanish

Better English performance on counting tasks. The vast majority of low-SES dual-language learners performed at ceiling in English for both counting tasks, meaning that they counted out loud to 10, and also correctly counted six objects in a row. Many children also performed at ceiling in Spanish, but there were some who performed noticeably worse in Spanish than in English. For example, a significant minority of children counted to 10 in English but only to six in Spanish. (Many of them recited the first six counting words in Spanish as a sort of nursery rhyme: *uno*, *dos*, *tres*, *cuatro*, *cinco*, *seis*.) In general, when children counted higher in one language than the other, the language in which they counted higher was English—presumably because English was the language of instruction in their Head Start classrooms, and most of their experience with counting happened at Head Start.

Cross-language consistency in number-knower levels and numeral identification. On the Give-N task, children generally scored at the same knower-level in both their languages. That is, a three-knower tested in English was still a three-knower when tested in Spanish and vice versa. Children's performance in the written numeral identification task also tended to be the same across languages, even if we exclude the group that appeared to be simply guessing. This suggests that these two tasks measure conceptual knowledge that transcends particular languages, that is, having mental symbols for sets of 1, 2, 3, and 4; understanding the principle of cardinality; and matching written numerals to particular number words. (For a related finding, see Wagner, Kimura, Cheung, & Barner, 2015.)

Floor effects in number-line tasks. It was hard for us to judge anything about the crosslinguistic consistency of performance on the scaffolded number line or classic number line tasks simply because most children performed so poorly on both of them. A child who always selected the fourth spot in the scaffolded number line task would get a score of 20. A child who always pointed to the middle of the line in the classic number line task would get a score of 2. Most children's scores hover around those marks on both tasks, in both languages, which suggest that our data contain strong floor effects. Among the small number of children who performed better, their performance was largely consistent across their two languages. There was no evidence that they performed better, as a group, in either English or Spanish.

Overall Conclusions of the Study

We drew four main conclusions from this study: (1) Being a dual-language learner does not adversely affect early number learning, but poverty certainly does. (2) Head Start provides crucial early math input to these students. (3) Future studies need not necessarily test young, low-SES DLLs in both of their languages in order to measure their number knowledge—testing only in the language of instruction may give equally valid results. (4) Interventions to improve low-SES students' math achievement should focus on symbolic number knowledge (spoken and written numbers), not the ANS.

Conclusion 1: Being a Dual-Language Learner Does Not Cause Educational Delays; Being Poor Does

Perhaps the most striking overall finding of our study was the huge gap between high- and low-SES children's knowledge of symbolic (i.e., spoken and written) numbers. This is not news. What we looked for but did not find was a big difference between monolinguals and dual-language learners at similar SES levels. This was most clear in the high-SES groups, where dual-language learners and monolinguals performed equally well on most measures. Of course, in the real world, one rarely finds groups that differ in only one way. For example, our high-SES dual-language learners were mostly the children of highly educated East Asian parents who had moved to the United States for graduate school or work—hardly a random sample of DLLs in the world or even in the United States. But the monolingual group had even more highly educated parents, and children in the two groups performed equally well.

These findings will be unsurprising to researchers, whose conversations about bilingualism and development are more often about the potential *benefits* of growing up bilingual. But discussions among policymakers are legitimately concerned with questions about how to serve students with limited English proficiency. To us, a take-home point from our study is the difference between preschool-age dual-language learners and older bilingual students who have completed some years of schooling in their first language. For older students, the challenge is to build on what they already know and to help them continue learning while at the same time adjusting to English as the language of instruction. But for dual-language learners (who by definition are still young children), they just do not have much of a prior knowledge base to import. Particularly in the case of the low-SES children in our sample, they seem to have had very little exposure to numbers or counting in any language.

Conclusion 2: Low-SES Kids Do Not Seem to Get the Experience With Counting and Numbers at Home That High-SES Kids Get; What They Know, They Learned at Head Start

When we designed this study, we were concerned that if we tested DLLs only in English, we might accidentally underestimate their knowledge. We assumed that early number knowledge is learned informally, through everyday interactions where caregivers use number words when talking with children, demonstrate counting to children, point out patterns and spatial relationships in children's environments, and so forth. We assumed that, because most of the literature on early number knowledge is based on studies with high-SES children, and that seems to be how high-SES children learn numbers.

We thought that the low-SES DLLs in our study, whose caregivers typically speak Spanish with them, would be more familiar with numbers and counting in Spanish than in English. (Just as their vocabularies were greater in Spanish, we expected their number knowledge to be greater as well.) So we decided to test children twice on all measures of spoken and written numbers—once in Spanish and once in English.

We were surprised to discover that the children performed as well or better in English than in Spanish. Rather than learning numbers at home, in Spanish, and translating those concepts for an English-speaking classroom, these children seemed to have learned the numbers in English—presumably *in* the English-speaking classroom—and then to varying degrees translated them (or not) into Spanish.

This was an unexpected gift to us as researchers. Most of the time when you measure children's knowledge, you cannot tell where the knowledge came from. And while no one should be surprised that 4-year olds from low-SES households have far less number knowledge than 4-year olds from high-SES households, comparing their performance in their two languages gives us some insight into the origin of the number knowledge they do have. Despite having larger vocabularies in Spanish than in English, the low-SES dual-language learners actually performed as well or better in English (the language of instruction in their Head Start programs) on all number measures, suggesting that they primarily learned about numbers and math at their Head Start programs rather than at home.

This seems like an important finding, and one that should be kept in mind when assessing the value of publicly funded preschool programs. If we only compare low-SES students to high-SES students, we will see a huge gap in their early math knowledge and school readiness. One interpretation of this gap, by people who oppose publicly funded preschool, could be that the Head Start programs attended by the low-SES students are not working, because they fail to close the gap between rich and poor students. But what we can infer from this language difference (i.e., from the fact that low-SES DLLs perform as well or better in English than in Spanish) is that whatever knowledge the children *do* have, they acquired in an English-language environment—presumably their Head Start classroom. In other words, if they did not have Head Start, they would be getting virtually no early math experience at all, placing them even further behind the high-SES kids.

There may be another implication of this finding, for education policymakers who are making choices about how to spend limited resources to provide educational services for low-income Hispanic preschoolers. Of course, we would like to see more resources devoted to these services overall. But assuming a fixed amount of money to spend, our findings suggest that getting English-language pre-K services to more children would be a better investment (at least in terms of improving math outcomes) than spending more money per child on bilingual services but reaching fewer children. More exposure to spoken and written numbers and counting games—in any language—seem to be what the children in our study needed. Although there are many compelling arguments in favor of bilingual preschool education (e.g., Castro, 2014; Castro, Garcia, & Markos, 2013; Garcia & Jensen, 2009; Hammer et al., 2014; Winsler et al., 2014), the added value of bilingual instruction in math seems small compared with the value of serving more children overall, even if the services are provided only in English.

Conclusion 3: A Methodological Note for Early Numeracy Researchers

Just as policymakers must decide whether to spend resources on bilingual vs English-language education, so must researchers decide whether to spend resources on bilingual testing of DLLs. Although there are many good reasons to conduct bilingual assessments of DLLs (Espinosa, 2014), we did not find enough differences in our participants' Spanish vs English performance on any number measure to justify the added time and expense of testing them twice. If we had tested them only in English, we would have saved a lot of time, increased the pool of research assistants we could recruit from (because they would not have to be bilingual), and ended up with virtually the same picture of their number knowledge as we got by testing them twice. Of course, there was no way to know that except by doing this study. And there are certainly other research questions (e.g., those relating to language and literacy development) for which it would make sense to take separate measurements in the child's two languages. But given these results, our future studies of early math development in this population of children will probably stick to English-language assessments.

Conclusion 4: The SES Gap in Math Achievement Is a Gap in Symbolic Number Knowledge (Spoken and Written Numbers) Much More Than ANS Acuity, and Interventions Should Focus on Improving Symbolic Number Knowledge

As discussed earlier in this chapter, it is difficult to accurately measure the ANS acuity of young children. But where we were able to measure ANS acuity, it did not differ very much between children from high- and low-SES households. It differed enough for us to measure, but not enough to matter much in everyday life (or in math class), and the difference seems very unlikely to be the source of the huge gaps we find in children's knowledge of spoken and written numbers already by age 4 (For related findings, see Geary & vanMarle, 2016; Merkley & Ansari, 2016; Raghubar & Barnes, 2017).

Given that spoken and written numbers are already the focus of most preschool math curricula, the message from our findings seems to be "if it ain't broke, don't fix it." Rather than trying to modify preschool math instruction to include more training of the ANS system (or of nonnumeracy skills such as shapes and patterns), low-SES students should be provided with as much experience as possible of counting and using number words. Once they have begun to master spoken numbers (i.e., once they have learned the first few number-word meanings and the cardinality principle), they will be much better able to use the visual representations such as written numerals and the number line.

Thus, in a sense, our conclusions can be summed up as an argument that low-SES DLLs would have more success in math if they had more opportunities to learn about and use numbers before starting kindergarten. Let us not spend resources on interventions to improve children's ANS acuity or even to provide bilingual preschool if it means that fewer children could attend preschool overall. Instead, if we want to narrow the math achievement gap, let us focus on making developmentally appropriate, play-based preschool experiences with numbers and counting available to as many low-SES DLLs as possible.

ACKNOWLEDGMENT

This research was supported by NSF grant DRL-0953521 to the first author.

REFERENCES

Abreu-Mendoza, R. A., Soto-Alba, E. E., & Arias-Trejo, N. (2013). Area vs. density: Influence of visual variables and cardinality knowledge in early number comparison. *Frontiers in Psychology, 4*. https://doi.org/10.3389/fpsyg.2013.00805.

Almoammer, A., Sullivan, J., Donlan, C., Marušič, F., Žaucer, R., O'Donnell, T., et al. (2013). Grammatical morphology as a source of early number word meanings. *Proceedings of the National Academy of Sciences of the United States of America, 110*(46), 18448–18453. https://doi.org/10.1073/pnas.1313652110.

Aunola, K., Leskinen, E., Lerkkanen, M.-K., & Nurmi, J.-E. (2004). Developmental dynamics of math performance from preschool to grade 2. *Journal of Educational Psychology, 96*(4), 699–713. https://doi.org/10.1037/0022-0663.96.4.699.

Barner, D., Libenson, A., Cheung, P., & Takasaki, M. (2009). Cross-linguistic relations between quantifiers and numerals in language acquisition: Evidence from Japanese. *Journal of Experimental Child Psychology, 103*(4), 421–440. https://doi.org/10.1016/j.jecp.2008.12.001.

Baroody, A. J., & Price, J. (1983). The development of the number-word sequence in the counting of three-year-olds. *Journal for Research in Mathematics Education*, *14*(5), 361. https://doi.org/10.2307/748681.

Bialystok, E. (2001). *Bilingualism in development: Language, literacy, and cognition.* Cambridge University Press.

Bialystok, E. (2009). Bilingualism: The good, the bad, and the indifferent. *Bilingualism: Language and Cognition*, *12*(1), 3–11. https://doi.org/10.1017/S1366728908003477.

Brannon, E. M., Abbott, S., & Lutz, D. J. (2004). Number bias for the discrimination of large visual sets in infancy. *Cognition*, *93*(2), B59–B68. https://doi.org/10.1016/j.cognition.2004.01.004.

Briars, D., & Siegler, R. S. (1984). A featural analysis of preschoolers' counting knowledge. *Developmental Psychology*, *20*(4), 607–618. https://doi.org/10.1037/0012-1649.20.4.607.

California Dept of Education. (2008). *California preschool learning foundations: 1* [government]. Retrieved from: http://www.cde.ca.gov/sp/cd/re/psfoundations.asp.

Carey, S. (2009). *The origin of concepts.* Oxford University Press.

Castro, D. C. (2014). The development and early care and education of dual language learners: Examining the state of knowledge. *Early Childhood Research Quarterly*, *29*(4), 693–698. https://doi.org/10.1016/j.ecresq.2014.08.003.

Castro, D. C., Garcia, E. E., & Markos, A. M. (2013). *Dual language learners: Research informing policy.* Chapel Hill, NC: The University of North Carolina, Frank Porter Graham Child Development Institute. Retrieved from: http://www.edweek.org/media/policypaper-05-10-13-learning-the-language.pdf.

Chen, Q., & Li, J. (2014). Association between individual differences in non-symbolic number acuity and math performance: A meta-analysis. *Acta Psychologica*, *148*, 163–172. https://doi.org/10.1016/j.actpsy.2014.01.016.

childstats.gov. (2015). *Childstats.gov—America's children 2015—Language spoken at home and difficulty speaking English.* Retrieved from: https://www.childstats.gov/americaschildren15/family5.asp.

Condry, K. F., & Spelke, E. S. (2008). The development of language and abstract concepts: The case of natural number. *Journal of Experimental Psychology: General*, *137*(1), 22–38. https://doi.org/10.1037/0096-3445.137.1.22.

Council, National Research. (2009). *Mathematics learning in early childhood: Paths toward excellence and equity.* Washington, DC: The National Academies Press. Retrieved from: http://www.nap.edu/catalog.php?record_id=12519.

Dehaene, S. (2011). *The number sense: How the mind creates mathematics.* Revised and Updated Edition. Oxford University Press.

Duncan, G. J., Dowsett, C. J., Claessens, A., Magnuson, K., Huston, A. C., Klebanov, P., et al. (2007). School readiness and later achievement. *Developmental Psychology*, *43*(6), 1428–1446. https://doi.org/10.1037/0012-1649.43.6.1428.

Duncan, G. J., & Murname, R. J. (2014). *Restoring opportunity the crisis of inequality and the challenge for American education.* Harvard Education Press.

Dunn, L. M., & Dunn, D. M. (2007). *Peabody picture vocabulary test* (4th ed.). San Antonio, TX: Pearson.

Dunn, L. M., Lugo, D. E., Padilla, E. R., & Dunn, L. M. (1986). *Test de Vocabulario en Imagenes Peabody (TVIP).*

Dyson, N. I., Jordan, N. C., & Glutting, J. (2013). A number sense intervention for low-income kindergartners at risk for mathematics difficulties. *Journal of Learning Disabilities*, *46*(2), 166–181. https://doi.org/10.1177/0022219411410233.

Espinosa, L. M. (2014). Perspectives on assessment of DLLs development & learning, PreK-third grade. In *Presented at the national research summit on the early care and education of dual language learners, Washington, DC.*

Fan, S. P., Liberman, Z., Keysar, B., & Kinzler, K. D. (2015). The exposure advantage: Early exposure to a multilingual environment promotes effective communication. *Psychological Science, 26*(7), 1090–1097. https://doi.org/10.1177/0956797615574699.

Fazio, L. K., Bailey, D. H., Thompson, C. A., & Siegler, R. S. (2014). Relations of different types of numerical magnitude representations to each other and to mathematics achievement. *Journal of Experimental Child Psychology, 123*, 53–72. https://doi.org/10.1016/j.jecp.2014.01.013.

Feigenson, L., Dehaene, S., & Spelke, E. (2004). Core systems of number. *Trends in Cognitive Sciences, 8*(7), 307–314. https://doi.org/10.1016/j.tics.2004.05.002.

Frye, D., Braisby, N., Lowe, J., Maroudas, C., & Nicholls, J. (1989). Young children's understanding of counting and cardinality. *Child Development, 60*(5), 1158–1171. https://doi.org/10.2307/1130790.

Fuson, K. C. (1988). *Children's counting and concepts of number: xiv*. New York, NY: Springer-Verlag Publishing.

Fuson, K. C., Smith, S. T., & Lo Cicero, A. M. (1997). Supporting Latino first graders' ten-structured thinking in urban classrooms. *Journal for Research in Mathematics Education, 28*(6), 738–766. https://doi.org/10.2307/749640.

Gallistel, C. R. (1989). Animal cognition: The representation of space, time and number. *Annual Review of Psychology, 40*(1), 155–189. https://doi.org/10.1146/annurev.ps.40.020189.001103.

Garcia, E., & Jensen, B. (2009). Early educational opportunities for children of hispanic origins. Social policy report. Vol. XXIII, Number II. *Society for research in child development.* Retrieved from: http://eric.ed.gov/?id=ED509217.

Geary, D. C., & vanMarle, K. (2016). Young children's core symbolic and nonsymbolic quantitative knowledge in the prediction of later mathematics achievement. *Developmental Psychology, 52*(12), 2130–2144. https://doi.org/10.1037/dev0000214.

Gelman, R. (1993). *Psychology of learning and motivation: Advances in research and theory.* Academic Press.

Gelman, R., & Gallistel, C. R. (1978). *The child's understanding of number.* Cambridge, MA: Harvard University Press.

Gennetian, L. A., Rodrigues, C., Hill, H. D., & Morris, P. (2015). *Income instability in the lives of Hispanic children* (No Publication #2015-47). National Research Center on Hispanich Children & Families. Retrieved from: http://www.childtrends.org/wp-content/uploads/2015/12/2015-47Hisp-Ctr-Income-Instability.pdf.

Goldman, M. C., Negen, J., & Sarnecka, B. W. (2014). Are bilingual children better at ignoring perceptually misleading information? A novel test. *Developmental Science.* https://doi.org/10.1111/desc.12175.

Griffin, S. A., Case, R., & Siegler, R. S. (1994). Rightstart: Providing the central conceptual prerequisites for first formal learning of arithmetic to students at risk for school failure. In *Classroom lessons: Integrating cognitive theory and classroom practice* (pp. 25–49). Cambridge, MA: The MIT Press.

Halberda, J., & Feigenson, L. (2008). Developmental change in the acuity of the "number sense": The approximate number system in 3-, 4-, 5-, and 6-year-olds and adults. *Developmental Psychology, 44*(5), 1457–1465. https://doi.org/10.1037/a0012682.

Halberda, J., Mazzocco, M. M. M., & Feigenson, L. (2008). Individual differences in non-verbal number acuity correlate with maths achievement. *Nature*, *455*(7213), 665–668. https://doi.org/10.1038/nature07246.

Hammer, C. S., Hoff, E., Uchikoshi, Y., Gillanders, C., Castro, D. C., & Sandilos, L. E. (2014). The language and literacy development of young dual language learners: A critical review. *Early Childhood Research Quarterly*, *29*(4), 715–733.

CA Head Start Assoc. (2017). 2014–2015 program statistics (Head Start/Early Head Start in California Data Report). California Head Start Association. Retrieved from: http://caheadstart.org/facts.html.

Hilchey, M. D., & Klein, R. M. (2011). Are there bilingual advantages on nonlinguistic interference tasks? Implications for the plasticity of executive control processes. *Psychonomic Bulletin & Review*, *18*(4), 625–658. https://doi.org/10.3758/s13423-011-0116-7.

Iglesias, C. (2012). *Investigating the status of early numeracy skills in bilingual dual language learner latino children attending head start and the association with parent demographic characteristics*. Graduate theses and dissertations. Retrieved from: http://scholarcommons.usf.edu/etd/4084.

Izard, V., Pica, P., Spelke, E. S., & Dehaene, S. (2008). Exact equality and successor function: Two key concepts on the path towards understanding exact numbers. *Philosophical Psychology*, *21*(4), 491–505. https://doi.org/10.1080/09515080802285354.

Jordan, N. C., Glutting, J., & Ramineni, C. (2010). The importance of number sense to mathematics achievement in first and third grades. *Learning and Individual Differences*, *20*(2), 82–88. https://doi.org/10.1016/j.lindif.2009.07.004.

Jordan, N. C., Huttenlocher, J., & Levine, S. C. (1992). Differential calculation abilities in young children from middle- and low-income families. *Developmental Psychology*, *28*(4), 644–653. https://doi.org/10.1037/0012-1649.28.4.644.

Jordan, N. C., Huttenlocher, J., & Levine, S. C. (1994). Assessing early arithmetic abilities: Effects of verbal and nonverbal response types on the calculation performance of middle- and low-income children. *Learning and Individual Differences*, *6*(4), 413–432. https://doi.org/10.1016/1041-6080(94)90003-5.

Jordan, N. C., Kaplan, D., Locuniak, M. N., & Ramineni, C. (2007). Predicting first-grade math achievement from developmental number sense trajectories. *Learning Disabilities Research & Practice*, *22*(1), 36–46.

Jordan, N. C., Kaplan, D., Nabors Oláh, L., & Locuniak, M. N. (2006). Number sense growth in kindergarten: A longitudinal investigation of children at risk for mathematics difficulties. *Child Development*, *77*(1), 153–175. https://doi.org/10.1111/j.1467-8624.2006.00862.x.

Jordan, N. C., Kaplan, D., Ramineni, C., & Locuniak, M. N. (2009). Early math matters: Kindergarten number competence and later mathematics outcomes. *Developmental Psychology*, *45*(3), 850–867. https://doi.org/10.1037/a0014939.

Jordan, N. C., & Levine, S. C. (2009). Socioeconomic variation, number competence, and mathematics learning difficulties in young children. *Developmental Disabilities Research Reviews*, *15*(1), 60–68. https://doi.org/10.1002/ddrr.46.

Kirk, G. E., Hunt, J. M., & Volkmar, F. (1975). Social class and preschool language skill: V. Cognitive and semantic mastery of number. *Genetic Psychology Monographs*. Retrieved from: http://psycnet.apa.org/psycinfo/1975-31403-001.

Le Corre, M., & Carey, S. (2007). One, two, three, four, nothing more: An investigation of the conceptual sources of the verbal counting principles. *Cognition*, *105*(2), 395–438. https://doi.org/10.1016/j.cognition.2006.10.005.

Le Corre, M., Li, P., Huang, B. H., Jia, G., & Carey, S. (2016). Numerical morphology supports early number word learning: Evidence from a comparison of young Mandarin and English learners. *Cognitive Psychology*, *88*, 162–186. https://doi.org/10.1016/j.cogpsych.2016.06.003.

Le Corre, M., Van de Walle, G., Brannon, E. M., & Carey, S. (2006). Re-visiting the competence/performance debate in the acquisition of the counting principles. *Cognitive Psychology*, *52*(2), 130–169. https://doi.org/10.1016/j.cogpsych.2005.07.002.

Lee, M. D., & Sarnecka, B. W. (2010). A model of knower-level behavior in number concept development. *Cognitive Science*, *34*(1), 51–67. https://doi.org/10.1111/j.1551-6709.2009.01063.x.

Lee, M. D., & Sarnecka, B. W. (2011). Number-knower levels in young children: Insights from Bayesian modeling. *Cognition*, *120*(3), 391–402. https://doi.org/10.1016/j.cognition.2010.10.003.

Li, P., Le Corre, M., Shui, R., Jia, G., & Carey, S. (2003). Effects of plural syntax on number word learning: A cross-linguistic study. In *Presented at the Boston University Conference on Language Development (BUCLD), Boston, MA*.

Lipton, J. S., & Spelke, E. S. (2003). Origins of number sense: Large-number discrimination in human infants. *Psychological Science*, *14*(5), 396–401. https://doi.org/10.1111/1467-9280.01453.

Lipton, J. S., & Spelke, E. S. (2004). Discrimination of large and small numerosities by human infants. *Infancy*, *5*(3), 271–290. https://doi.org/10.1207/s15327078in0503_2.

Locuniak, M. N., & Jordan, N. C. (2008). Using kindergarten number sense to predict calculation fluency in second grade. *Journal of Learning Disabilities*, *41*(5), 451–459. https://doi.org/10.1177/0022219408321126.

Lopez, M. H., & Velasco, G. (2011). *Childhood poverty among Hispanics sets record, leads nation*. Washington, DC: Pew Hispanic Center. Retrieved from: http://permanencyenhancementproject.org/wp-content/uploads/2012/09/147-Childhood-Poverty-Among-Hispanics-Sets-Record-Lead-Nation.pdf.

Mazzocco, M. M. M., Feigenson, L., & Halberda, J. (2011a). Impaired acuity of the approximate number system underlies mathematical learning disability (Dyscalculia). *Child Development*, *82*(4), 1224–1237. https://doi.org/10.1111/j.1467-8624.2011.01608.x.

Mazzocco, M. M. M., Feigenson, L., & Halberda, J. (2011b). Preschoolers' precision of the approximate number system predicts later school mathematics performance. *PLoS ONE*. *6*(9), e23749. https://doi.org/10.1371/journal.pone.0023749.

Merkley, R., & Ansari, D. (2016). Why numerical symbols count in the development of mathematical skills: Evidence from brain and behavior. *Current Opinion in Behavioral Sciences*, *10*, 14–20. https://doi.org/10.1016/j.cobeha.2016.04.006.

Morton, J. B., & Harper, S. N. (2007). What did Simon say? Revisiting the bilingual advantage. *Developmental Science*, *10*(6), 719–726. https://doi.org/10.1111/j.1467-7687.2007.00623.x.

Mussolin, C., Nys, J., Leybaert, J., & Content, A. (2012). Relationships between approximate number system acuity and early symbolic number abilities. *Trends in Neuroscience and Education*, *1*(1), 21–31. https://doi.org/10.1016/j.tine.2012.09.003.

Negen, J., & Sarnecka, B. W. (2012). Number-concept acquisition and general vocabulary development. *Child Development*, *83*(6), 2019–2027. https://doi.org/10.1111/j.1467-8624.2012.01815.x.

Negen, J., & Sarnecka, B. W. (2014). Is there really a link between exact-number knowledge and approximate number system acuity in young children? *The British Journal of Developmental Psychology*. https://doi.org/10.1111/bjdp.12071.

Paap, K. R., & Greenberg, Z. I. (2013). There is no coherent evidence for a bilingual advantage in executive processing. *Cognitive Psychology*, *66*(2), 232–258. https://doi.org/10.1016/j.cogpsych.2012.12.002.

Piantadosi, S. T., Jara-Ettinger, J., & Gibson, E. (2014). Children's learning of number words in an indigenous farming-foraging group. *Developmental Science*, *17*(4), 553–563. https://doi.org/10.1111/desc.12078.

Purpura, D. J., & Lonigan, C. J. (2015). Early numeracy assessment: The development of the preschool early numeracy scales. *Early Education and Development*, *26*(2), 286–313. https://doi.org/10.1080/10409289.2015.991084.

Raghubar, K. P., & Barnes, M. A. (2017). Early numeracy skills in preschool-aged children: A review of neurocognitive findings and implications for assessment and intervention. *The Clinical Neuropsychologist*, *31*(2), 329–351. https://doi.org/10.1080/13854046.2016.1259387.

Ramani, G. B., & Siegler, R. S. (2008). Promoting broad and stable improvements in low-income children's numerical knowledge through playing number board games. *Child Development*, *79*(2), 375–394.

Reich, R. B. (2015). *Saving capitalism: For the many, not the few* (1st ed.). New York, NY: Knopf.

Rousselle, L., & Noël, M.-P. (2007). Basic numerical skills in children with mathematics learning disabilities: A comparison of symbolic vs non-symbolic number magnitude processing. *Cognition*, *102*(3), 361–395. https://doi.org/10.1016/j.cognition.2006.01.005.

Sarnecka, B. W., & Carey, S. (2008). How counting represents number: What children must learn and when they learn it. *Cognition*, *108*(3), 662–674. https://doi.org/10.1016/j.cognition.2008.05.007.

Sarnecka, B. W., & Gelman, S. A. (2004). Six does not just mean a lot: Preschoolers see number words as specific. *Cognition*, *92*(3), 329–352. https://doi.org/10.1016/j.cognition.2003.10.001.

Sarnecka, B. W., Kamenskaya, V. G., Yamana, Y., Ogura, T., & Yudovina, Y. B. (2007). From grammatical number to exact numbers: Early meanings of "one", "two", and "three" in English, Russian, and Japanese. *Cognitive Psychology*, *55*(2), 136–168. https://doi.org/10.1016/j.cogpsych.2006.09.001.

Sarnecka, B. W., & Lee, M. D. (2009). Levels of number knowledge during early childhood. *Journal of Experimental Child Psychology*, *103*(3), 325–337. https://doi.org/10.1016/j.jecp.2009.02.007.

Sarnecka, B. W., & Wright, C. E. (2013). The idea of an exact number: Children's understanding of cardinality and equinumerosity. *Cognitive Science*. https://doi.org/10.1111/cogs.12043.

Schaeffer, B., Eggleston, V. H., & Scott, J. L. (1974). Number development in young children. *Cognitive Psychology*, *6*, 357–379.

Shusterman, A., Slusser, E., Halberda, J., & Odic, D. (2016). Acquisition of the cardinal principle coincides with improvement in approximate number system acuity in preschoolers. *PLoS ONE*, *11*(4), e0153072. https://doi.org/10.1371/journal.pone.0153072.

Slaughter, V., Itakura, S., Kutsuki, A., & Siegal, M. (2011). Learning to count begins in infancy: Evidence from 18 month olds' visual preferences. *Proceedings of the Royal Society B: Biological Sciences*, *278*(1720), 2979–2984. https://doi.org/10.1098/rspb.2010.2602.

Slusser, E. B., Ditta, A., & Sarnecka, B. W. (2013). Connecting numbers to discrete quantification: A step in the child's construction of integer concepts. *Cognition, 129*, 31–41. https://doi.org/10.1016/j.cognition.2013.05.011.

Slusser, E. B., & Sarnecka, B. W. (2011). Find the picture of eight turtles: A link between children's counting and their knowledge of number word semantics. *Journal of Experimental Child Psychology, 110*(1), 38–51. https://doi.org/10.1016/j.jecp.2011.03.006.

Starkey, P., Klein, A., & Wakeley, A. (2004). Enhancing young children's mathematical knowledge through a pre-kindergarten mathematics intervention. *Early Childhood Research Quarterly, 19*(1), 99–120. https://doi.org/10.1016/j.ecresq.2004.01.002.

Stepler, R., & Brown, A. (2015). Statistical Portrait of Hispanics in the United States, (1980–2013). Retrieved from: http://www.pewhispanic.org/2015/05/12/statistical-portrait-of-hispanics-in-the-united-states-1980-2013-trends/.

Thomas, A. J., & Sarnecka, B. W. (2015). Exploring the relation between people's theories of intelligence and beliefs about brain development. *Frontiers in Psychology, 6*. https://doi.org/10.3389/fpsyg.2015.00921.

U.S. DHHS. (2017). *Head start program facts fiscal year 2015*. U.S. DHHS, Administration for Children and Families. Retrieved from: https://eclkc.ohs.acf.hhs.gov/hslc/data/factsheets/2015-hs-program-factsheet.html.

van Marle, K., Chu, F. W., Li, Y., & Geary, D. C. (2014). Acuity of the approximate number system and preschoolers' quantitative development. *Developmental Science, 17*(4), 492–505. https://doi.org/10.1111/desc.12143.

Wagner, J. B., & Johnson, S. C. (2011). An association between understanding cardinality and analog magnitude representations in preschoolers. *Cognition, 119*(1), 10–22. https://doi.org/10.1016/j.cognition.2010.11.014.

Wagner, K., Kimura, K., Cheung, P., & Barner, D. (2015). Why is number word learning hard? Evidence from bilingual learners. *Cognitive Psychology, 83*, 1–21. https://doi.org/10.1016/j.cogpsych.2015.08.006.

Wagner, S. H., & Walters, J. (1982). A longitudinal analysis of early number concepts: From numbers to number. *Action and thought: From sensorimotor schemes to symbolic operations*. (pp. 137–161).

Wiley, T. G. (2009). *The education of language minority immigrants in the United States*. Multilingual Matters.

Winsler, A., Burchinal, M. R., Tien, H.-C., Peisner-Feinberg, E., Espinosa, L., Castro, D. C., et al. (2014). Early development among dual language learners: The roles of language use at home, maternal immigration, country of origin, and socio-demographic variables. *Early Childhood Research Quarterly, 29*(4), 750–764. https://doi.org/10.1016/j.ecresq.2014.02.008.

Wynn, K. (1990). Children's understanding of counting. *Cognition, 36*(2), 155–193. https://doi.org/10.1016/0010-0277(90)90003-3.

Wynn, K. (1992). Children's acquisition of the number words and the counting system. *Cognitive Psychology, 24*(2), 220–251. https://doi.org/10.1016/0010-0285(92)90008-P.

Xu, F. (2003). Numerosity discrimination in infants: Evidence for two systems of representations. *Cognition, 89*(1), B15–B25. https://doi.org/10.1016/S0010-0277(03)00050-7.

Xu, F., & Arriaga, R. I. (2007). Number discrimination in 10-month-old infants. *British Journal of Developmental Psychology, 25*(1), 103–108. https://doi.org/10.1348/026151005X90704.

Xu, F., & Spelke, E. S. (2000). Large number discrimination in 6-month-old infants. *Cognition, 74*(1), B1–B11. https://doi.org/10.1016/S0010-0277(99)00066-9.

Xu, F., Spelke, E. S., & Goddard, S. (2005). Number sense in human infants. *Developmental Science*, *8*(1), 88–101. https://doi.org/10.1111/j.1467-7687.2005.00395.x.

Xue, Y., Atkins-Burnett, S., & Moiduddin, E. (2012). Informing the Performance-Based Contract Between First 5 LA and LAUP: Child Progress in the 2011– 2012 Program Year: Final Report. *Los Angeles: Mathematica Policy Research.* Retrieved from https://pdfs.semanticscholar.org/091f/df8a7548bc40d75534496830b29b9e22876a.pdf on June 28, 2017.

Chapter 10

How Culturally Predominant Reading Direction and Number Word Construction Influence Numerical Cognition and Mathematics Performance

Silke M. Göbel
University of York, York, United Kingdom

INTRODUCTION

At first glance, processing of Arabic digits might seem to be a perfect example of an abstract task that is carried out without the influence of any linguistic and cultural factors. As it turns out, this is not the case. First, all aspects of Arabic digit acquisition (e.g., recognizing, naming, and acquiring their meaning) are a cultural achievement (Skwarchuk, Sowinski, & LeFevre, 2014). Just like letter knowledge (Muter, Hulme, Snowling, & Stevenson, 2004), Arabic digit knowledge needs to be taught, and this teaching happens within a particular cultural context. Second, in order to learn the meaning of Arabic digits, they must be mapped onto spoken number words. Languages, however, differ from each other in various aspects of number word structure. Clearly, linguistic and cultural factors affect the acquisition of basic number skills in young children, but their influence leaves traces even in adult processing of Arabic digits.

In this chapter, I will present our work on how one cultural factor, reading direction, and one linguistic factor, number word construction, influence numerical processing including counting, number generation, number comparison, and arithmetic (for other culturally mediated factors, see Chapter 11 by Shaki and Fischer). Furthermore, I will investigate whether these factors might contribute to individual differences in mathematical performance and development. While the extent to which children and adults are influenced by the culturally predominant reading direction does not seem to be related

Mathematical Cognition and Learning, Vol. 4. https://doi.org/10.1016/B978-0-12-812574-8.00010-9

to their overall mathematical performance, number word construction and the ability to transcode (i.e., translate) spoken number words into Arabic digits are likely to be significant predictors of growth in mathematical performance in the early primary school.

EFFECTS OF READING DIRECTION

In this first section, I will focus on one cultural factor: reading direction and its relation to spatial representations of number. There is a large body of evidence showing an association between representations of number and space (Fischer & Shaki, 2014; Göbel, Shaki, & Fischer, 2011; Hubbard, Piazza, Pinel, & Dehaene, 2005; Nuerk, Patro, et al., 2015; Shaki & Fischer, 2012; van Dijck, Gevers, Lafosse, & Fias, 2011; Wood, Willmes, Nuerk, & Fischer, 2008). Typically, Western participants associate small numbers with the left side of space and larger numbers with the right side of space. For example, in the classic spatial-numerical association of response codes (SNARC) task (Dehaene, Bossini, & Giraux, 1993), participants are faster to judge the parity (i.e., odd or even) of a single Arabic digit with a left-hand button press than a right-hand button press when the number is small. In contrast, they are quicker to respond to larger numbers with a right-hand button press than with a left-hand button press. Traditionally, this has been taken as evidence that numbers are represented along a mental number line with numbers increasing in magnitude along that line from left to right. Further support that representations of number and space are linked come, for example, from the finding that patients with unilateral brain damage to the right hemisphere who neglect the left side of space overestimate the midpoint of a numerical interval (i.e., neglect the left side of the numerical interval, Zorzi, Priftis, & Umiltà, 2002). Conversely, numerical size can also shift spatial attention: in a spatial cueing task, participants have been found to detect targets faster on the left side after small numbers cues and on the right side after large numbers cues, despite the number cues being nonpredictive of target location (Fischer, Castel, Dodd, & Pratt, 2003; but see also Fattorini, Pinto, Rotondaro, & Doricchi, 2015).

Although Western participants generally show an association of left space with small numbers and right space with large numbers, this association seems to be flexible and can be changed easily by recent spatial experience. For example, when participants performed a magnitude comparison task and had to decide whether the Arabic digit displayed was longer or shorter than 6 cm on a ruler, they were faster to respond to smaller numbers with the left hand and to larger numbers with the right hand (Bächtold, Baumüller, & Brugger, 1998). However, when they were instructed to think about the same numbers as hours on a clock face and to decide whether it was earlier or later than 6 o'clock, they were faster to respond to smaller numbers with the right hand and larger numbers with the left hand. Several other experiments

(Fischer, Mills, & Shaki, 2010; Fischer, Shaki, & Cruise, 2009; Ristic, Wright, & Kingstone, 2006; Shaki & Fischer, 2008) have shown that recent spatial experience shapes the direction of the mental number line. Although these studies demonstrate clearly that spatial reference frames and spatial experience can modulate the association between numbers and space, they do not contribute to our understanding of the origin of this association.

Recent findings suggest that there might be a biological predisposition for an association of the left side with small numerosity. A general preference for increasing magnitudes from left to right space has been found in both preverbal infants (Bulf, de Hevia, & Macchi Cassia, 2015; de Hevia, Girelli, Addabbo, & Macchi Cassia, 2014) and nonhuman animals (Adachi, 2014; Rugani, Kelly, Szelest, Regolin, & Vallortigara, 2010; Rugani, Vallortigara, Priftis, & Regolin, 2015). For example, Bulf et al. (2015) showed that centrally presented numerosities (e.g., collections of dots) bias the visual attention of 8- to 9-month-olds in a directional manner, such that the infants were faster to fixate on targets appearing on the left after small numerosities and faster to fixate on targets appearing on the right after larger numerosities were presented centrally. Rugani et al. (2015) reported that 3-day-old chicks trained with an intermediate number of dots subsequently spontaneously associated a smaller numerosity with the left and a larger numerosity with the right side of space (but see Shaki & Fischer, 2015).

A biological account needs to make at least three assumptions. First, there must be an early biological, possibly innate, affinity between numerosity and space. This might occur through neuronal recycling: numerosity coding in the brain might have emerged later in evolution from the brain systems that support spatial coding, allowing number representations to "piggyback" on evolutionarily older networks for representing and processing spatial information (Dehaene & Cohen, 2007). Second, hemispheric brain asymmetries might then produce an attentional bias toward the left side of space. This assumption is consistent with neuropsychological evidence (Heilman & Van Den Abell, 1980; Kinsbourne, 1970). Finally, this attentional preference toward the left side of space then needs to be combined with a preference for increasing over decreasing sequences of numerosity in infants (Macchi Cassia, Picozzi, Girelli, & de Hevia, 2012) to establish an association of small numerosities with the left side of space and larger numerosities with the right.

A purely biological account predicts that the mental number line of most individuals should be going from left to right. However, associations of right space with small numbers have been reported consistently in individuals from right-to-left reading cultures. For example, Dehaene et al. (1993) reported that the extent to which adults coming from a right-to-left reading culture (Iran) showed the classic left-to-right SNARC effect depended on the length of time they had spent in a left-to-right reading culture (France). Comparing the direction of the SNARC effect directly in adults from different reading cultures,

Shaki, Fischer, and Petrusic (2009) found that Canadian participants (left-to-right reading culture) showed the classic left-to-right SNARC effect, while Arab participants—in line with their culturally predominant reading direction (right to left)—showed a reverse SNARC effect, associating small numbers with the right and larger numbers with the left side.

These results suggest a cultural modification of the potentially inherent number-space relation: the association between number and space might be shaped by the culturally predominant reading direction (Shaki & Fischer, 2008; Shaki et al., 2009). This mechanism seems plausible because not only does eye movement behavior in reading differ significantly across different writing systems (Deutsch & Rayner, 1999), but reading direction also influences the preferred visual scanning direction of nontext material (Abed, 1991; Rinaldi, Di Luca, Henik, & Girelli, 2014). During reading acquisition, many hours are spent scanning text using horizontal eye movements from left to right or right to left depending on the culturally predominant reading direction. This extensive directional horizontal tuning through eye movements might lead to a directional bias to expect increases in number magnitude (and other continuous quantities, e.g., time) to be in line with the acquired, culturally predominant reading direction. Indeed, there is evidence that the culturally predominant reading direction modulates the association not only between number and space but also between time and space (Bonato, Zorzi, & Umiltà, 2012; Casasanto & Bottini, 2010).

Counting Direction

Counting is one of children's earliest explicitly numerical activities. While the order in which items are counted is irrelevant with respect to the final outcome of the counting activity (the order irrelevance principle, Gelman & Gallistel, 1986), there is some evidence that the preferred directionality of counting is culturally shaped. For example, Kamawar et al. (2010) reported that 5- to 11-year-old Canadian children believed the order in which items were counted to be important: most children saw left-to-right counting as the "correct" direction of counting (see also Briars & Siegler, 1984). This direction is consistent with their culturally predominant reading direction (from left to right).

Horizontal Counting

In a series of experiments, my colleagues and I subsequently explored whether the culturally predominant reading direction influences the preferred counting direction (Shaki, Fischer, & Göbel, 2012). First, we investigated horizontal counting. The participants, both children and adults, sat at a table where four objects (plastic coins) were already placed in front of them along an invisible horizontal line on a mat (see Fig. 1). They were then asked to point to each object and to count them aloud. Most participants counted the

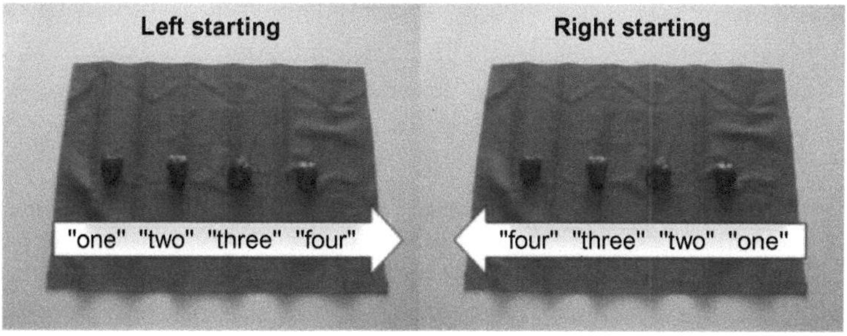

FIG. 1 Horizontal counting—experimental setup and coding.

objects either from left to right or from right to left. The very few individuals who counted differently were excluded from further analysis.

The majority of English-speaking adults and English-speaking 7- to 10-year-old children counted from left to right. In contrast, Arabic-speaking adults and Arabic-speaking 7- to 10-year-old children were significantly more likely to count from right to left than from left to right. These results clearly suggest that the amount of direct experience with a particular reading direction shapes the counting direction.

Even more convincingly, illiterates who had recently emigrated into Israel from Northern Ethiopia and who were unable to read simple sentences in their mother tongue (Amharic) or in Hebrew were equally likely to count from left to right as from right to left. This finding is incompatible with the idea of an early biological predisposition for a left-to-right mental number line (Bulf et al., 2015; Rugani et al., 2015) and clearly demonstrates the effect of reading direction on counting direction.

Consistent with the idea that reading experience shapes counting direction, the percentage of English-speaking children who count left to right increases with age, while the percentage of Arabic-speaking children who count left to right decreases with age. However, in this study, we already found significant preferences for a counting direction in line with the culturally dominant reading direction for children as young as 3 years of age. These findings suggest that a person's own reading experience might not be the only driving factor. Culturally mediated experiences with print such as being read to by parents seem to be shaping this effect as well (Göbel, Fischer, & Shaki, 2014).

Vertical Counting

While in most Western cultures the primary reading direction is from left to right, they also share a common secondary reading direction: text on a page is typically read from top to bottom (secondary reading direction). Does the vertical reading direction influence counting direction too? In order to investigate this, in a new set of experiments, we changed the display from horizontal to vertical. Participants now saw four objects placed on a mat along an invisible vertical line. Strictly speaking, the objects were displayed vertically in the horizontal plane, that is, just like text on a page in a book lying on a table can be at the top of the page or the bottom. (*Note*: this is along *y*-axis in a three-dimensional coordinate system and NOT along the *z*-axis.) When 3- to 5-year-old English-speaking children were asked to point to the objects and count them, the majority counted from the bottom to the top (Göbel, 2015); in contrast, English-speaking adults showed the opposite pattern with a significant preference to count from the top to the bottom in line with their secondary reading direction. This points to an influence of direct own experience with the secondary reading direction.

We also tested children and adults from Hong Kong, because the reading direction in Hong Kong is mixed. They mainly read text from left to right but are also exposed to a *primary* reading and writing direction from the top to the bottom. For example, in 2012, 100 books for children and 100 books for adults were picked randomly from the shelves of the Hong Kong Central Library. Eighty percent of the books for children used a left-to-right reading direction, 17% a top-to-bottom reading direction, and 3% had mixed reading directions. For adult books, the corresponding percentages were 46% left to right, 42% top to bottom, and 12% mixed (Göbel, 2015). When we tested 3- to 5-year-old children and adults from Hong Kong on their preferred vertical counting direction, both groups showed a significant preference for counting from top to bottom over bottom to top (Fig. 2).

How can this difference between English-speaking children and children from Hong Kong be explained, given that there was no significant difference between the adult groups? Children in Hong Kong start reading and writing much earlier than children in the United Kingdom (Curriculum Development Council, 2006), and some of this early reading and writing experience is vertical. So why did the majority of English-speaking children in the United Kingdom count bottom to top? They might not have shown a

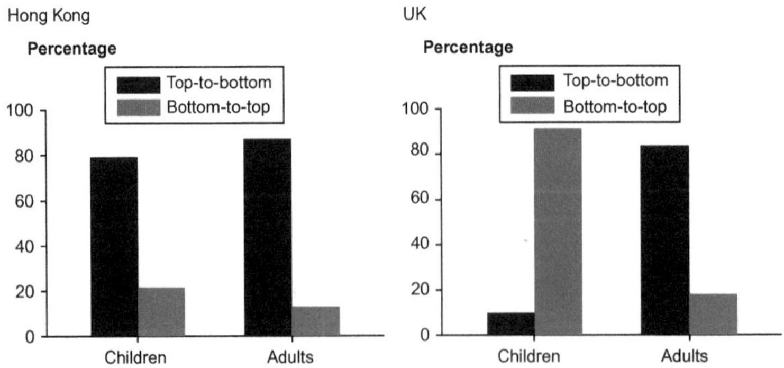

FIG. 2 Vertical counting—percentage of participants counting from top to bottom (black bars) and bottom to top (gray bars).

significant preference to count from top to bottom because of their lack of experience with this secondary reading direction. But then, why did they show a preference at all? This finding could be explained by the idea that initially, magnitude representations might be physically grounded (Fischer, 2012; Gevers, Lammertyn, Notebaert, Verguts, & Fias, 2006; Lakoff & Núñez, 2000). Experience of magnitude in the physical world might lead to an initial association of up with higher and down with lower: taller trees extend more upward in the vertical dimension, more juice in the glass is often associated with a higher level in the glass, and adults (e.g., older people) are taller. Children might start out with a physically grounded association between magnitude and vertical space. This association would then be reversed with increasing (direct) experience with the secondary reading direction (top to bottom). This explanation assumes that the experience of young children with the vertical dimension in the physical world (i.e., along the *z*-axis) influences their counting behavior in the *y*-axis.

Our results to date also suggest that different experimental paradigms showing associations between number and space might not all be driven by the same underlying mechanism (van Dijck et al., 2011). While our studies on counting direction show clear effects of both culturally predominant primary and secondary reading directions, there is convincing evidence that the secondary reading direction does not influence the vertical SNARC effect (e.g., the finding that in parity judgment tasks, faster responses are made with a top button for larger numbers and with a bottom button for smaller numbers) in Western participants (Gevers et al., 2006; Hartmann, Gashaj, Stahnke, & Mast, 2014; Holmes & Lourenco, 2012; Schwarz & Keus, 2004; Shaki & Fischer, 2012). In addition, the evidence is mixed for whether even the primary vertical reading direction influences the direction of the vertical SNARC effect. Japanese participants, whose primary reading direction is from top to bottom, nevertheless were faster to respond to small numbers with a bottom response and larger numbers with a top response (Ito & Hatta, 2004), showing the same association as Western participants and opposite to what would be expected given the Japanese reading direction. However, Hung, Hung, Tzeng, and Wu (2008) found that for Taiwanese participants, it depended on the type of numerical stimuli used. For Chinese number words, which are more often read in text from top to bottom, they were faster to respond to small numbers with a top and large numbers with a bottom response. These findings suggest that the vertical SNARC effect is only influenced by reading direction under certain conditions, for example, when a reading context is evoked. In contrast, our findings show a clear influence of reading direction on vertical counting direction. It is thus likely that tasks such as counting that test explicit associations between number and space might at least be partly influenced by different factors, other than implicit associations between number and space, as tested, for example, in the SNARC effect (Nuerk, Patro, et al., 2015).

Horizontal and Vertical Counting Combined

As the real world is three-dimensional, the locations of objects frequently vary relative to one another in both horizontal and vertical positions at the same time. We thus started wondering what happens in displays with objects along both horizontal and vertical dimensions. Is one dimension more dominant? In order to test this, we presented a new group of adults with a 6×6 grid of circles and asked them to count aloud the number of circles while pointing to each one (Göbel, 2015).

Most adults started counting on the top left circle. However, significantly more adults from Hong Kong than adults from the United Kingdom started counting on the top right circle. This can be explained by the mixed reading experience of adults from Hong Kong. In a second step, we tested directly whether recent reading experience can change counting direction. Adults from Hong Kong were split into two groups. The horizontal group had to read a text printed horizontally with a reading direction from left to right within each line and from the top to the bottom line on the page. The vertical group had to read exactly the same text, but it was printed vertically and had to be read from top to bottom in each column and from the right-most to the left-most column on the page. Comprehension questions were also printed horizontally or vertically and had to be answered in horizontal or vertical writing. All participants were students from Hong Kong who were living in the United Kingdom at the time of testing. After reading the text and answering the questions, they were asked to count the 6×6 array. While overall the majority of both groups still started counting at the top left circle, the group of participants who had read the vertical text was significantly more likely than the group of participants who had read the horizontal text to start at the top right circle (after reading the vertical text—35.7%, after reading the horizontal text—4.4%).

In sum, reading direction influences counting direction. As early as 3 years of age, children have counting preferences. At least for the horizontal direction, these directional preferences are already in line with the culturally predominant reading direction. The finding that this preference can be manipulated by experience and recent reading direction suggests that the mechanism underlying the association between reading direction and counting direction is flexible and adaptive.

Counting Direction and Mathematical Performance

Is the extent to which children respond to the influence of experiential and cultural cues in any way related to their broader numerical skills and mathematical development? While none of the studies reported here were designed to answer this question explicitly, additional results are suggestive. In an unpublished study of English-speaking preschool children, we tested 48 3-year olds, 87 4-year olds, and 64 5-year olds on their horizontal counting preference (as described above). We also tested them on three additional numerical

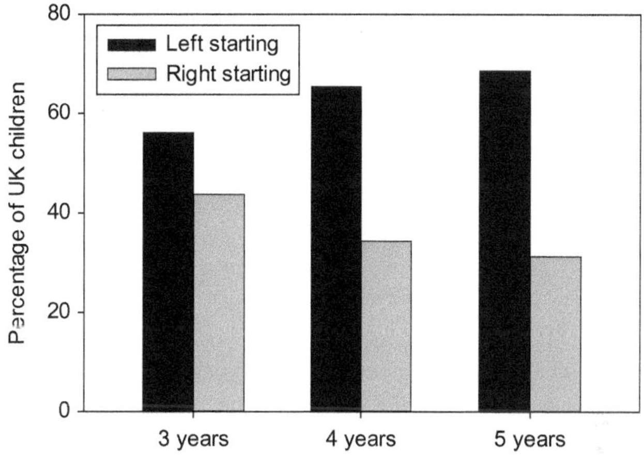

FIG. 3 Horizontal counting—percentage of UK children by age counting left to right (black bars) and right to left (gray bars).

tasks: verbal counting, digit naming, and digit recognition. In the verbal counting task, children were asked to count as high and as quickly as they could without making any mistakes. The highest number reached in 40 s (minus any mistakes) was recorded. For digit naming, the children were presented with eight cards, each having a different Arabic digit printed on it (1–8). The experimenter pointed to each card in turn, and the children were asked to name the digit (maximum score = 8). In the digit recognition task, four of those cards were placed on the table in front of the children who were then asked to point to the Arabic digit that corresponded to the number word given verbally by the experimenter ("one," "two," "seven," and "eight," maximum score = 4).

As can be seen in Fig. 3, we replicated our previous findings: the youngest children (and all age groups) exhibited a significant preference to count from left to right. This preference increased significantly with age. Next, we analyzed children's verbal counting, digit naming, and digit recognition separately for the children who counted left to right versus those who counted right to left. Not surprisingly, digit naming, digit recognition, and verbal counting performance increased significantly with age. There was no strong empirical evidence that children who counted left to right outperformed the children who counted right to left (Fig. 4). There was no significant main effect of counting direction group and no significant interaction between the effects of counting direction group and age. However, as the direct group comparisons for digit naming and digit recognition approached significance, future studies might want to revisit this question.

To date, no strong evidence exists that the children who count in line with the culturally predominant reading direction differ in numerical skills from

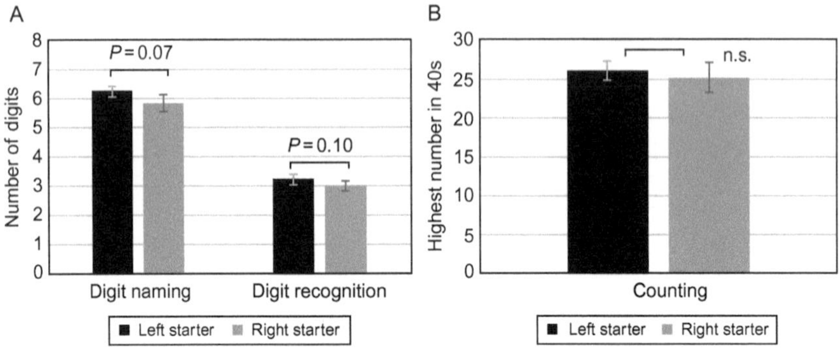

FIG. 4 (A) Digit naming, digit recognition, and (B) counting performance separately for children who started counting from the left (black bars) versus children starting counting from the right (gray bars).

those who do not. In addition, this factor is unlikely to be a powerful and reliable predictor for mathematical development for at least three reasons. First, as we have shown, it can be changed by recent reading experience (see also Göbel et al., 2014). Second, it is a binary variable (most children count left to right or right to left) and thus limited in its explanatory power. Third, in our study, nearly all 6-year-old children counted from left to right; so at least from age 6 onward, this factor is unlikely to have any predictive power for future mathematical development.

Number Generation

The role of reading direction in number processing has also been studied with respect to its influence on the so-called number generation task. Asking participants to generate a sequence of random numbers has frequently been used as an executive function task (Brugger, 1997; Towse & Neil, 1998). While it has been shown repeatedly how difficult it is for participants to generate truly random sequences (Wagenaar, 1972), another interesting finding emerged when the frequency with which each number was named was investigated: adults showed a small number bias (Di Bono & Zorzi, 2013; Loetscher & Brugger, 2007; Towse, Loetscher, & Brugger, 2014). They were more likely to name smaller numbers than larger numbers (relative to the interval they were operating in). This small number bias could be due to smaller numbers being more frequently used in everyday contexts (Dehaene & Mehler, 1992). However, Loetscher and Brugger (2009) reported a significant correlation between the size of individuals' small number bias and their tendency to err toward smaller numbers in number bisection, a task in which the numerical midpoint between two other numbers has to be determined—such as 4 is the numerical midpoint between 2 and 6. This finding suggests that number generation might activate a mental number line. Loetscher, Schwarz,

Schubiger, and Brugger (2008) asked Western adults from a left-to-right reading culture to generate numbers randomly picked from the numbers 1 to 30. To explicitly activate space-number relations, participants were asked to perform head turns to the left side and to the right side. Participants named significantly more smaller numbers after left than after right head turns. This finding is consistent with the activation of a left-to-right mental number line in Western participants during number generation.

Number Generation and Reading Direction

The findings described above beg the question: what outcomes would be expected in number generation tasks for people from a right-to-left reading culture? Given that their mental number line goes from right to left, they should show the opposite effect when turning their head to the left vs right side while generating numbers. In addition, as their number line starts at the right side, might they show a large number bias (see Loetscher & Brugger, 2007) instead of a small number bias? We investigated this question with a slightly altered design (see Fig. 5). Instead of actively turning their heads, participants were passively lying on the floor, either on their left or on their right side with closed eyes while naming 40 numbers between 1 and 50 to the sound of a metronome, at the rate of one number every 2 s.

We tested children and adults from a left-to-right reading culture (British adults—$N=54$, mean age = 20.13 years; British children—$N=81$, mean age = 8.19 years) and from a right-to-left reading culture (Arab adults—$N=60$, mean age = 24.38 years; Arab children—$N=52$, mean age = 7.73 years). In order to explain the task, we told adults and children to imagine a jar containing 50 marbles, each marble with a unique number ranging from 1 to 50. We then told them to imagine with each metronome beat taking out one marble from the jar, to read out loud the number on the marble, and then to put the marble back. We highlighted that all the 50 marbles numbered 1–50 were back in the jar, when at the next beat of the metronome, they were again asked to take out a marble from the jar. This could either be the same marble they just took out or another marble. Once we had given them these instructions, we started the metronome and asked them to name

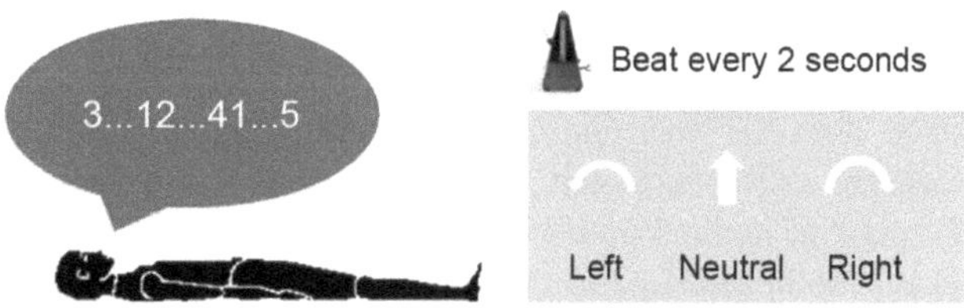

FIG. 5 Illustration of the procedure for the number generation task. Please note, the participant is shown in neutral position (starting position). The participant's body was moved onto their left or right side before number generation started.

a number from 1 to 50 at each beat of the metronome and recorded each number given by the participant.

Although there was no significant difference between the adults and the children in terms of the mean numbers named in each position, there was a significant interaction between body position and reading direction with respect to number generation. Arab participants named on average smaller numbers when they were lying on their right side than when they were lying on their left side. British participants showed the opposite effect: they produced significantly larger number when lying on their right than on their left side.

Interestingly, overall, there was a small number bias, with participants more likely to name smaller than larger numbers, and this effect was stronger in the Arab participants. This suggests that the overall small number bias that has repeatedly been reported (Grade, Lefèvre, & Pesenti, 2013; Loetscher & Brugger, 2007, 2009; Loetscher et al., 2008) is unlikely to be related to reading direction.

Number Generation and Mathematical Performance

In the study just described, there were, however, individual differences in the extent to which body position influenced number generation. Is it possible that people who are more sensitive to body position are also better at mathematics? While investigating individual differences was not the aim of this study, we also included a measure of mathematical performance for the adults. All adults completed the numerical operations subtest of the wide-range achievement test (WRAT 4, Wilkinson & Robertson, 2006). There was no significant correlation between individuals' differences in average size of the numbers generated when they were lying on their left vs right side and their raw score on the numerical operations subtest ($N=114$, $r=0.03$). So at least in our study, people whose size of generated numbers was strongly influenced by body position were not performing significantly better or worse on the mathematical achievement test than those whose number generation was less influenced by body position.

Summary of the Effects of Reading Direction

To sum up this first section, the cultural factor of reading direction clearly shapes the preferred counting direction, influences number generation, and is related to at least the horizontal SNARC effect. Future studies will have to clarify whether experience with reading direction is the source of some associations between number and space or whether these associations are preexisting and get modulated by this cultural factor. Our data so far suggest that neither the extent to which an individual's association of number and space is influenced by reading direction nor the extent to which individuals show an association between number and space is a reliable predictor of their mathematical abilities. This is in line with several other studies that failed to find a significant relationship between the extent to which participants show an association between number and space and their mathematical abilities (e.g., Cipora & Nuerk, 2013; Schneider, Grabner, & Paetsch, 2009).

However, there is clear evidence that networks involved in number processing and arithmetic overlap with networks involved in spatial processing (Hubbard et al., 2005; Knops, Thirion, Hubbard, Michel, & Dehaene, 2009) and that even mathematical experts strongly rely on brain circuits for space and number (Amalric & Dehaene, 2016). In addition, some studies have reported significant associations between the size of an individual's SNARC effect and her or his mathematical abilities (Cipora et al., 2016; Hoffmann, Mussolin, Martin, & Schiltz, 2014). However, our studies and other research (Bächtold et al., 1998; Fischer et al., 2009; Ristic et al., 2006) have emphasized the flexible nature of these associations; they are easily changed by recent experience and thus tend to be of low reliability (Cipora & Göbel, 2013; Viarouge, Hubbard, & McCandliss, 2014). It is thus unlikely that measures of the cultural influence of reading direction on spatial-numerical associations or even measures of the strength of the spatial-numerical associations themselves will be robust predictors of children's mathematical development.

EFFECTS OF NUMBER WORD CONSTRUCTION

In the second part of this chapter, I will describe the effect of one aspect of number word construction on numerical processing and arithmetic: number word inversion. First, I will present a brief overview of the relationship between language and number processing. The extent to which the linguistic properties of a language influence numerical cognition has intrigued researchers for many years, and views have ranged from the idea that the numerical system is largely independent of the language system (e.g., Brysbaert, Fias, & Noel, 1998) to the proposal of a very close relationship between language and numerical cognition (Hurford, 1987). Finding clear-cut answers to this question is complicated by the fact that groups who speak different languages mostly live in different countries. Consequently, in addition to the desired differences in the spoken language, there are inevitably differences between those groups in cultural experiences such as teaching practices, parental involvement, attitudes, and beliefs (e.g., LeFevre, Clarke, & Stringer, 2002; Towse & Saxton, 1997) that can modulate numerical performance (see Chapter 3 by LeFevre et al., Chapter 4 by Okamoto, and Chapter 5 by Opfer et al.). Several ingenious researchers have thus used different language groups within one culture (e.g., English and Welsh speakers in Wales (Dowker, Bala, & Lloyd, 2008), French and Flemish speakers in

Belgium (Imbo, Vanden Bulcke, De Brauwer, & Fias, 2014; Noël & Seron, 1997; Seron & Fayol, 1994), Chinese and English speakers in the United States (Geary, Bow-Thomas, Liu, & Siegler, 1996), and French and English speakers in Canada (LeFevre et al., 2002)). However, even within one culture, there is evidence that groups that differ in the language they speak might also differ on other factors (e.g., parents' beliefs and behaviors) that are related to children's numeracy performance (LeFevre et al., 2002). For an excellent in-depth investigation about the relative contribution of linguistic and cultural factors influencing children's early numeracy skills, see Chapter 3 by LeFevre et al.

Irrespective of these important caveats, clear evidence is mounting that linguistic factors affect numerical cognition independently of cultural factors. For example, children from a language background with singular/plural marking (e.g., English and Russian) learn the meaning of the number word "one" faster than children from language backgrounds without singular/plural marking (e.g., Japanese—Sarnecka, Kamenskaya, Yamana, Ogura, & Yudovina, 2007; Mandarin—Le Corre, Li, Huang, Jia, & Carey, 2016), even if the children from the language background without singular/plural marking are exposed more frequently to number words (Le Corre et al., 2016). This demonstrates clearly the importance of linguistic structure on number learning (see Chapter 4 by Okamoto, Chapter 6 by Donlan, Chapter 9 by Sarnecka et al., and Chapter 13 by Bender and Beller). In addition to grammatical markers, Dowker (2005) proposed four other characteristics of languages that might affect number processing: (1) whether the language includes number words at all and if there is an upper limit, (2) whether there is a written number system, (3) the regularity of the spoken number system, and (4) the degree and consistency of mapping between the spoken and the written number system.

Research on tribes speaking languages with a limited number of exact number words, for example, the Munduruku (Pica, Lemer, Izard, & Dehaene, 2004) who only have exact number words for up to five, has clearly shown that performance on exact numerical tasks, but not approximate tasks, is limited by their exact number word vocabulary (Butterworth, Reeve, Reynolds, & Lloyd, 2008; Frank, Everett, Fedorenko, & Gibson, 2008; Gordon, 2004). This is in line with earlier research in adult bilinguals showing that after they were trained on new arithmetic facts, they were better at a later test in the language they were trained in only for exact arithmetic tasks. For approximate tasks, the language in which they were tested did not influence their performance (Spelke & Tsivkin, 2001).

The regularity of the spoken number word system has received a lot of attention. The construction of number words in Asian languages, for example, is highly regular. Once the number words for one to ten have been acquired, number words up to ninety-nine can be constructed by combining the first 10 number words; for example, 32 is constructed as three-ten-two. Children speaking languages that have a regular number word structure are typically better at verbal counting, reading and comparing numbers, place-value understanding, and arithmetic (Aunio, Aubrey, Godfrey, Pan, & Liu, 2008; Dowker et al., 2008; Geary et al., 1996; LeFevre et al., 2002; Chapter 3 by LeFevre et al.; Miller, Smith, Zhu, & Zhang, 1995; Miura, Okamoto, Kim, Steere, & Fayol, 1993; Chapter 4 by Okamoto). One criticism of these studies however is that the children from the more regular language background frequently had much more early numeracy experience as well (e.g., Towse & Saxton, 1997). However, recent studies have shown unambiguously that while early numeracy experiences are important for numerical development, the structure of the number naming system is important too and regularity of the number naming systems is a bonus. Cankaya, LeFevre, and Dunbar (2014), for example, showed that young children when exposed systematically to the counting sequence in a game learn this sequence more quickly when it is presented in language with a more regular number naming system (e.g., Turkish compared with English, see Chapter 3 by LeFevre et al.).

Several studies have shown that the consistency of the mapping between spoken and written number forms influences number processing too. Early studies compared the processing of French-speaking participants from France and Belgium who differ in their way of constructing larger decade words (e.g., seventy in France, soixante-dix (60 + 10), in Belgium, septante; eighty in France, quatre-vingt (4 × 20), in Belgium, octante; and ninety in France, quatre-vingt-dix (4 × 20 + 10), in Belgium, nonante). Seron and Fayol (1994) found that second-grade children in France made more errors in Arabic number production than Belgian second-grade children (see also Noël & Seron, 1997). In a more recent study, Colomé, Laka, and Sebastián-Gallés (2010) provided evidence that the mapping between the number word structure and the Arabic form influences performance in adults even when the input and output formats used in the task were purely in Arabic notation. They tested Basque speakers (in Basque, base 20 is used in number words, e.g., 35 is twenty-fifteen) and Spanish/Italian speakers (both languages use a base ten system) and found that Basque speakers were significantly faster to respond to addition problems in the form of 20 + 15 than in the form of 25 + 10 than Spanish/Italian speakers.

Number Word Inversion

Another example of inconsistent mapping between number words and the corresponding Arabic digits can be found in languages with number word inversion (e.g., German and Dutch). In languages with inverted number words, the number word starts with the unit followed by the decade (e.g., zwei-und-dreißig, two and thirty). This is inconsistent with the order of the digits in the Arabic form (e.g., 32). In contrast, in languages with noninverted number words, such as English, the sequence in the number words for double-digit numbers is decade first followed by unit (e.g., thirty-two), just as in the Arabic notation (e.g., 32). Several studies have shown that 7-year-old children from languages with number word inversion make many inversion errors in number transcoding (e.g., writing 23 for two and thirty; e.g., Zuber, Pixner, Moeller, & Nuerk, 2009). This error type is rare for children from languages without number word inversion (Imbo et al., 2014; see also Chapter 4 by Okamoto).

Number Word Inversion and the Carry Effect in Addition

Can number word inversion also affect symbolic arithmetic? In order to investigate this question, we tested Austrian and Italian children aged 7–9 years of age on a computerized addition task with a two-alternative-forced-choice response (e.g., "26+3" followed by "29–27"; Göbel, Moeller, Pixner, Kaufmann, & Nuerk, 2014). The children had to choose the correct response as quickly as possible. Half of the items were sums that required a carry operation (e.g., 26+5), half did not (e.g., 32+5). On no-carry trials, children can select the correct answer just by adding the units, because the decade digit stays the same for the correct answer. In contrast, on carry trials, the unit sum is 10 or larger so the decade digit has to be increased. This additional processing step is known to increase reaction times (Geary, Widaman, & Little, 1986) and tax working memory (Fürst & Hitch, 2000). In line with the literature (Deschuyteneer, de Rammelaere, & Fias, 2005; Fürst & Hitch, 2000), we expected increased response times on carry trials compared with no-carry trials.

We were particularly interested in whether the structure of the number words (inverted in the German-speaking children versus noninverted in the Italian-speaking children) affects the size of the carry effect in symbolic written arithmetic. If number words are activated during the processing of the sum, the carry operation should be more difficult for inverted than noninverted number words because the inversion of the number word with respect to the number sequence in the Arabic digit string should make the required place-value integration and decade place holding/updating more difficult.

Overall, as expected, for both language groups, we found a significant increase in response times on carry trials compared with no-carry trials, replicating earlier findings. Most importantly, the size of the carry effect was significantly larger for the German-speaking children, that is, for participants with a language with inverted number words (see also Lonnemann & Yan, 2015). This means the cost of a carry trial is higher for children speaking a language with inverted number words than for children using a language with noninverted number words. This shows clearly that number word structure influences written symbolic arithmetic, at least in children.

Carry Effect and Mathematical Performance

Could the size of the carry effect be related to overall mathematical performance? At the time of testing, we did not administer standardized tests of children's mathematical performance. However, for a subset of these children, their teachers' mathematics grade at the end of year 3 (2 years later; Austrian, $N=47$ and Italian, $N=71$) was available. For those children, the size of the carry effect in the addition task in year 1 was not significantly correlated with their mathematics grade in year 3 ($r=-0.02$, $P=0.81$, and $N=118$). Likewise, the relationship between the carry effect and mathematics grade was not significant when analyzed separately for Austrian ($r=0.05$, $P=0.73$, and $N=47$) and Italian ($r=-0.12$, $P=0.31$, and $N=71$) children.

Effects of Reading Direction and Number Word Construction Can interact

One could argue that children might be more likely to still activate the number words when seeing Arabic digits, but that in adults, Arabic number processing might proceed without activation of a verbal code. However, there is evidence for effects of the structure of number words on the processing of Arabic digits in adults too (e.g., Colomé et al., 2010; Nuerk, Weger, & Willmes, 2005).

Further evidence that adults still activate the number words when processing Arabic digits came also from one of our recent studies in which we showed that number word inversion and reading direction interacted in adults (Moeller, Shaki, Göbel, & Nuerk, 2015). The idea behind this was that when seeing multidigit Arabic numerals, the preferred reading direction of these Arabic digits is in line with the cultural reading direction—from left to right for Western adults. As discussed in the previous section, number word inversion might lead to more interference with double-digit Arabic number processing because, while reading the Arabic digit string from left to right and thus processing the decade digit first, an inverted number word starts with the unit digit. The effect of number word inversion on Arabic digit processing might therefore originate from the incongruity between the timing when the decade digit is read (first) and its position in the number word (last).

If this proposal is correct, then, for cultures reading right to left, speaking a language with number word inversion should actually be advantageous relative to having no number word inversion. We tested this in a study using four groups of adult participants: German speaking, English speaking, Hebrew speaking, and Arabic speaking. German and Arabic use number word inversion. English and Hebrew number words are not inverted. We thus expected that when processing Arabic double digits, number word inversion for people from a left-to-right reading culture (German) would be disadvantageous, while number word inversion for people from a right-to-left reading culture (Arabic) should be advantageous—relative to languages that are read in the same direction but have noninverted number words.

We asked participants to perform a comparison task on pairs of double-digit Arabic numerals presented on a computer screen. Please note that there is no need to explicitly activate number words to perform this task. There were three trial types: within decade pairs, compatible pairs, and incompatible pairs. For within decade pairs, the decision had to be made based on the unit information (e.g., 62 and 67); for compatible and incompatible pairs, the decision could in principle be made based entirely on the decade information. On compatible trials, the information from the unit comparison and the decade comparison was consistent (e.g., 42 and 67, $4<6$ and $2<7$). In contrast, on incompatible trials, the comparison of units and the comparison of decades led to contrasting decisions (e.g., 47 and 62, $4<6$ but $7>2$).

Despite the fact that on both compatible and incompatible trials the decision could be made based solely on the decade digit, in previous studies, participants were significantly slower on incompatible than on compatible pairs; this is the so-called unit-decade compatibility effect (Nuerk, Weger, & Willmes, 2001). For left-to-right reading cultures, number word inversion influences the size of this effect: it was larger in German (inverted number words) than in English (noninverted number words); that is, the cost of processing incompatible pairs is larger in languages with number word inversion for left-to-right reading cultures (Nuerk et al., 2005; see also Pixner, Moeller, Hermanova, Nuerk, & Kaufmann, 2011).

The novel contribution of our study was that in addition to two left-to-right reading groups (German and English), we also tested two right-to-left reading groups (Arabic and Hebrew). The other factor was number word construction: participants were from a language background either with number word inversion (German and Arabic) or without number word inversion (English and Hebrew).

Following the logic explained above, our hypothesis was that the compatibility effect will be increased when the sequencing within the number words (determined by the inversion factor) did not match the sequencing in the Arabic digit strings (determined by the reading direction). Consequently, for the left-to-right reading direction, we predicted a larger compatibility effect for the language with number word inversion (German) than for the language without number word inversion (English). In contrast, for the right-to-left reading groups, we predicted a smaller compatibility for the language with number word inversion (Arabic) than the language without number inversion (Hebrew).

And this is exactly what we found (see Fig. 6): there was a significant interaction between reading direction and number word inversion. Although for the left-to-right reading direction the compatibility effect was higher for the language with inverted number words than for language without number word inversion, the converse was found for the right-to-left reading direction. Here, the language with number word inversion showed a smaller compatibility effect than the language without inversion. This outcome led us to conclude that it is not the number word inversion per se that is producing the interference. It crucially depends on the relationship between the reading direction and number word inversion. Clearly, both factors influence symbolic

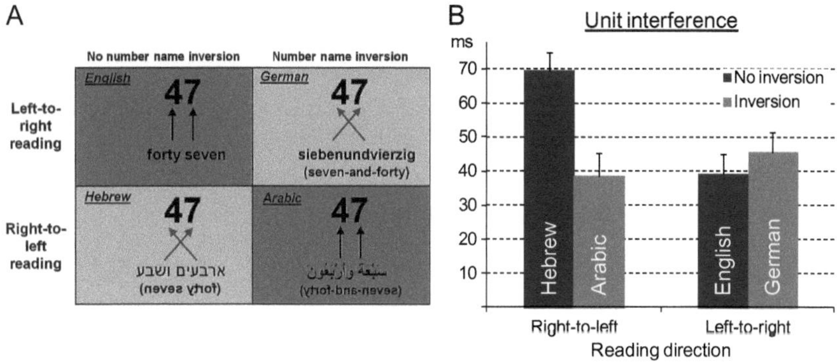

FIG. 6 Reading direction by number word inversion design and results. *images are adapted with permission from Elsevier for Figs. 1 and 2C from Moeller, K., Shaki, S., Göbel, S. M., & Nuerk, H.-C. (2015). Language influences number processing—A quadrilingual study.* Cognition, 136, *150–155.*

number comparison. The results of this study also provide further evidence that even adults activate number words when processing Arabic digits.

Compatibility Effect and Mathematical Performance

Could the size of the compatibility effect be related to individual differences in mathematical performance? In an unpublished pilot study, we tested 48 English-speaking adults on double-digit number comparison. As expected, they exhibited a significant compatibility effect. Participants were also tested on a standardized mathematics test (numerical operations subtest of the WRAT 4; Wilkinson & Robertson, 2006). In this study, the size of their compatibility effect was not significantly related to their performance on the mathematics test ($N=48$, $r=-0.13$, and $P=0.38$).

However, there is some evidence that the size children's compatibility effect might be related to their mathematical performance. In a longitudinal study by Moeller, Pixner, Zuber, Kaufmann, and Nuerk (2011), Austrian 7-year olds were tested on double-digit number comparison. Two years later, the same children were tested on a computerized addition task. Children who had a larger compatibility effect at time 1 (i.e., they showed a greater difference between compatible and incompatible trials) made significantly more mistakes in the addition task 2 years later.

These findings are intriguing, because they hint at the importance of early place-value integration for later arithmetic. However, this study was carried out with Austrian children, that is, with children whose language (German) uses number word inversion. Consequently, on the basis of this study, we cannot definitively conclude that there is a significant relationship between the compatibility effect and arithmetic specifically because of the inversion property of the children's language, nor can we definitely conclude that the compatibility effect is a significant predictor of later arithmetic performance, independent of number word inversion. A longitudinal study in a language without number word inversion is needed in order to decide between these two possibilities.

Summary of Effects of Number Word Construction

As the studies discussed in this section show clearly, the degree and consistency of the mapping between number words and Arabic digits affect number processing and number comparison, not to mention addition performance even when only Arabic digits are presented, in adults and children. There is some evidence that at least for young children, the ability to and the speed with which one maps between number words and Arabic digits might be related to mathematical performance (e.g., Mundy & Gilmore, 2009). One of our own longitudinal studies provided additional evidence that mapping between the verbal and Arabic forms might be making an important contribution to individual differences in mathematical achievement (Göbel, Watson, Lervåg, & Hulme, 2014). We tested a large sample of 6-year-old British children ($N = 165$) on a range of number comparison and arithmetic tasks when they were in year 1 of primary school. They were retested again 11 months later in year 2. Using structural equation modeling, we identified significant predictors of growth in arithmetic in these children over those 11 months. Not surprisingly, their arithmetic performance at time 1 was a significant predictor of their arithmetic growth over the next year. However, contrary to previous findings (for reviews, see De Smedt, Noël, Gilmore, & Ansari, 2013; Schneider et al., 2016), none of our symbolic or nonsymbolic comparison measures explained a significant amount of variance in arithmetic growth. But in addition to arithmetic performance at time 1, a task we called number identification emerged as a significant longitudinal predictor. Taken together, number identification and arithmetic performance at time 1 predicted 86% of the variance in growth of arithmetic over the next 11 months. This is impressive given that the number identification task consisted of just eight items. In this task, children heard an auditorily presented number word (e.g., "six") and had to choose the correct corresponding Arabic digit or digit string from four or five options (e.g., 8 6 3 9). Distractors were either visually similar, inverted number words, or typical transcoding errors.

Another example of a more difficult item in this task would be hearing "seventy-six" and having to choose the correct answer out of the following options: 17, 76, 6, 706, 67. This task measures at least two component skills: Arabic digit knowledge and place-value understanding. First, it is conceivable that in early development, Arabic digit knowledge might play a similar role for mathematical development as letter knowledge does for reading development (Muter et al., 2004). Second, as a good understanding of place value is particularly important for multidigit processing (Nuerk, Willmes, & Fischer, 2015), the more difficult items in this task clearly require such knowledge to transcode between number words and Arabic digits quickly and accurately.

Future research will have to shed light on the relative contribution of these two skills to arithmetic knowledge and competence. Although it seems very plausible that both skills are essential in this regard, their comparative importance may well differ at different points in development.

In an analysis of a subset of the sample, we were also able to show that number identification is still a significant predictor of arithmetic at time 2 when children's counting performance is included as a predictor. This finding demonstrates that the predictive power of the task was not just due to a general verbal/linguistic component, as this component would be shared with counting performance. It seems to be the ability to identify the corresponding Arabic digit string in response to a number word that is driving the predictive power.

CONCLUSIONS AND FUTURE DIRECTIONS

In summary, it is clear that the way in which children and adults process Arabic digits is influenced by several culture-specific factors. In this chapter, I have presented evidence for how one cultural factor (reading direction) and one linguistic factor (number word construction) influence numerical processing (counting, number generation, number comparison, and arithmetic). However, the fact that reading direction and number word construction shape numerical processing does not automatically mean that these factors influence individual differences in mathematical performance and development. For example, in our data, there is no evidence that the extent to which children and adults are influenced by the culturally predominant reading direction is related to their mathematical performance. In contrast, results from longitudinal studies suggest that number word construction and the ability to transcode spoken number words into Arabic digits are significant predictors of growth in mathematical performance in the first years of primary school.

Clearly, many other linguistic and cultural factors shape the acquisition of number processing and arithmetic, even in adults. For example, one important linguistic factor for early number word understanding is whether a child is exposed to a language that uses singular/plural markings (Sarnecka et al., 2007). In contrast, one crucial cultural factor for early numeracy development is clearly the frequency with which children at home are involved in numeracy-related activities (Chapter 9 by Sarnecka et al.; Skwarchuk et al., 2014).

Future research will continue to document cultural and linguistic factors that influence numerical cognition. However, it will be crucial to identify which of these factors contribute significantly to numerical development and/or individual differences in mathematical skill, as this work could open up unique opportunities for teaching and intervention. Although some cultural and linguistic factors can be changed relatively easily to optimize numerical development, other factors are so ingrained in a particular language or culture that it will be essential to find the best way to ameliorate any disadvantageous effects through instructional interventions.

ACKNOWLEDGMENTS

The experiments reported in this chapter were carried out in collaboration with Martin Fischer, Liane Kaufmann, Korbinian Moeller, Hans-Christoph Nuerk, Silvia Pixner, and Samuel Shaki. Silke Göbel's work was supported by a British Academy/Leverhulme Small Research Grant (SG121544). Thanks to Allison Cheung, Yinwen Ko, Kelsey Lam, Carmen Lau, Jasmine Lei, Carolin Maier, Rachel Meyrick, Courtney Poole, Hannah Roome, and Ellen Teder for their assistance with data collection and coding. Thanks to Liz Kirk, Harriet Over, and Jelena Mirkovic for comments on the first version of the talk on which this chapter is based and Graham Hitch for feedback on a previous draft of this chapter.

REFERENCES

Abed, F. (1991). Cultural influences on visual scanning patterns. *Journal of Cross-Cultural Psychology*, *22*(4), 525–534.

Adachi, I. (2014). Spontaneous spatial mapping of learned sequence in chimpanzees: Evidence for a SNARC-like effect. *PLoS ONE*, *9*(3). e90373, https://doi.org/10.1371/journal.pone.0090373.

Amalric, M., & Dehaene, S. (2016). Origins of the brain networks for advanced mathematics in expert mathematicians. *Proceedings of the National Academy of Sciences of the United States of America*, *113*(18), 4909–4917. https://doi.org/10.1073/pnas.1603205113.

Aunio, P., Aubrey, C., Godfrey, R., Pan, Y., & Liu, Y. (2008). Children's early numeracy in England, Finland and People's Republic of China. *International Journal of Early Years Education*, *16*(3), 203–221. https://doi.org/10.1080/09669760802343881.

Bächtold, D., Baumüller, M., & Brugger, P. (1998). Stimulus-response compatibility in representational space. *Neuropsychologia*, *36*(8), 731–735.

Bonato, M., Zorzi, M., & Umiltà, C. (2012). When time is space: Evidence for a mental time line. *Neuroscience and Biobehavioral Reviews*, *36*(10), 2257–2273. https://doi.org/10.1016/j.neubiorev.2012.08.007.

Briars, D., & Siegler, R. S. (1984). A featural analysis of preschoolers' counting knowledge. *Developmental Psychology*, *20*(4), 607.

Brugger, P. (1997). Variables that influence the generation of random sequences: An update. *Perceptual and Motor Skills*, *84*(2), 627–661. https://doi.org/10.2466/pms.1997.84.2.627.

Brysbaert, M., Fias, W., & Noel, M. -P. (1998). The Whorfian hypothesis and numerical cognition: is "twenty-four" processed in the same way as "four-and-twenty?". *Cognition*, *66*(1), 51–77.

Bulf, H., de Hevia, M. D., & Macchi Cassia, V. (2015). Small on the left, large on the right: Numbers orient visual attention onto space in preverbal infants. *Developmental Science*. https://doi.org/10.1111/desc.12315.

Butterworth, B., Reeve, R., Reynolds, F., & Lloyd, D. (2008). Numerical thought with and without words: Evidence from indigenous Australian children. *Proceedings of the National*

Academy of Sciences of the United States of America, *105*(35), 13179–13184. https://doi.org/10.1073/pnas.0806045105.

Cankaya, O., LeFevre, J. -A., & Dunbar, K. (2014). The role of number naming systems and numeracy experiences in children's rote counting: Evidence from Turkish and Canadian children. *Learning and Individual Differences*, *32*, 238–245. https://doi.org/10.1016/j.lindif.2014.03.016.

Casasanto, D., & Bottini, R. (2010). Can mirror-reading reverse the flow of time? *Spatial cognition VII*. Springer Berlin Heidelberg (pp. 335–345). https://doi.org/10.1007/978-3-642-14749-4_28.

Cipora, K., & Göbel, S. M. (2013). Number—space associations: Just how reliable is the SNARC effect. In *XXXIst European Workshop on Cognitive Neuropsychology. Bressanone: Italy*.

Cipora, K., Hohol, M., Nuerk, H. -C., Willmes, K., Brożek, B., Kucharzyk, B., et al. (2016). Professional mathematicians differ from controls in their spatial-numerical associations. *Psychological Research*, *80*(4), 710–726. https://doi.org/10.1007/s00426-015-0677-6.

Cipora, K., & Nuerk, H. -C. (2013). Is the SNARC effect related to the level of mathematics? No systematic relationship observed despite more power, more repetitions, and more direct assessment of arithmetic skill. *The Quarterly Journal of Experimental Psychology*, *66*(10), 1974–1991. https://doi.org/10.1080/17470218.2013.772215.

Colomé, À., Laka, I., & Sebastián-Gallés, N. (2010). Language effects in addition: How you say it counts. *The Quarterly Journal of Experimental Psychology*, *63*(5), 965–983. https://doi.org/10.1080/17470210903134377.

Curriculum Development Council. (2006). *Guide to the pre-primary curriculum*. Hong Kong: Curriculum Development Council.

de Hevia, M. D., Girelli, L., Addabbo, M., & Macchi Cassia, V. (2014). Human infants' preference for left-to-right oriented increasing numerical sequences. *PLoS ONE*, *9*(5), e96412. https://doi.org/10.1371/journal.pone.0096412.g005.

De Smedt, B., Noël, M. -P., Gilmore, C., & Ansari, D. (2013). The relationship between symbolic and non-symbolic numerical magnitude processing skills and the typical and atypical development of mathematics: A review of evidence from brain and behavior. *Trends in Neuroscience & Education*, *2*, 48–55.

Dehaene, S., Bossini, S., & Giraux, P. (1993). The mental representation of parity and number magnitude. *Journal of Experimental Psychology. General*, *122*, 371–396.

Dehaene, S., & Cohen, L. (2007). Cultural recycling of cortical maps. *Neuron*, *56*(2), 384–398. https://doi.org/10.1016/j.neuron.2007.10.004.

Dehaene, S., & Mehler, J. (1992). Cross-linguistic regularities in the frequency of number words. *Cognition*, *43*(1), 1–29.

Deschuyteneer, M., de Rammelaere, S., & Fias, W. (2005). The addition of two-digit numbers: Exploring carry versus no-carry problems. *Psychology Science Quarterly*, *47*(1), 74–83.

Deutsch, A., & Rayner, K. (1999). Initial fixation location effects in reading hebrew words. *Language and Cognitive Processes*, *14*(4), 393–421. https://doi.org/10.1080/016909699386284.

Di Bono, M. G., & Zorzi, M. (2013). The spatial representation of numerical and non-numerical ordered sequences: Insights from a random generation task. *Quarterly Journal of Experimental Psychology*, *66*(12), 2348–2362. https://doi.org/10.1080/17470218.2013.779730.

Dowker, A. (2005). *Individual differences in arithmetic: Implications for psychology, neuroscience and education*. Hove, East Sussex: Taylor & Francis.

Dowker, A., Bala, S., & Lloyd, D. (2008). Linguistic influences on mathematical development: How important is the transparency of the counting system? *Philosophical Psychology*, *21*(4), 523–538. https://doi.org/10.1080/09515080802285511.

Fattorini, E., Pinto, M., Rotondaro, F., & Doricchi, F. (2015). Perceiving numbers does not cause automatic shifts of spatial attention. *Cortex: A Journal Devoted to the Study of the Nervous System and Behavior*, *73*, 298–316. https://doi.org/10.1016/j.cortex.2015.09.007.

Fischer, M. H. (2012). A hierarchical view of grounded, embodied, and situated numerical cognition. *Cognitive Processing*, *13*(Suppl 1), S161–S164. https://doi.org/10.1007/s10339-012-0477-5.

Fischer, M. H., Castel, A. D., Dodd, M. D., & Pratt, J. (2003). Perceiving numbers causes spatial shifts of attention. *Nature Neuroscience*, *6*(6), 555–556. https://doi.org/10.1038/nn1066.

Fischer, M. H., Mills, R. A., & Shaki, S. (2010/4). How to cook a SNARC: Number placement in text rapidly changes spatial–numerical associations. *Brain and Cognition*, *72*(3), 333–336. https://doi.org/10.1016/j.bandc.2009.10.010.

Fischer, M. H., & Shaki, S. (2014). Spatial associations in numerical cognition—From single digits to arithmetic. *The Quarterly Journal of Experimental Psychology*, *67*(8), 1461–1483. https://doi.org/10.1080/17470218.2014.927515.

Fischer, M. H., Shaki, S., & Cruise, A. (2009). It takes just one word to quash a SNARC. *Experimental Psychology*, *56*(5), 361–366. https://doi.org/10.1027/1618-3169.56.5.361.

Frank, M. C., Everett, D. L., Fedorenko, E., & Gibson, E. (2008). Number as a cognitive technology: Evidence from Pirahã language and cognition. *Cognition*, *108*(3), 819–824. https://doi.org/10.1016/j.cognition.2008.04.007.

Fürst, A. J., & Hitch, G. J. (2000). Separate roles for executive and phonological components of working memory in mental arithmetic. *Memory & Cognition*, *28*(5), 774–782.

Geary, D. C., Bow-Thomas, C. C., Liu, F., & Siegler, R. S. (1996). Development of arithmetical competencies in Chinese and American children: Influence of age, language, and schooling. *Child Development*, *67*(5), 2022–2044. https://doi.org/10.2307/1131607.

Geary, D. C., Widaman, K. F., & Little, T. D. (1986). Cognitive addition and multiplication: Evidence for a single memory network. *Memory & Cognition*, *14*(6), 478–487.

Gelman, R., & Gallistel, C. R. (1986). *The child's understanding of number*. Cambridge, MA: Harvard University Press.

Gevers, W., Lammertyn, J., Notebaert, W., Verguts, T., & Fias, W. (2006). Automatic response activation of implicit spatial information: Evidence from the SNARC effect. *Acta Psychologica*, *122*(3), 221–233. https://doi.org/10.1016/j.actpsy.2005.11.004.

Göbel, S. M. (2015). Up or down? Reading direction influences vertical counting direction in the horizontal plane—a cross-cultural comparison. *Frontiers in Psychology*, *6*, 228. https://doi.org/10.3389/fpsyg.2015.00228.

Göbel, S. M., Fischer, M. H., & Shaki, S. (2014). Experience counts: Transmitting culture through space. In *Poster and Oral Presentation at the 32. European Workshop on Cognitive Neuropsychology, Bressanone*.

Göbel, S. M., Moeller, K., Pixner, S., Kaufmann, L., & Nuerk, H. -C. (2014). Language affects symbolic arithmetic in children: The case of number word inversion. *Journal of Experimental Child Psychology*, *119*, 17–25. https://doi.org/10.1016/j.jecp.2013.10.001.

Göbel, S. M., Shaki, S., & Fischer, M. H. (2011). The cultural number line: A review of cultural and linguistic influences on the development of number processing. *Journal of Cross-Cultural Psychology*, *42*(4), 543–565. https://doi.org/10.1177/0022022111406251.

Göbel, S. M., Watson, S. E., Lervåg, A., & Hulme, C. (2014). Children's arithmetic development: It is number knowledge, not the approximate number sense, that counts. *Psychological Science*, *25*(3), 789–798. https://doi.org/10.1177/0956797613516471.

Gordon, P. (2004). Numerical cognition without words: Evidence from Amazonia. *Science*, *306*(5695), 496–499. https://doi.org/10.1126/science.1094492.

Grade, S., Lefèvre, N., & Pesenti, M. (2013). Influence of gaze observation on random number generation. *Experimental Psychology*, *60*(2), 122–130. https://doi.org/10.1027/1618-3169/a000178.

Hartmann, M., Gashaj, V., Stahnke, A., & Mast, F. W. (2014). There is more than "more is up": Hand and foot responses reverse the vertical association of number magnitudes. *Journal of Experimental Psychology. Human Perception and Performance*, *40*(4), 1401. https://doi.org/10.1037/a0036686.

Heilman, K. M., & Van Den Abell, T. (1980). Right hemisphere dominance for attention: The mechanism underlying hemispheric asymmetries of inattention (neglect). *Neurology*, *30*(3), 327–330.

Hoffmann, D., Mussolin, C., Martin, R., & Schiltz, C. (2014). The impact of mathematical proficiency on the number-space association. *PloS ONE*. *9*(1), e85048. https://doi.org/10.1371/journal.pone.0085048.

Holmes, K. J., & Lourenco, S. F. (2012). Orienting numbers in mental space: Horizontal organization trumps vertical. *Quarterly Journal of Experimental Psychology*, *65*(6), 1044–1051. https://doi.org/10.1080/17470218.2012.685079.

Hubbard, E. M., Piazza, M., Pinel, P., & Dehaene, S. (2005). Interactions between number and space in parietal cortex. *Nature Reviews. Neuroscience*, *6*(6), 435–448. https://doi.org/10.1038/nrn1684.

Hung, Y. -H., Hung, D. L., Tzeng, O. J. L., & Wu, D. H. (2008). Flexible spatial mapping of different notations of numbers in Chinese readers. *Cognition*, *106*(3), 1441–1450. https://doi.org/10.1016/j.cognition.2007.04.017.

Hurford, J. R. (1987). *Language and number: The emergence of a cognitive system*. Oxford: Blackwell.

Imbo, I., Vanden Bulcke, C., De Brauwer, J., & Fias, W. (2014). Sixty-four or four-and-sixty? The influence of language and working memory on childrens number transcoding. *Frontiers in Psychology*, *5*, 313. https://doi.org/10.3389/fpsyg.2014.00313.

Ito, Y., & Hatta, T. (2004). Spatial structure of quantitative representation of numbers: Evidence from the SNARC effect. *Memory & Cognition*, *32*(4), 662–673.

Kamawar, D., LeFevre, J. -A., Bisanz, J., Fast, L., Skwarchuk, S. -L., Smith-Chant, B., et al. (2010). Knowledge of counting principles: How relevant is order irrelevance? *Journal of Experimental Child Psychology*, *105*(1-2), 138–145. https://doi.org/10.1016/j.jecp.2009.08.004.

Kinsbourne, M. (1970). The cerebral basis of lateral asymmetries in attention. *Acta Psychologica*, *33*, 193–201.

Knops, A., Thirion, B., Hubbard, E. M., Michel, V., & Dehaene, S. (2009). Recruitment of an area involved in eye movements during mental arithmetic. *Science*, *324*(5934), 1583–1585. https://doi.org/10.1126/science.1171599.

Lakoff, G., & Núñez, R. E. (2000). *Where mathematics comes from: How the embodied mind brings mathematics into being*. New York, NY: Basic Books.

Le Corre, M., Li, P., Huang, B. H., Jia, G., & Carey, S. (2016). Numerical morphology supports early number word learning: Evidence from a comparison of young Mandarin and English learners. *Cognitive Psychology*, *88*, 162–186. https://doi.org/10.1016/j.cogpsych.2016.06.003.

LeFevre, J. -A., Clarke, T., & Stringer, A. P. (2002). Influences of language and parental involvement on the development of counting skills: Comparisons of French- and English-speaking Canadian children. *Early Child Development and Care*, *172*(3), 283–300. https://doi.org/10.1080/03004430212127.

Loetscher, T., & Brugger, P. (2007). Exploring number space by random digit generation. *Experimental Brain Research*, *180*(4), 655–665. https://doi.org/10.1007/s00221-007-0889-0.

Loetscher, T., & Brugger, P. (2009). Random number generation in neglect patients reveals enhanced response stereotypy, but no neglect in number space. *Neuropsychologia*, *47*(1), 276–279. https://doi.org/10.1016/j.neuropsychologia.2008.08.005.

Loetscher, T., Schwarz, U., Schubiger, M., & Brugger, P. (2008). Head turns bias the brain's internal random generator. *Current Biology*, *18*(2), R60–R62. https://doi.org/10.1016/j.cub.2007.11.015.

Lonnemann, J., & Yan, S. (2015). Does number word inversion affect arithmetic processes in adults? *Trends in Neuroscience and Education*, *4*(1–2), 1–5. https://doi.org/10.1016/j.tine.2015.01.002.

Macchi Cassia, V., Picozzi, M., Girelli, L., & de Hevia, M. D. (2012). Increasing magnitude counts more: Asymmetrical processing of ordinality in 4-month-old infants. *Cognition*, *124*(2), 183–193. https://doi.org/10.1016/j.cognition.2012.05.004.

Miller, K. F., Smith, C. M., Zhu, J., & Zhang, H. (1995). Preschool origins of cross-national differences in mathematical competence: The role of number-naming systems. *Psychological Science*, *6*(1), 56–60. https://doi.org/10.1111/j.1467-9280.1995.tb00305.x.

Miura, I. T., Okamoto, Y., Kim, C. C., Steere, M., & Fayol, M. (1993). First graders' cognitive representation of number and understanding of place value: Cross-national comparisons: France, Japan, Korea, Sweden, and the United States. *Journal of Educational Psychology*, *85*(1), 24.

Moeller, K., Pixner, S., Zuber, J., Kaufmann, L., & Nuerk, H. -C. (2011). Early place-value understanding as a precursor for later arithmetic performance—A longitudinal study on numerical development. *Research in Developmental Disabilities*, *32*, 1837–1851. https://doi.org/10.1016/j.ridd.2011.03.012.

Moeller, K., Shaki, S., Göbel, S. M., & Nuerk, H. -C. (2015). Language influences number processing—a quadrilingual study. *Cognition*, *136*, 150–155.

Mundy, E., & Gilmore, C. K. (2009). Children's mapping between symbolic and nonsymbolic representations of number. *Journal of Experimental Child Psychology*, *103*(4), 490–502. https://doi.org/10.1016/j.jecp.2009.02.003.

Muter, V., Hulme, C., Snowling, M. J., & Stevenson, J. (2004). Phonemes, rimes, vocabulary, and grammatical skills as foundations of early reading development: Evidence from a longitudinal study. *Developmental Psychology*, *40*(5), 665–681. https://doi.org/10.1037/0012-1649.40.5.665.

Noël, M. -P., & Seron, X. (1997). On the existence of intermediate representations in numerical processing. *Journal of Experimental Psychology. Learning, Memory, and Cognition*, *23*(3), 697. https://doi.org/10.1037/0278-7393.23.3.697.

Nuerk, H. -C., Patro, K., Cress, U., Schild, U., Friedrich, C. K., & Göbel, S. M. (2015). How space-number associations may be created in preliterate children: Six distinct mechanisms. *Frontiers in Psychology*, *6*, 215. https://doi.org/10.3389/fpsyg.2015.00215.

Nuerk, H. -C., Weger, U., & Willmes, K. (2001). Decade breaks in the mental number line? Putting the tens and units back in different bins. *Cognition*, *82*(1), B25–B33.

Nuerk, H. -C., Weger, U., & Willmes, K. (2005). Language effects in magnitude comparison: Small, but not irrelevant. *Brain and Language*, *92*(3), 262–277. https://doi.org/10.1016/j.bandl.2004.06.107.

Nuerk, H. -C., Willmes, K., & Fischer, M. H. (2015). Multi-digit number processing. *Zeitschrift Für Psychologie*, https://doi.org/10.1027/2151-2604/a000040.

Pica, P., Lemer, C., Izard, V., & Dehaene, S. (2004). Exact and approximate arithmetic in an Amazonian indigene group. *Science, 306*(5695), 499–503. https://doi.org/10.1126/science.1102085.

Pixner, S., Moeller, K., Hermanova, V., Nuerk, H. C., & Kaufmann, L. (2011). Whorf reloaded: Language effects on nonverbal number processing in first grade - A trilingual study. *Journal of Experimental Child Psychology, 108*(2), 371–382. https://doi.org/10.1016/j.jecp.2010.09.002.

Rinaldi, L., Di Luca, S., Henik, A., & Girelli, L. (2014). Reading direction shifts visuospatial attention: An interactive account of attentional biases. *Acta Psychologica, 151*, 98–105. https://doi.org/10.1016/j.actpsy.2014.05.018.

Ristic, J., Wright, A., & Kingstone, A. (2006). The number line effect reflects top-down control. *Psychonomic Bulletin & Review, 13*(5), 862–868.

Rugani, R., Kelly, D. M., Szelest, I., Regolin, L., & Vallortigara, G. (2010). Is it only humans that count from left to right? *Biology Letters, 6*(3), 290–292. https://doi.org/10.1037/0097-7403.33.1.21.

Rugani, R., Vallortigara, G., Priftis, K., & Regolin, L. (2015). Animal cognition. Number-space mapping in the newborn chick resembles humans' mental number line. *Science, 347*(6221), 534–536. https://doi.org/10.1126/science.aaa1379.

Sarnecka, B. W., Kamenskaya, V. G., Yamana, Y., Ogura, T., & Yudovina, Y. B. (2007). From grammatical number to exact numbers: early meanings of "one", "two", and "three" in English, Russian, and Japanese. *Cognitive Psychology, 55*(2), 136–168. https://doi.org/10.1016/j.cogpsych.2006.09.001.

Schneider, M., Beeres, K., Coban, L., Merz, S., Susan Schmidt, S., Stricker, J., et al. (2016). Associations of non-symbolic and symbolic numerical magnitude processing with mathematical competence: A meta-analysis. *Developmental Science*. https://doi.org/10.1111/desc.12372/pdf.

Schneider, M., Grabner, R. H., & Paetsch, J. (2009). Mental number line, number line estimation, and mathematical achievement: Their interrelations in grades 5 and 6. *Journal of Educational Psychology, 101*(2), 359–372. https://doi.org/10.1037/a0013840.

Schwarz, W., & Keus, I. M. (2004). Moving the eyes along the mental number line: Comparing SNARC effects with saccadic and manual responses. *Perception & Psychophysics, 66*(4), 651–664.

Seron, X., & Fayol, M. (1994). Number transcoding in children: A functional analysis. *The British Journal of Developmental Psychology, 12*(3), 281–300. https://doi.org/10.1111/j.2044-835X.1994.tb00635.x.

Shaki, S., & Fischer, M. H. (2008). Reading space into numbers—a cross-linguistic comparison of the SNARC effect. *Cognition, 108*(2), 590–599.

Shaki, S., & Fischer, M. H. (2012). Multiple spatial mappings in numerical cognition. *Journal of Experimental Psychology. Human Perception and Performance, 38*(3), 804–809. https://doi.org/10.1037/a0027562.

Shaki, S., & Fischer, M. H. (2015). Newborn chicks need no number tricks. Commentary: Number-space mapping in the newborn chick resembles humans' mental number line. *Frontiers in Human Neuroscience, 9*, 451. https://doi.org/10.3389/fnhum.2015.00451.

Shaki, S., Fischer, M. H., & Göbel, S. M. (2012). Direction counts: A comparative study of spatially directional counting biases in cultures with different reading directions. *Journal of Experimental Child Psychology, 112*(2), 275–281. https://doi.org/10.1016/j.jecp.2011.12.005.

Shaki, S., Fischer, M. H., & Petrusic, W. M. (2009). Reading habits for both words and numbers contribute to the SNARC effect. *Psychonomic Bulletin & Review*, *16*(2), 328–331. https://doi.org/10.3758/PBR.16.2.328.

Skwarchuk, S. -L., Sowinski, C., & LeFevre, J. -A. (2014). Formal and informal home learning activities in relation to children's early numeracy and literacy skills: the development of a home numeracy model. *Journal of Experimental Child Psychology*, *121*, 63–84. https://doi.org/10.1016/j.jecp.2013.11.006.

Spelke, E. S., & Tsivkin, S. (2001). Language and number: A bilingual training study. *Cognition*, *78*(1), 45–88.

Towse, J. N., Loetscher, T., & Brugger, P. (2014). Not all numbers are equal: Preferences and biases among children and adults when generating random sequences. *Frontiers in Psychology*, *5*, 19. https://doi.org/10.3389/fpsyg.2014.00019.

Towse, J. N., & Neil, D. (1998). Analyzing human random generation behavior: A review of methods used and a computer program for describing performance. *Behavior Research Methods, Instruments, & Computers: A Journal of the Psychonomic Society, Inc*, *30*(4), 583–591. https://doi.org/10.3758/BF03209475.

Towse, J. N., & Saxton, M. (1997). Linguistic influences on children's number concepts: Methodological and theoretical considerations. *Journal of Experimental Child Psychology*, *66*(3), 362–375. https://doi.org/10.1006/jecp.1997.2389.

van Dijck, J. -P., Gevers, W., Lafosse, C., & Fias, W. (2011). The heterogeneous nature of number–space interactions. *Frontiers in Human Neuroscience*, *5*. https://doi.org/10.3389/fnhum.2011.00182.

Viarouge, A., Hubbard, E. M., & McCandliss, B. D. (2014). The cognitive mechanisms of the SNARC effect: An individual differences approach. *PLoS ONE*. *9*(4), e95756. https://doi.org/10.1371/journal.pone.0095756.

Wagenaar, W. A. (1972). Generation of random sequences by human subjects: A critical survey of literature. *Psychological Bulletin*, *77*(1), 65.

Wilkinson, G. S., & Robertson, G. J. (2006). WRAT 4: Wide range achievement test; Professional manual. Psychological Assessment Resources, Incorporated.

Wood, G., Willmes, K., Nuerk, H. -C., & Fischer, M. H. (2008). On the cognitive link between space and number: A meta-analysis of the SNARC effect. *Psychology Science Quarterly*, *50*(4), 489.

Zorzi, M., Priftis, K., & Umiltà, C. (2002). Brain damage: Neglect disrupts the mental number line. *Nature*, *417*(6885), 138–139. https://doi.org/10.1038/417138a.

Zuber, J., Pixner, S., Moeller, K., & Nuerk, H. -C. (2009). On the language specificity of basic number processing: Transcoding in a language with inversion and its relation to working memory capacity. *Journal of Experimental Child Psychology*, *102*(1), 60–77. https://doi.org/10.1016/j.jecp.2008.04.003.

Chapter 11

Language, Culture, and Space: Reconstructing Spatial-Numerical Associations

Samuel Shaki* and Martin H. Fischer†
**Ariel University, Ariel, Israel*
†*University of Potsdam, Potsdam, Germany*

Cognitive scientists aim to explain the prevalence of a series of performance signatures that characterize human numerical cognition, namely, the size effect (better performance with small compared with large numbers), the distance effect (better discrimination between numbers that are farther apart in numerical magnitude compared with numbers that are closer together in numerical magnitude), and the spatial-numerical association of response codes (SNARC) effect, a tendency to associate small numbers with left space and larger numbers with right space. We focus here on the SNARC effect and its origin in the context of language and culture. The SNARC effect was discovered by measuring the speed of lateralized button presses in French adults who classified single digits as odd or even (the parity judgment task; Dehaene, Bossini, & Giraux, 1993). They responded faster with their left-side button to smaller numbers, such as 1 or 2, and faster with their right-side button to larger numbers, such as 8 or 9.

The SNARC effect itself is well established and has been replicated with different number formats (symbolic and nonsymbolic, visual, and auditory), with modified task instructions (e.g., magnitude comparison and magnitude classification), and with different responding effectors (eyes, hands, and head; for review, see Fischer & Shaki, 2014). Even in the original report, SNARC was found to vary with the predominant reading direction of the adult participants, implying that habitual directional activity shapes the number-space mapping and suggesting a culturally mediated phenomenon. However, research with children has since determined that the onset of SNARC occurs early in development: as we will review below, children at ever younger ages exhibit a directional spatial-numerical bias when tested with age-appropriate

Mathematical Cognition and Learning, Vol. 4. https://doi.org/10.1016/B978-0-12-812574-8.00011-0

assessment methods (for review, see Patro, Nuerk, Cress, & Haman, 2014). The early emergence of this bias reduces the potential period of influence of directional language processing on the development of the SNARC signature and instead suggests that earlier culture-specific factors and spatial experiences and even evolutionary factors may shape the directionality of SNARC. Here, we systematize the various language-related and cultural influences on SNARC as follows: first, we consider two verbal components of language-related behavior: reading and listening. Next, we investigate nonverbal components of language-related behavior such as gestures and finger counting. Finally, we comment on the possible contribution of evolved predispositions on the emergence of SNARC.

VERBAL COMPONENTS OF LANGUAGE-RELATED SPATIAL-NUMERICAL ASSOCIATIONS

We all learn to enumerate spatially or temporally distributed objects by reciting a sequence of number words, thus assigning to each member of the set a unique name. The one-to-one correspondence between number words and objects through counting constitutes fundamental knowledge about quantitative aspects of our reality (Gallistel & Gelman, 1992). Through this assignment process, we learn that all numerosities can be spatially and temporally ordered and that "earlier" and "later" assignments reflect "smaller" and "larger" sets. Although the assignment of count labels to simultaneously available objects is, in principle, order invariant, there is a systematic preference to start counting on the left or right edges of the set, both in children and in adults. This behavior likely reflects a memory strategy to eliminate the need to keep track of which item has already been counted and which remains to be counted (Trick & Pylyshyn, 1994). More interestingly, the starting side for counting is culturally dependent, covarying with the predominant reading direction (Shaki, Fischer, & Göbel, 2012). Why do small numbers go with one side of space and larger numbers with the other side of space? What role, if any, do culturally acquired spatial-directional habits, such as reading and finger counting, play in the development of this performance signature? In this section, we report a series of studies that have investigated the role of reading experiences.

Do Space-Number Associations Require Directional Harmony?

In a comparison of three groups of adult readers from different cultures, Shaki, Fischer, and Petrusic (2009) documented that reading direction went hand in hand with SNARC direction. The first group was Canadians who read both text and Arabic digits from left to right, and the second group was Palestinians who read both text and (East Arabic) numerals ٠ ١ ٢ ٣ ٤ ٥ ٦ ٧ ٨ from right to left. The third group was Israelis who read Hebrew text from right to left but Arabic digits from left to right. All participants made speeded parity judgments to classify digits from 1 to 9 (except 5) as odd or even. Before we present the results, the reader should understand that in the standard analysis, a typical SNARC effect is depicted as a decreasing linear function that summarizes how the difference score "right-hand reaction time minus left-hand reaction time" changes across increasing number magnitude. As can be seen in Fig. 1, the results were clear: Canadians showed the typical SNARC pattern of faster responses to small numbers with the left-side key and to larger numbers with the right-side key. In contrast, Palestinians showed a reverse SNARC with faster responses to large numbers with the left-side key and to smaller numbers with the right-side key. Finally, and most interesting, the Israeli group exhibited no spatial-numerical association (SNA), either at the group level or the individual level. This was attributed to their conflicting directional habits for reading text (from right to left) and reading numbers (from left to right). These data led us to propose, at the time, that the long-standing reading habits might lead to the specific direction of the SNARC effect, thus indicating a powerful influence of cultural practice on number representation. In other words, we thought that "harmony" between directional habits for text and numbers was necessary for the SNARC effect to emerge.

This view was reinforced by a replication of our findings, reported by Rashidi-Ranjbar, Goudarzvand, Jahangiri, Brugger, and Loetscher (2014), who studied number representations in Farsi readers who also read words from right to left but numbers from left to right. Their participants generated random numbers while their heads turned alternatingly left or right, a method pioneered by Loetscher, Schwarz, Schubiger, and Brugger (2008) and known to modulate the availability of numbers in a production task. Specifically, turning one's head to the left leads to the generation of more small than large

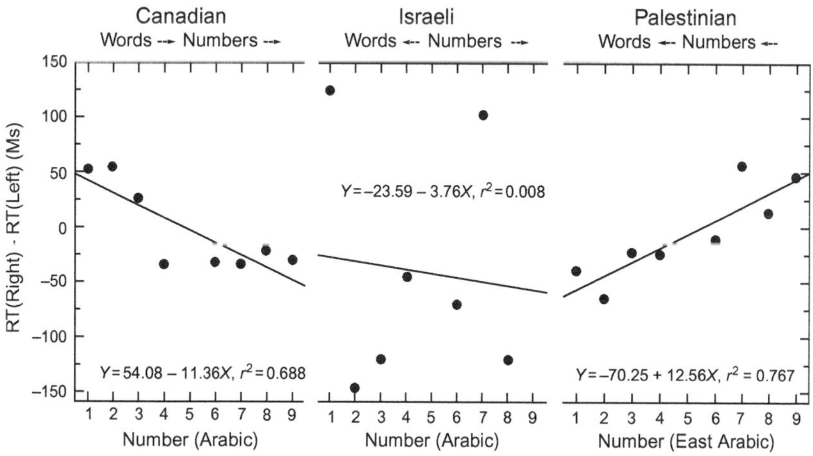

FIG. 1 Mean SNARC effect (RT(Right)–RT(Left) in Ms) as a function of digit presented in the parity task with Canadian, Israeli, and Palestinian participants. Direction of reading words and numbers for each group are indicated by *arrows. (Reprinted with permission from Shaki, S., Fischer, M. H., & Petrusic, W. M. (2009). Reading habits for both words and numbers contribute to the SNARC effect.* Psychonomic Bulletin and Review, 16*(2), 328–331.)*

numbers, and vice versa for right-side head turns. In the study by Rashidi-Ranjbar and colleagues, however, there was no such effect. This null result is consistent with the view that conflicting directional reading habits for words and numbers cancel each other and that no directional mapping will appear for numbers as a consequence of this hypothesized lack of directional harmony.

Another follow-up study with Hebrew readers (Shaki & Fischer, 2012) also reinforced the harmony account explanation for the lack of SNARC along the horizontal dimension in Israeli adults. In this study, a single group of Hebrew readers was tested in two different response mappings with the parity judgment task: in one condition, the response keys were placed left and right from the participant (horizontal condition); in the other condition, the keyboard was rotated on the tabletop, and the response keys were near and far from the participant. While there was no reliable SNARC effect in the horizontal condition, the same participants showed a reliable SNARC effect in the orthogonal condition (i.e., faster responses when smaller numbers were classified on a key in the near position). This observation suggested that the cancelation of SNARC only occurs when the responses are made within the same dimension that exhibits the conflict. In other words, placing the response keys orthogonal to the dimension of habitual scanning removed the conflict, established "harmony," and thereby allowed the SNARC effect to emerge.

So far, it seemed to us that all people, regardless of their culturally mediated directional habits, do associate numbers with space, even when they sometimes cannot map them along the horizontal dimension. However, a subsequent study (Fischer & Shaki, 2016) indicated that the horizontal mapping itself may be universal after all, even when its cognitive consequences are not measurable with the speeded classification approach. Cognitive representations can be assessed with the implicit association test (IAT; Greenwald, McGhee, & Schwartz, 1998). The IAT normally involves two interleaved speeded classification tasks: in one task, participants decide the category membership of sequentially presented stimuli (e.g., is the stimulus male vs female?); in the other, they classify stimulus attributes according to their valence (e.g., is the stimulus good vs bad?), always using two lateralized response keys. Variation of the response mappings for the two interleaved dimensions, such as mapping "good" and "female" on the same vs on opposite keys, leads to systematically slower responses for conflicting dimensions, thus revealing implicit attitudes (cf. Lane, Banaji, Nosek, & Greenwald, 2007). In the nonspatial version of this test (Nosek & Banaji, 2001), participants respond with a single nonlateralized button in only half of the trials, based on a conjunction rule for the two dimensions under investigation (go/no-go task). An example would be "respond only if the stimulus is either female or good."

When applied to the study of number representations, participants see random sequences of digits and left- or right-facing objects. One of several

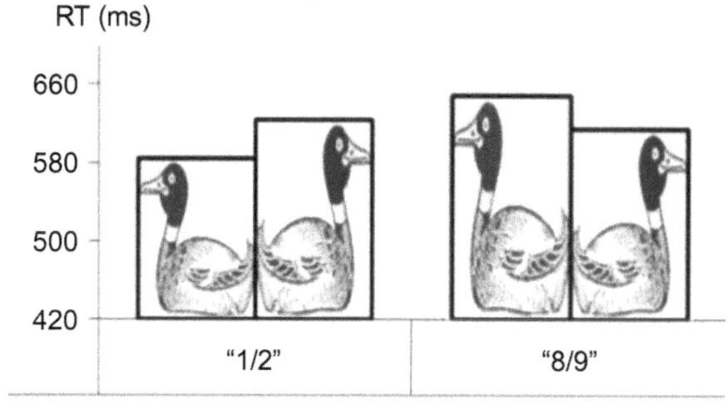

FIG. 2 Results of the nonspatial SNA assessment. *Inserts* illustrate the response-relevant direction of the objects (these were left- or right-facing ducks or cars). *(Reprinted with permission from Fischer, M. H., & Shaki, S. (2016). Measuring spatial-numerical associations: Evidence for a purely conceptual link. Psychological Research, 80, 109–112.)*

blockwise go instructions could be "Respond only when the stimulus is either smaller than 5 or oriented to the left side." As can be seen in Fig. 2, this new task has recently indicated a normal SNARC effect, thereby showing that numbers are spatially associated at the conceptual level alone, without mediated spatial activation in the number stimuli or the responses (Fischer & Shaki, 2016). Importantly, participants in the last study were Israelis who would not normally show a horizontal SNARC effect in classification tasks (but see below for debate regarding a link between their object reading habits and the direction of their SNARC effect).

At this point, we find it necessary to introduce a terminological clarification (see also Fischer & Brugger, 2011): since there is no longer a selection and preparation of response codes in the go/no-go version of the IAT task, we prefer to speak of "SNAs" more generally from now on, to accommodate the fact that such mappings can be studied with a variety of methods, many of which impose no response selection. We reserve the term "SNARC effect" for the more familiar speeded assessment of number parity or magnitude in a two-alternative forced choice task.

To summarize, our series of cross-cultural studies showed the importance of long-standing directional reading habits with both text and numbers for the spatial-numerical mapping. The associated development of theoretical positions also showed us how the measurement method strongly influences the obtained pattern of results.

Longstanding Cultural Habits or Recent Experience?

Taking into account that we found an effect of long-standing reading direction habit on SNAs, we can ask what the evidence is for a role of recent experience in shaping SNAs. A set of three studies into the relation between SNAs and the reading process itself suggested to us a considerable influence of recent experiences. In our first study, bilingual Russian-Hebrew readers performed the parity task before and after reading either a short Russian or a short

Hebrew paragraph (Shaki & Fischer, 2008). After reading the left-to-right Cyrillic Russian script, the typical SNARC effect persisted, but after reading the right-to-left Hebrew script, a significantly reduced SNARC effect was found (but no reversal). This finding suggests that long-standing and laboriously acquired reading habits do not necessarily exert a persistent influence on SNAs.

It is worth noting that this study included an auditory control condition. Specifically, a separate group of bilingual Russian-Hebrew adults was presented with auditory versions of the same text passages in the two languages before being assessed on their SNARC effect. Importantly, there was no change of SNARC from merely listening to the texts. This outcome suggests that the modification of SNARC from a brief reading episode in the first experiment was due to the visual presentation modality of texts with its associated directional scanning activity for the eyes. In a follow-up study, bilingual Russian-Hebrew readers continuously made parity judgments while alternatingly seeing Arabic digits and number words (Fischer, Shaki, & Cruise, 2009). Number words were shown randomly in either Cyrillic or Hebrew script, thus imposing different reading directions on our participants. This method allowed us to measure the SNARC effect both for the number words and for the Arabic digits that immediately followed them. Interestingly, the left-to right SNARC effect appeared within the Russian context (both for number words and for the subsequent Arabic digits) but vanished in the Hebrew context, from one second to the next and within the same participants. This striking observation indicates that after reading a single word, the directionality of SNAs can be changed in accordance with the most recent visual scanning direction. Again, the study contained an auditory control experiment to assess the effect of listening to number words on the SNARC effect of a fresh group of Russian-Hebrew bilingual adults. This time, we found strong regular SNARC effects regardless of the language in which the number words had been presented. However, the SNARC effect was significantly stronger in the Russian language, presumably because this was the mother tongue of our participants.

The third study in this context explored the role of the spatial positions of visually presented numbers within a single language and showed that SNARC changes even during reading. Fischer, Mills, and Shaki (2010) asked separate groups of English and Hebrew readers to read a set of 20 short cooking recipes, one per page, in their native language. Cooking recipes have the interesting feature that they naturally contain many numbers. Each language group was further subdivided according to the following manipulation of digit positions on the screen: For each recipe, the digits 1 through 4 were on the left side of the page and digits 6 through 9 on the right side for one group, whereas digits 1 through 4 were on the right side of the page and digits 6 through 9 on the left side for another group. This created SNARC-congruent and SNARC-incongruent reading conditions, respectively. For example, a sentence would state something like "Please use 8 eggs and take great care to boil them for exactly 2 min." The other version of this sentence would say "Please use 2 eggs and take great care to boil them for exactly 8 min." Of course, these recipes hardly resulted in edible meals, but they allowed us to manipulate the association between a number's magnitude and its place on the screen. The SNARC was measured before and after reading these recipes. As a result of random assignment, both English groups had similar typical SNARC effects before reading the recipes, and both Hebrew groups initially showed the typical absence of SNARC. Importantly, after reading these recipes for just 20 min (and answering two comprehension questions for each recipe), the SNARC-incongruent English group no longer mapped numbers onto space, at least as indicated by the absence of a reliable SNARC effect. In contrast, the SNARC effect was unchanged in the SNARC-congruent English group. For the Hebrew group, the posttest in the "incongruent" (small numbers on the right side) group showed a rare but reliable horizontal number-to-space mapping in Hebrew readers. We interpreted this nice result as indicating that a brief experience with small numbers in right-side locations was sufficient to remove a previously existing conflict between the reading directions for numbers and words, in line with our original "harmony" account.

To summarize this section, the results of the three just described studies converge in showing that the influence of culturally mediated reading habits on SNARC is not as persistent and long-lived as previously thought. Nevertheless, it must be acknowledged that reading direction does exert an influence on SNARC and other SNAs, but in a situation-dependent fashion and in conjunction with other influences. Moreover, it is clear that the Hebrew language context alone is not sufficient to eliminate (or even reverse) the SNARC effect compared with the typical Western pattern. In addition, the visual modality and its associated directional scanning activity is an important prerequisite for this cultural modulation of SNARC. From this language-based review, it is clear that it is worth looking at how number-related spatial biases are acquired. This leads us to look at developmental studies of SNAs with preliterate children in order to identify other contributing factors.

NON-VERBAL LANGUAGE-RELATED BEHAVIOR AND SPATIAL-NUMERICAL ASSOCIATIONS

Language communication includes not only oral and written channels but also gesturing and other bodily signals. The body has traditionally been used to represent numerosities through finger and body part counting (Bender & Beller, 2012, Ifrah, 1981; Saxe, 2012, Chapter 12 by Saxe and Chapter 13 by Bender and Beller), and this universal strategy is further shaped by cultural constraints. Additionally, we use our bodies to direct the attention of others across space. Here, we review some of this evidence in a numerical context in order to provide a further perspective on the origin and development of SNAs.

Reading Observation

How does the reading activity shape SNAs? Does it have to be active reading or is passively being read to enough? Moreover, is the observation of other spatially directional activities outside of reading equally effective in shaping our own SNAs? We can look at recent developmental studies for answers. At the behavioral level, reading consists not merely of the systematic movement of one's eyes across a page of text. It also involves, at least in beginning readers, a slow movement of the finger across the line of text to help one's eyes keep their place. This same behavior can, in turn, also be observed by future young readers as they are being entertained by their caregivers before being able to read by themselves; that is, people reading to them also often track the read text using their fingers. For example, storybook reading is a pervasive activity in around half of the homes, preschools, and daycares of literate societies (Montag, Jones, & Smith, 2015). Children also often flip through books without being able to read from them (Tolchinsky, 2003). This reading observation account is beginning to receive attention as a likely mechanism shaping SNAs in preliterate children.

Göbel, McCrink, Fischer, & Shaki (in revision; see also Chapter 10 by Göbel) investigated the role of reading-related enculturation for shaping the directionality of the preliterate association between number and space. They focused on counting behavior in 3- to 5-year-old preliterate children. After observing someone reading from storybooks (left-to-right or right-to-left storybook), children changed their counting direction in line with the observed direction of reading. This same result was not observed in a subsequent study with visuospatial directional movement of an object. Moreover, the authors reported that book illustrations commonly exhibit directionality that conforms to the direction of a culture's written language. They proposed that shared book reading activates spatiotemporal representations of order in young children, which in turn affect their spatial representation of numbers. Shared book reading with young children might be a pervasive and potentially powerful phenomenon through which spatial-numerical relations become culturally tuned.

Even outside of the reading situation, there is evidence for asymmetrical enculturation of infants. McCrink, Caldera, and Shaki (2016) recorded the gestural activity of caregivers trying to draw their toddler's attention to a screen on which objects or symbols were presented. American caregivers were more likely than Israeli caregivers to use left-to-right gestures, supporting the asymmetrical exposure of infants to directional cultural activities. Other evidence for culture-specific action schemas was recently reviewed by Maass, Suitner, and Deconchy (2014).

Together, these early influences might shape the direction of SNAs in infants. As a result, a general preference for left-to-right SNAs can already be found in preverbal infants (Bulf, Hevia, & Cassia, 2015; de Hevia,

Girelli, Addabbo, & Cassia, 2014). For example, Bulf et al. (2015) showed that centrally presented numerosities bias visual attention of 8- to 9-month-olds in a directional manner. This effect was specific to numerical cues such as clusters of dots and not present for small or large shapes. Using nonsymbolic stimuli, Patro and Haman (2012) found a left-to-right SNA in 3- to 4-year-old Polish children: the children were faster to respond to small nonsymbolic numerosities presented on the left side and larger nonsymbolic numerosities on the right side of space. Furthermore, 3-year-old American and British children already show a preference to count horizontally aligned objects from left to right (Briars & Siegler, 1984; Opfer, Thompson, & Furlong, 2010; Shaki et al., 2012).

Finger Counting

Fingers are almost universally involved in the initial acquisition of number concepts through culturally mediated finger counting (cf. Butterworth, 1999). Thus, in most cultures, children spontaneously discover and use their fingers to establish one-to-one correspondences between sets of objects and their numerosities (Chapter 13 by Bender and Beller). A surprising exception to this principle of embodied number processing may be certain indigenous Australian tribes who have not been observed using their fingers for counting (cf. Butterworth & Reeve, 2008; Butterworth, Reeve, Reynolds, & Lloyd, 2008).

The cultural variety of finger-counting behaviors is evident from a historical review (Ifrah, 1981) and from a recent Internet-based questionnaire (Lindemann, Alipour, & Fischer, 2011). Historically, there seems to have been a preference to start counting on the left hand in most medieval Western societies (Bede, 725; cited in Bloom, 2001), thus establishing an association of small numbers with left space that appears to be independent of handedness (cf. Fischer, 2008). This is still true in the contemporary English-speaking societies examined by Lindemann et al. (2011), including the United States, Canada, and the United Kingdom. This bias dissipates as one travels eastward across Europe and reverses into a right-start preference in Mideastern Iran. While it is unclear which factors contributed to the conspicuous left-starting preference in the English-speaking West, the reversal in the East can be attributed to the change in writing direction and other cultural rules for (or rather against) the use of one's left hand.

Do finger-counting activities shape the SNAs? We first review some evidence for a role of finger activities and finger-counting activities in numerical cognition and then turn to a discussion of possible links between finger counting and the directionality of SNAs.

Finger activities, as measured by finger gnosis scores (the ability to distinguish tactile stimulation patterns on the fingers), were found to be a strong longitudinal predictor for various mathematical tests in grade 1 children

(Fayol, Barrouillet, & Marinthe, 1998; Noël, 2005; Penner-Wilger et al., 2009). However, recent studies demonstrated that finger gnosis seems to predict only a small part of variance in mathematical competence (Long et al., 2016; Wasner, Nuerk, Martignon, Roesch, & Moeller, 2016).

How do finger-counting activities shape numerical cognition? One surprising result has been reported by Domahs, Moeller, Huber, Willmes, and Nuerk (2010) who showed that the size effect, an increase in decision latencies for larger compared with small numbers, depends on finger-counting habits. The authors presented pairs of digits in the range from 1 to 20 with a fixed numerical distance of 2 to either Chinese or German adults and recorded their decision speed in a magnitude comparison task with two response buttons. This is the typical assessment situation for SNARC, and there was indeed a SNARC-like effect in both groups: smaller numbers were classified faster with the left hand, and larger numbers were classified faster with the right hand. Moreover, in both groups, the decision times increased with number size (the typical size effect). Importantly, there was an additional group-specific cognitive penalty once a visually presented number in a pair exceeded the limit of what the participants in that group could habitually count on the fingers of one hand. Thus, the number pair 4–6 was classified slower than expected on the basis of number size alone, but only for Germans who switch to their second hand when counting above 5. Chinese participants, who can habitually count up to 9 on one hand, did not show this penalty.

One obvious thought is to predict a link between the starting hand for finger counting and the directionality of SNAs. Specifically, one might expect that left starters show a clear SNARC effect while right starters show a diluted or even reversed SNARC effect. This prediction was examined by Fischer (2008) who showed that the SNARC effect is modulated by finger-counting habits: in a reanalysis of speeded parity judgments from Scottish adults who habitually started counting on the fingers of the left vs right hand, he found a reliable SNARC effect for the group of 53 left starters but not for the group of 47 right starters, thus suggesting that the habitual association of small numbers (1–5) with the left hand and larger numbers (6–10) with the right hand mediates the SNARC effect and might even be the origin of this association in children. However, this result was based on a post hoc analysis of data from previous studies and is in conflict with more recent evidence for a reverse pattern (stronger SNARC effect in Italian right starters compared with left starters, Fabbri, 2013). A more recent set of studies by Wasner, Moeller, Fischer, and Nuerk (2014, 2015) clarified that this link is dependent on the method of assessment, with a larger proportion of left starters obtained when answering with written responses compared with pantomime responses (show me how you count).

An important issue under debate when discussing the role of finger counting for SNAs is the reference frame involved in the spatial mapping of numbers. Conflicting evidence has accumulated for the relative contributions of both hand space and external space in spatial-numerical mappings. Brozzoli et al. (2008) stimulated the right thumb or pinkie of their participants after showing them a small or large number cue. Detection speed for tactile stimuli was compared for prone and supine hand orientations, which varied across blocks but within participants. Following a small number, tactile stimuli to the leftmost finger were detected faster, regardless of hand orientation. Similarly, following a large number, tactile stimuli to the rightmost finger were detected faster, again regardless of hand orientation. This result provided evidence for a strong contribution of external and no hand-based spatial associations of numbers. However, Brozzoli et al.'s (2008) task minimized the role of hand-based representations by requiring merely a rough discrimination between the two outmost fingers of the hand, by requiring no response selection, and by recording foot responses, thus biasing the outcome against the involvement of any hand-based reference frames.

Riello and Rusconi (2011) directly examined the relative contributions of external and hand-based reference frames to number processing. They measured the speed of parity judgments in Italian adults who counted from thumb to pinkie on both hands when counting 1–5 or 6–10. By assigning the two response buttons to the index and middle fingers of either hand and assessing both hands separately in either prone or supine orientations, the authors created four experimental blocks: congruent relations between both reference frames for the left supinated or right pronated hand and incongruent relations between both reference frames for the left pronated or right supinated hand. Their results showed a reliable SNARC effect only in the congruent conditions, thus demonstrating that hand space and external space must be aligned to jointly induce the SNARC effect.

In summary, there are currently conflicting and therefore inconclusive observations available about the relationship between finger counting and SNAs: opposite SNARC patterns were found for left and right starters. In addition, the method of assessment (written or pantomime responses) and the reference frame (prone or supine orientations) were found to influence the complicated link between finger counting and SNAs. A recent study by Knudsen, Fischer, and Aschersleben (2015), however, shed some light on this controversial debate: they showed a clear preference for finger counting from right to left in 6-year-old German children. The same children, however, had a preference to count objects from left to right (such as the reading direction of their culture). This evidence speaks against an origin of SNAs in finger counting and indicates that the culturally predominant reading direction seems to be a better predictor of SNA direction than finger-counting direction.

Object Counting

As we reviewed above, there is a lack of clarity about the contribution of finger counting to the direction of SNAs. Now, we turn to another possible involvement of fingers (or hands) in counting, namely, object counting. Does the direction from which we start to count off objects shape the SNAs? The above report by Knudsen et al. (2015) indeed implies the possibility of such a link, as German children preferred to count objects from left to right, in accordance with the left-to-right directionality of SNAs in adults from the same culture (see also Briars & Siegler, 1984; Opfer et al., 2010; Van Galen & Reitsma, 2008).

Surprisingly, so far, only a single study systematically explored the cross-cultural directional habits of object counting: Shaki et al. (2012, Experiment 1) recorded directional counting preferences in adults with different reading experiences (left to right, right to left, mixed for different materials, and illiterate) and observed a clear relationship between reading and counting directions. Specifically, they asked their participants to count a series of four coins placed horizontally in front of them and found that, in adults reading both text and numbers from left to right, there was a 90% preference to count objects from left to right. In contrast, for adults reading text and numbers from right to left, this left-start preference was reduced to under 20%. Illiterates and mixed readers had no clear preference. These results were obtained despite the fact that participants normally started counting with their dominant hand.

Shaki et al. (2012, Experiment 2) reported a comparison of directional preferences for object counting in preschoolers and school children in order to evaluate the forming influence of formal schooling on SNAs. They found that preschoolers from a left-to-right reading culture at 4.4 years of age already exhibited a culture-consistent counting bias that was further enhanced through around 2 years of schooling (at age 8.8 years). The mirror pattern was observed in right-to-left reading children. Finally, the children with mixed reading habits demonstrated that the developmental trajectory of object counting depends on the harmony between the direction of reading text and reading numbers: preschool children show a clear preference to count from right to left, in accordance with reading text. However, with age, this right-to-left counting bias became weaker with acquiring conflicting reading habits for numbers.

It must be noted that we did not measure the direction of SNAs within these groups of children in the previous study. However, the direction of object counting in the adult groups is compatible with the direction of SNARC effects found previously for left-to-right, right-to-left, and mixed reading direction habits (Shaki et al., 2009). This one-to-one match between the direction of object counting and the direction of SNAs (at least on the cultural group level) speaks in clear favor of a link between these two directional biases or even in favor of a possible modulation of SNAs by object-counting experiences in early childhood.

Finally, in order to test if the direction of object counting modulates the direction of SNAs, we applied the go/no-go version of IAT (Fischer & Shaki, 2016; see above) while also documenting the object-counting direction of adult participants. We found a clear modulation of SNAs depends on the directional counting habits of our participants: left-to-right counters associated small numbers with leftward facing objects and large numbers with rightward facing objects. The reverse pattern was found for right-to-left counters, implying a clear link between object-counting direction and SNAs (Fischer & Shaki, in revision).

In summary, while there is extensive evidence for a link between adults' numerical cognition and hand use, we should distinguish between finger counting and object counting. Regarding finger-counting preferences, the evidence for a causal link between finger use and numerical abilities is hard to come by. This state of affairs may reflect the highly controlled access to number knowledge that characterizes the developing mind, while for adults, this access occurs in an automatic fashion that enables multiple sources of cognitive support to contaminate the behavioral evidence. One such source is the signature of embodied cognition during knowledge retrieval, that is, the permanent coactivation of those sensory and motor features that were present during the acquisition of the relevant knowledge. This makes finger counting a prime example for embodied numerical cognition (Fischer & Brugger, 2011). Regarding object counting, there is consistent evidence from both children and adults that points to a close link between directional preferences and cultural factors. Specifically, object counting and other SNAs seem closely related in a bidirectional manner, thus suggesting a developmental continuum that reflects our need for harmony in the way we express spatial-numerical relationships.

EVOLUTIONARY PREDISPOSITIONS TO SNARC

Having so far given considerable weight to cultural factors in our analysis of the origin of SNAs, we wish to end this chapter with a brief reflection on potential evolutionary influences. After all, numbers are symbols that were invented across many different cultures for the sole purpose of representing magnitudes and quantities. This does not, however, imply a negligible role for the impact of nature on numerical cognition. In fact, there is considerable evidence that magnitude and quantity processing is shaped by our evolutionary heritage (Geary, Berch, & Mann Koepke, 2015). Both the size effect (better performance with small compared with large quantities) and the distance effect (better discrimination between quantities that are less similar) are present in nonhuman animals. The situation is less clear with regard to the SNARC effect. Perhaps supporting an evolutionary account of its origin, the SNA is already measurable in humans from the age of 7 months (Bulf et al., 2015; de Hevia et al., 2014). However, even this short time interval leaves open the possibility of spatial experience as a contributing factor. Specifically, infants may have learned from observing directional preferences of adults within their cultural surroundings (Göbel et al., in revision; McCrink et al., 2017).

Drucker and Brannon (2014) provided initial evidence for a SNA in monkeys who preferentially located the fourth of five positions in a horizontal array on their right side, consistent with the cultural preferences of their caretakers. Similarly, Adachi (2014) reported that chimpanzees were faster to point to digits on a screen when they were horizontally congruent rather than incongruent with a mental number line (for methodological concerns and further discussions, see Rugani & de Hevia, 2016). Perhaps the strongest piece of evidence in support of an evolutionarily inherited SNA is the recent observation that newborn chicks tend to associate relatively small numerosities with left space and relatively larger numerosities with right space (Rugani, Vallortigara, Priftis, & Regolin, 2015). In this case, it is unlikely that cultural factors are responsible for the observed SNARC (but see Shaki & Fischer, 2015). Instead, the authors attributed their finding to hemispheric asymmetries whereby the right hemisphere is dominant in attending to visuospatial and numerical information that, in turn, leads to an implementation of number-space mappings in inferior parietal cortex.

This interpretation assumes both an inherited number sense and a universal predisposition to map such number knowledge onto space in a left-to-right orientation for increasing magnitudes. Arguing against this interpretation, Shaki and Fischer (2015) noted several nonnumerical spatial associations that can parsimoniously account for the choices of chicks without involving SNAs (such as area, perimeter, density of the displayed dots, or the complexity of the patterns). In turn, Rugani, Vallortigara, Priftis, and Regolin (2016) replied that the concerns raised were addressed by further control experiments and attributed the ubiquity of SNAs to the fact that they characterize the processing of all quantities. Hence, while the findings of Rugani et al. (2015) cannot be considered as ultimate evidence in favor of an evolutionary origin of SNAs, they do underline the importance of numbers as representatives for quantities.

SUMMARY AND CONCLUSIONS

In this chapter, we reviewed verbal and nonverbal components of language-related and culturally mediated SNAs. Indeed, several recent reports of SNAs in nonhuman animals and the development of creative and age-appropriate new tasks for measuring SNAs in humans from the age of 7 months seem to support an evolutionary account of its origin. However, the culturally consistent bias of SNAs with reading direction found already in preschool children indicates that these associations are a largely cultural phenomenon.

When considering the various components of SNAs discussed in this chapter, the following picture (see Fig. 3) emerges: building on possible evolutionary predispositions to associate quantities with space in a direction-specific manner, there is evidence for systematic nonverbal, gestural activity of caregivers that influences infants' SNAs through passive observation at first. This initial bias is later cultivated by active counting of objects and one's own fingers in a preferred direction, which in turn is shaped by cultural immersion. Later on, language-related verbal experiences from reading text and numbers

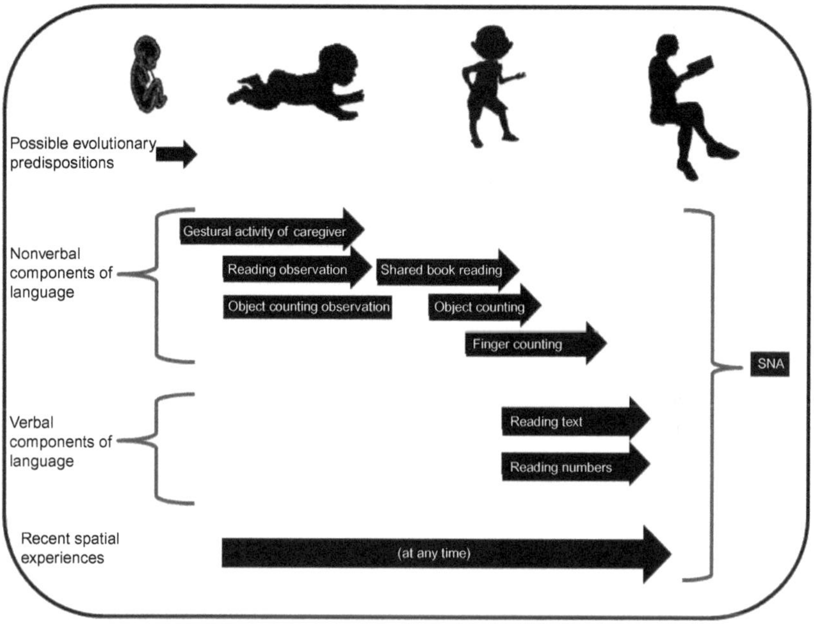

FIG. 3 The *top row* shows a schematic depiction of human development from fetus to adulthood. The *arrows* below represent various factors that may contribute to spatial-numerical associations, separately for verbal and nonverbal components of language, as well as for evolutionary predispositions and recent experiences.

come into play, and this rich repertoire of SNAs together determines the observed effect, possibly giving more weight to more recent experiences.

Surprisingly, there are currently inconclusive observations available about the relationship between SNAs and finger counting, a ubiquitous and spontaneous physical activity that is typically involved in the initial acquisition of number words and their associated concepts. Instead, a clear link seems to exist between object-counting direction and the direction of SNAs.

Finally, we provided solid evidence that reading direction habits for both numbers and text separately shape the directionality of SNAs: while consistent directional habits for reading text and numbers create internal harmony of spatial associations and can thus strengthen the resulting SNAs, conflicting reading directions for numbers and text can weaken (or at least make it difficult to measure) the effect. Nevertheless, long-standing verbal and nonverbal components of language-related directional behaviors are not alone in shaping these SNAs; recent directional sensory-motor experience can temporarily modulate the direction and strength of the SNAs, thereby making them a convenient tool to study the interface between sensory-motor experiences and cognitive representations (cf. Fischer, 2012). Future tasks in this endeavor include a determination of the relative contributions of the different ingredients of SNAs and their ontogenetic time course.

REFERENCES

Adachi, I. (2014). Spontaneous spatial mapping of learned sequence in chimpanzees: Evidence for a SNARC-like effect. *PLoS ONE*, *9*(3), e90373.

Bender, A., & Beller, S. (2012). Nature and culture of finger counting: Diversity and representational effects of an embodied cognitive tool. *Cognition*, *124*, 156–182.

Bloom, J. M. (2001). *The history and impact of paper in the Islamic world.* New Haven, CT: Yale University Press.

Briars, D., & Siegler, R. S. (1984). A featural analysis of preschoolers' counting knowledge. *Developmental Psychology*, *20*, 607–618.

Brozzoli, C., Ishihara, M., Göbel, S. M., Salemme, R., Rossetti, Y., & Farnè, A. (2008). Touch perception reveals the dominance of spatial over digital representation of numbers. *Proceedings of the National Academy of Sciences of the United States of America*, *105*(14), 5644–5648.

Bulf, H., Hevia, M. D., & Cassia, V. M. (2015). Small on the left, large on the right: Numbers orient visual attention onto space in preverbal infants. *Developmental Science*, *19*, 394–401.

Butterworth, B. (1999). *The mathematical brain.* London: Macmillan, pp. 217–250.

Butterworth, B., & Reeve, R. (2008). Verbal counting and spatial strategies in numerical tasks: Evidence from indigenous Australia. *Philosophical Psychology*, *21*(4), 443–457.

Butterworth, B., Reeve, R., Reynolds, F., & Lloyd, D. (2008). Numerical thought with and without words: Evidence from indigenous Australian children. *Proceedings of the National Academy of Sciences of the United States of America*, *105*(35), 13179–13184.

de Hevia, M. D., Girelli, L., Addabbo, M., & Cassia, V. M. (2014). Human infants' preference for left-to-right oriented increasing numerical sequences. *PLoS ONE*, *9*(5), e96412.

Dehaene, S., Bossini, S., & Giraux, P. (1993). The mental representation of parity and number magnitude. *Journal of Experimental Psychology: General*, *122*, 371–396.

Domahs, F., Moeller, K., Huber, S., Willmes, K., & Nuerk, H.-C. (2010). Embodied numerosity: Implicit hand-based representations influence symbolic number processing across cultures. *Cognition*, *116*, 251–266.

Drucker, C. B., & Brannon, E. M. (2014). Rhesus monkeys (Macaca mulatta) map number onto space. *Cognition*, *132*, 57–67.

Fabbri, M. (2013). Finger counting habits and spatial-numerical association in horizontal and vertical orientations. *Journal of Cognition and Culture*, *13*, 95–110.

Fayol, M., Barrouillet, P., & Marinthe, C. (1998). Predicting arithmetical achievement from neuro-psychological performance: A longitudinal study. *Cognition*, *68*, B63–B70.

Fischer, M. H. (2008). Finger counting habits modulate spatial-numerical associations. *Cortex*, *44*, 386–392.

Fischer, M. H. (2012). A hierarchical view of grounded, embodied and situated numerical cognition. *Cognitive Processing*, *13*, S161–S164. https://doi.org/10.1007/s10339-012-0477-5.

Fischer, M. H., & Brugger, P. (2011). When digits help digits: Spatial-numerical associations point to finger counting as prime example of embodied cognition. *Frontiers in Psychology*, *2*, 260. https://doi.org/10.3389/fpsyg.2011.00260.

Fischer, M. H., Mills, R. A., & Shaki, S. (2010). How to cook a SNARC: Number placement in text rapidly changes spatial-numerical associations. *Brain and Cognition*, *72*, 333–336.

Fischer, M. H., & Shaki, S. (2014). Spatial associations in numerical cognition: From single digits to arithmetic. *Quarterly Journal of Experimental Psychology*, *67*(8), 1461–1483.

Fischer, M. H., & Shaki, S. (2016). Measuring spatial-numerical associations: Evidence for a purely conceptual link. *Psychological Research*, *80*, 109–112.

Fischer, M. H., & Shaki, S. (2017). Implicit Spatial-Numerical Associations: Negative numbers and the role of counting direction. *Journal of Experimental Psychology: Human Perception and Performance*, *43*(4), 639–643.

Fischer, M. H., Shaki, S., & Cruise, A. (2009). It takes only one word to quash the SNARC. *Experimental Psychology*, *56*(5), 361–366.

Gallistel, C. R., & Gelman, R. (1992). Preverbal and verbal counting and computation. *Cognition*, *44*(1), 43–74.

Geary, D. C., Berch, D. B., & Mann Koepke, K. (Eds.), (2015). *Evolutionary origins and early development of number processing*. San Diego, CA: Elsevier/Academic Press.

Göbel, S. M., McCrink, K., Fischer, M. H., & Shaki, S. (in revision). Experience with Storybooks counts: Transmitting Culture through Space. *Journal of Experimental Child Psychology.*

Greenwald, A. G., McGhee, D. E., & Schwartz, J. L. K. (1998). Measuring individual differences in implicit cognition: The implicit association test. *Journal of Personality and Social Psychology*, *74*, 1464–1480.

Ifrah, G. (1981). *Histoire universelle des chiffres*. Paris: Seghers.

Knudsen, B., Fischer, M. H., & Aschersleben, G. (2015). Development of spatial preferences for counting and picture naming. *Psychological Research*, *79*(6), 939–949.

Lane, K. A., Banaji, M. R., Nosek, B. A., & Greenwald, A. G. (2007). Understanding and using the implicit association test: IV. Implicit measures of attitudes. In B. Wittenbrink & N. S. Schwarz (Eds.), *Implicit measures of attitudes: Procedures and controversies* (pp. 59–102). New York: Guilford Press.

Lindemann, O., Alipour, A., & Fischer, M. H. (2011). Finger counting habits in Middle-Eastern and Western individuals: An online survey. *Journal of Cross-Cultural Psychology*, *42*, 566–578.

Loetscher, T., Schwarz, U., Schubiger, M., & Brugger, P. (2008). Head turns bias the brain's internal random generator. *Current Biology*, *18*, R60–R62.

Long, I., Malone, S. A., Tolan, A., Burgoyne, K., Heron-Delaney, M., Witteveen, K., et al. (2016). The cognitive foundations of early arithmetic skills: It is counting and number judgment, but not finger gnosis, that count. *Journal of Experimental Child Psychology*, *152*, 327–334.

Maass, A., Suitner, C., & Deconchy, J. P. (2014). *Living in an asymmetrical world: How writing direction affects thought and action*. Hove: Psychology Press. https://doi.org/10.4324/9781315780054.

McCrink, K., Caldera, C., & Shaki, S. (2017). The early construction of spatial attention: Culture, space, and gesture in parent-child interactions. *Child Development*. https://doi.org/10.1111/cdev.12781.

Montag, J. L., Jones, M. N., & Smith, L. B. (2015). The words children hear picture books and the statistics for language learning. *Psychological Science*, *26*(9), 1489–1496.

Noël, M.-P. (2005). Finger gnosia: A predictor of numerical abilities in children? *Child Neuropsychology*, *11*, 413–430.

Nosek, B. A., & Banaji, M. R. (2001). The Go/No-go association task. *Social Cognition*, *19*(6), 625–664.

Opfer, J. E., Thompson, C. A., & Furlong, E. E. (2010). Early development of spatial numeric associations: Evidence from spatial and quantitative performance of preschoolers. *Developmental Science*, *13*, 761–771.

Patro, K., & Haman, M. (2012). The spatial–numerical congruity effect in preschoolers. *Journal of Experimental Child Psychology*, *111*(3), 534–542.

Patro, K., Nuerk, H.-C., Cress, U., & Haman, M. (2014). How number space relationships are assessed before formal schooling: A taxonomy proposal. *Frontiers in Psychology*, *5*, 419. https://doi.org/10.3389/fpsyg.2014.00419.

Penner-Wilger, M., Fast, L., LeFevre, J., Smith-Chant, B. L., Skwarchuk, S., Kamawar, D., et al. (2009). Subitizing, finger gnosis, and the representation of number. In *Proceedings of the 31st annual Cognitive Science Society* (pp. 520–525), Austin, TX: Cognitive Science Society.

Rashidi-Ranjbar, N., Goudarzvand, M., Jahangiri, S., Brugger, P., & Loetscher, T. (2014). No horizontal numerical mapping in a culture with mixed-reading habits. *Frontiers in Human Neuroscience*, *8*, 72. https://doi.org/10.3389/fnhum.2014.00072.

Riello, M., & Rusconi, E. (2011). Unimanual SNARC Effect: Hand matters. *Frontiers in Psychology*, *2*, 372. https://doi.org/10.3389/fpsyg.2011.00372.

Rugani, R., & de Hevia, M. (2016). Number-space associations without language: Evidence from preverbal human infants and non-human animal species. *Psychonomic Bulletin and Review*. https://doi.org/10.3758/s13423-016-1126-2.

Rugani, R., Vallortigara, G., Priftis, K., & Regolin, L. (2015). Number-space mapping in the newborn chick resembles humans' mental number line. *Science*, *347*, 534–536.

Rugani, R., Vallortigara, G., Priftis, K., & Regolin, L. (2016). Response: Newborn chicks need no number tricks. Commentary: Number-space mapping in the newborn chick resembles humans' mental number line. *Frontiers in Human Neuroscience*, *10*, 31. https://doi.org/10.3389/fnhum.2016.00031.

Saxe, G. B. (2012). *Cultural development of mathematical ideas: Papua New Guinea studies*. Cambridge: University Press.

Shaki, S., & Fischer, M. H. (2008). Reading space into numbers—A cross-linguistic comparison of the SNARC effect. *Cognition*, *108*(2), 590–599.

Shaki, S., & Fischer, M. H. (2012). Multiple spatial mappings in numerical cognition. *Journal of Experimental Psychology: Human Perception and Performance*, *38*(3), 804–809.

Shaki, S., & Fischer, M. H. (2015). Newborn chicks need no number tricks. Commentary: Number-space mapping in the newborn chick resembles humans' mental number line. *Frontiers in Human Neuroscience*, *9*, 451. https://doi.org/10.3389/fnhum.2015.00451.

Shaki, S., Fischer, M. H., & Göbel, S. M. (2012). Direction counts: A comparative study of spatially directional counting biases in cultures with different reading directions. *Journal of Experimental Child Psychology*, *112*(2), 275–281.

Shaki, S., Fischer, M. H., & Petrusic, W. M. (2009). Reading habits for both words and numbers contribute to the SNARC effect. *Psychonomic Bulletin and Review*, *16*(2), 328–331.

Tolchinsky, L. (2003). *The cradle of culture and what children know about writing and numbers before being*. Mahwah, New Jersey: Lawrence Erlbaum Associates.

Trick, L. M., & Pylyshyn, Z. W. (1994). Why are small and large numbers enumerated differently? A limited-capacity preattentive stage in vision. *Psychological Review*, *101*, 80–102.

Van Galen, M. S., & Reitsma, P. (2008). Developing access to number magnitude: A study of the SNARC effect in 7- to 9-year-olds. *Journal of Experimental Child Psychology*, *101*, 99–113.

Wasner, M., Moeller, K., Fischer, M. H., & Nuerk, H. C. (2014). Aspects of situated cognition in embodied numerosity: The case of finger counting habits. *Cognitive Processing*, *15*, 317–328.

Wasner, M., Moeller, K., Fischer, M. H., & Nuerk, H. C. (2015). Ordinality, cardinality and 1-to-1 correspondence in finger-based numerical representations. *Journal of Cognitive Psychology*, *27*(4), 426–441.

Wasner, M., Nuerk, H. C., Martignon, L., Roesch, S., & Moeller, K. (2016). Finger gnosis predicts a unique but small part of variance in initial arithmetic performance. *Journal of Experimental Child Psychology*, *146*, 1–16.

Chapter 12

Cultural Processes in Elementary Mathematics: Studies in a Remote Papua New Guinea Community

Geoffrey B. Saxe
University of California, Berkeley, Berkeley, CA, United States

Most scholars would agree that culture (and language) figures prominently into children's development of numerical ideas. In this chapter, I sketch two methodological approaches for studying relations between culture, language, and numerical understandings (see Saxe & de Kirby, 2014). One, referred to as a "dichotomous approach," is mainstream and commonly used in the psychological sciences to investigate ways that culture and language may influence people's cognition. The other, an "intrinsic relations" approach, is less frequently used. It focuses on people's participation in daily activities and the complex relations between cognitive and cultural processes that emerge through their participation. In this chapter, I illustrate my use of the intrinsic relations approach, with a sketch of four field visits over a 36-year period with the Oksapmin of Papua New Guinea. My aim is to illustrate how the approach can be used to provide insights into culture-cognition relations not possible with dichotomous approaches. In the chapter's end, I point to a rapprochement: how some empirical techniques associated with the dichotomous approach can be productively recruited to advance the intrinsic relations' focus on the reproduction and alteration of numerical ideas in communities.

DICHOTOMOUS APPROACHES TO ANALYZING RELATIONS BETWEEN CULTURE AND THE DEVELOPMENT OF NUMERICAL UNDERSTANDINGS

With a "dichotomous" approach, researchers use empirical techniques that sharply differentiate cognition from culture, treating elements of cognition

Mathematical Cognition and Learning, Vol. 4. https://doi.org/10.1016/B978-0-12-812574-8.00012-2

as "in the heads" of individuals, and culture as the "surround," distinct from their cognitive processes.[1] In empirical research, the bifurcation supports the operationalization of cognition as one or more dependent variables (such as cognitive structures and numerical skills) and culture as one or more independent factors that produce effects in the cognitions of individuals (such as language group and parenting style). With such operational definitions of culture and cognition, researchers' analytic task is to identify predictors of individuals' cognitions, the dependent variables, as a function of cultural elements, the independent variables.

The dichotomous approach to studying culture-cognition relations is prominent in both cross-cultural and within-cultural studies of culture-cognition relations. In cross-cultural studies, researchers compare whether selected cultural and linguistic features of the surround differ by cultural group and then examine whether the presence or absence of these cultural features are associated with the cognitive development of individuals. The methodological approach is depicted graphically in Fig. 1(i) in which features of culture (e.g., type of numeration system used and kind of child rearing style) are represented as points on the circle's circumference and individuals are represented as small circles within the cultural surround. The approach leads to the evaluation of claims that particular features (or groups of features) of culture affect a cognitive-developmental process.

In within-cultural studies, as illustrated in Fig. 1(ii), researchers measure the extent of exposure of individuals to an element of the surround like the extent of children's interactions with caregivers about number (as represented by boldness/lightness of arrows emanating from a single point (*a*) on the

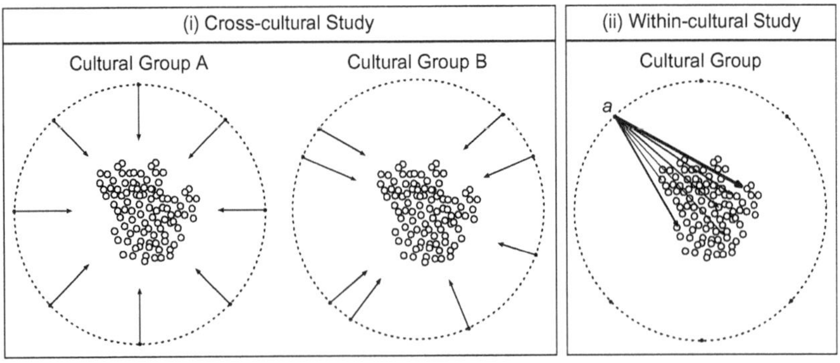

FIG. 1 (i) Cross-cultural comparison of two groups (Cultural Groups A and B) with elements of cultural surround represented as bolded points on circumference; (ii) within-cultural study of a group with relative intensity of cultural element emanating from a point to a subpopulation, represented by thickness of arrow.

1. Similar distinctions have been presented by other writings about relations between cultural and developmental processes (e.g., Cole, 1996; Rogoff, 2003; Vygotsky, 1986).

circle's circumference); they then correlate the extent of exposure with measures of number development for individuals, like children's counting fluency. Thus, whether in cross-cultural or within-cultural studies, investigators examine to what extent cognition can be explained by cultural/linguistic variables that are constitutive of the surround.

INTRINSIC RELATIONS APPROACHES TO ANALYZING RELATIONS BETWEEN CULTURE AND THE DEVELOPMENT OF NUMERICAL UNDERSTANDINGS

In an "intrinsic relations" approach, researchers take the position that culture and cognition are processes, coconstituted as people participate in daily activities. From an intrinsic relations perspective, to treat culture and cognition as distinct and to relate them as independent and dependent variables remove from analysis the dynamics of how they are coproduced and complexly interwoven in activities of individuals. It is this focus on culture and cognition as processes in motion and the dynamic interplay between them that is prominent in my research in Oksapmin communities.

The particular intrinsic relations approach I take has both cultural and developmental aspects. An orienting assumption is that, in their activities, individuals often unwittingly reproduce and alter *culture*, including norms, conventions, values, and language, as they engage with problems that emerge in daily life; at the same time, in those activities, individuals elaborate their own cognitive processes as they make sense of and develop approaches to addressing emergent problems. Below, I sketch cultural and cognitive-developmental aspects of the framework (see Saxe, 2014, for an extended exposition).

Cultural Dimension of the Framework: A Focus on Collective Practices

The cultural dimension of the framework elevates collective practices as a principal arena for analysis. By collective practices, I mean recurring socially patterned activities in which norms, social positions, values, and kinds of social interactions are reproduced and altered through the actions of participating individuals. In this regard, collective practices are microcosms of cultural processes. Some examples of collective practices that I have studied include children selling candy on the streets of an urban center in Brazil (Saxe, 1991), parents' interactive engagement with their children with number games like Candyland (Saxe, Guberman, & Gearhart, 1987), students engaged with mathematics lessons in upper elementary classrooms (Saxe, de Kirby, Kang, Le, & Schneider, 2015), and peer play of an educational game (Saxe & Guberman, 1998).

In the Oksapmin case, consider a collective practice situated in trade stores involving customer-store clerk transactions. In such transactions, Oksapmin participants inevitably (and often unwittingly) divide the cognitive work in relation to social positions (like "customer" and "clerk"), and their goals emerge and shift in relation to these roles and values (currency and commodities), conventions (e.g., the value of particular currency tokens), and other representational forms, like the Oksapmin 27-body-part counting system. To illustrate, consider the exchange that occurred in an Oksapmin trade store *c.*1978 captured in Fig. 2.

Through participation in trade store exchanges, people do more than accomplish what they want with respect to obtaining goods or completing transactions. They unwittingly reproduce and alter the institution of the trade store in the community, including social positions, conventions, values, and patterns of interaction. To illustrate, in the process of an exchange captured in Fig. 2, one can see an Oksapmin clerk standing behind a counter facing customers, a customary position that he reproduces in the transaction, one that positions himself with the authority often attributed to him by customers. Moments prior to the scene captured in the photos, the customers had requested cans of fish from the clerk, the clerk had retrieved them, and the customers presented payment. In the sequence depicted in Fig. 2A–I, one customer retrieves payment from a small can that he keeps in his string bag and requests additional cans (Fig. 2A–D). In Fig. 2E the clerk returns with additional cans, and in Fig. 2F–I, payment is made and change is returned.

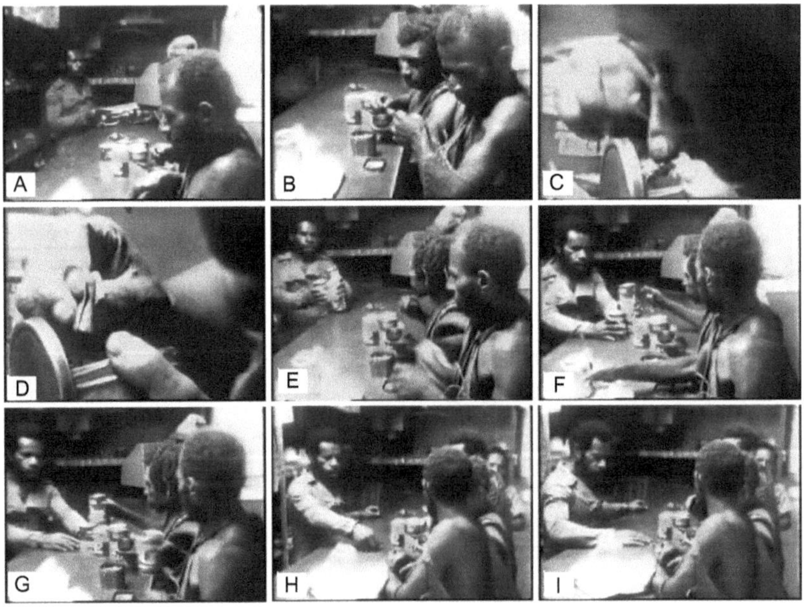

FIG. 2 Two men making a purchase of tins of fish in the mission trade store from an Oksapmin trade store clerk, 1978. Images A-D: Customer retrieves cash for payment and requests additional cans for purchase. Image E: Clerk returns with cans. Images F-I: Payment is made and change is returned.

Like any collective practice, the structure of the activity portrayed in the images of Fig. 2 has no permanence, other than its reproduction in the activity of the participants as they work to complete an exchange of goods for cash. Similarly, the artifacts that are constitutive of emergent numerical problems in exchanges, the cash and commodities, have no inherent numerical or exchange value. Artifacts take on value as they are used and interpreted in the recurrent collective problems that constitute the exchange. Noteworthy is that the recreated structure and meanings attributed to artifacts by participants are more or less predictable, often rooted in precedent, rendering them interpretable by one another. Further, alterations in the practice's structure (for instance, to accommodate a customer's request to pay later) or the properties of the collective problems (for instance, the government's issue of a new currency token) may lead to alterations and a back-and-forth between participants that renders new ways of coordinating actions and the emergence of new precedents that support communication in subsequent interactions.

In trade store exchanges in 1978 like the one depicted in Fig. 2, I observed different kinds of recurrent numerical problems in exchanges. For example, recurrent problems were created as a customer anticipated (or did not anticipate) the cost of an item and evaluated the amount of currency in his possession or when clerk and customer negotiated a cash-for-commodities transaction. As people structured and accomplished collective problems, they generated local goals in their activities, drawing upon cultural forms like the body system, currency tokens, or names for tokens to serve various cognitive functions. These functions included using tokens to pay for goods or using the body system to quantify tokens.

To sum up, in collective practices, individual and collective activities are reciprocally related: (a) Collective activity affords social purpose and meaning to the activities of individuals, and (b) individual activity is constitutive of the collective. It is the interplay between collective and individual activity that I take to be an arena for the construction and perpetual reconstruction of a local "culture"—a local culture in which social positions, values, norms, and even mathematics are constituted and reconstituted. The focus on collective practices brings into focus individuals' engagement with recurring problems reproduced and altered in collective life.

Developmental Aspects of the Framework: A Focus on the Interplay Between Representational Forms and the Functions They Serve in Activity

My focus is also developmental in an important sense. In collective practices like economic exchange in a trade store, individuals make use of cultural forms to serve cognitive functions, and relations between forms and functions shift over developmental time. Consider, for example, the Oksapmin 27-body-part count system depicted in Fig. 3, a cultural form that people

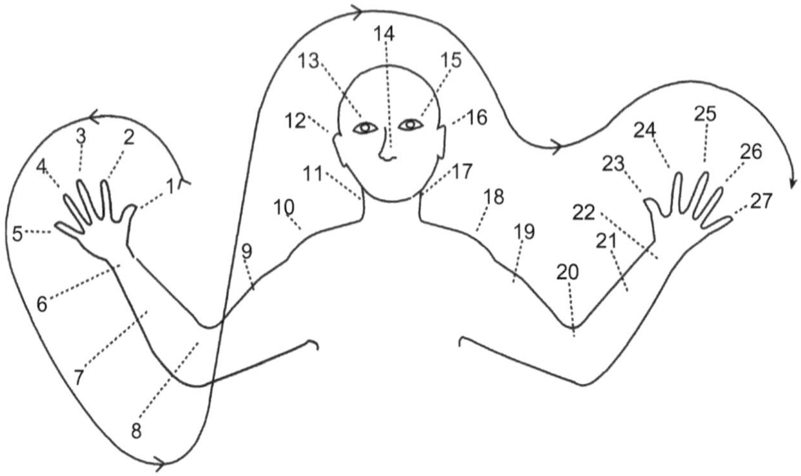

FIG. 3 Oksapmin 27-body-part counting system (for video display, see http://culturecognition.com/video/oksapmin-27-body-part-counting-system).

traditionally used to serve cognitive functions of representing cardinal and ordinal numbers. To count as Oksapmin do, one begins from the little finger of one hand and counts up the periphery of the body and down the other side. Conventional positions on the body are used to produce numerical representations. Thus, the position of the nose is the 14th body part in the conventional sequence, and the position and word form represents the equivalent of "fourteen."

At the crux of the developmental aspect of the framework is a focus on three strands of form-function relations: (1) A microgenetic strand is constituted in momentary activity through which individuals transform cultural forms into means to accomplish emerging goals as they conceptualize and structure solutions to problems, (2) an *ontogenetic* strand is constituted through individuals' elaboration and repurposing of forms as they structure new functions for them over their own developments, and (3) a sociogenetic strand is constituted by the massively distributed reproductions and alterations of form-function relations in the social history of communities as individuals interact and communicate with one another.

I argue that each of these genetic[2] strands have roots in the same activities. To illustrate, consider again an interaction in a trade store. To quantify a group of 10 toea coins in order to purchase a good, a trade store owner may produce a count of the coins with the body system, using the system to serve the function of cardinal representation in public display to the customer. The

2. I use "genetic" in the sense of "genesis"—developmental origins and transformation through time (Werner, 1948; Werner & Kaplan, 1963); such usage contrasts with the use of "genetic" to refer to genes or heredity.

process of quantification is a microgenetic activity. The owner creates a correspondence between a conventional sequence of body parts and the target of quantification, the coins. In this act, the owner sums the coins, a summation that is not in the coins or in the body-part form, but produced through activity. The production of a representation is a very short-term process, one in which a representational vehicle (e.g., the body parts) and an object of representation (e.g., the coins) are schematized in relation to one another in the development of a quantitative representation.

The very same microgenetic act is also a moment in the process of ontogenesis. The trade store owner, when a young child, presumably did not engage in actions of quantification and did not understand the role of correspondences in quantification. Indeed, the trade store owner's act of quantification is a point in his own ontogenetic history, one in which he came to use body-part names and positions to serve quantitative functions.

Finally, in the act of using the counting system during the transaction, the trade store owner is also engaged in a sociogenetic strand of development. Indeed, the body system, with its many conventions and norms of use, could not have always existed in its current form. Rather, the body system itself must have undergone a genesis as precursory forms were used and altered in people's communication with one another. Further, its structure does not exist apart from human activity—it exists only by virtue of its continued public use in communications and problem solving by individuals. Thus, the trade store owner, in his use of body parts in communications, is both reproducing the system and sustaining its conventions and structure, as well as adjusting the system to his particular task at hand, thus introducing alterations. Through his use and the massively distributed use by others, the body part form continues as a recognizable representation in the Oksapmin world; of course, alterations that individuals produce, whether the trade store owner and/or the many other Oksapmin individuals, may be taken up by others, leading to shifts in the structure of the form itself. I will have occasion to point to such alterations that have been taken up in the community over the course of the chapter.

Fig. 4 depicts the interplay between form and function in relation to the three kinds of genesis. The schematic focuses on three individuals (I_1, I_2, and I_3) in past, present, and prospective activities in a collective practice, say economic exchange. At the center is an individual engaged in a microgenetic act, using the body system (I_{2b}), an act that occurs in the back-and-forth with the constructions of others (I_{1b} and I_{3b}). In this process, the actors are reproducing forms and introducing alterations. Thus, the very same microgenetic acts are moments in sociogenetic processes involving the reproduction and alteration of forms of representation in the community. Moreover, the figure also reflects that these acts are moments in each individual's own ontogenetic trajectories (I_{2a}, I_{2b}, and I_{2c}) as he or she draws upon sociogenetic developments in relation to his or her own prior understandings, thereby creating adaptations of form-function relations to local circumstances.

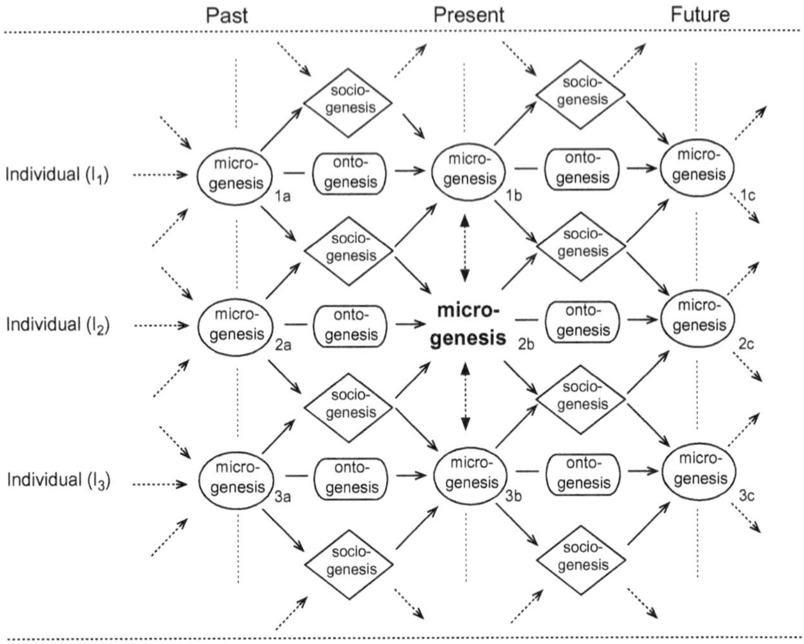

FIG. 4 The interplay between genetic processes in development.

THE INTERPLAY BETWEEN CULTURAL AND DEVELOPMENTAL PROCESSES IN MATHEMATICAL UNDERSTANDINGS: A PROGRAM OF RESEARCH IN REMOTE OKSAPMIN COMMUNITIES

In the pages to follow, I illustrate my cultural-developmental approach through studies in the Oksapmin community. My overarching objective is to show how, through their participation in collective practices of daily life, people are at once sustaining and altering those practices as they reproduce and alter cultural forms and representation and the functions they serve in processes of micro-, socio-, and ontogenesis. I begin by sketching some relevant Oksapmin background with a focus on the body system. I then describe some of my work in Oksapmin communities that occurred over the course of four field visits that span a 36-year stretch of recent history: 1978, 1980, 2001, and 2014.

Oksapmin Communities

The Oksapmin people reside in a mountain region of Papua New Guinea, east of the border with Irian Jaya. They are one of a number of cultural groups referred to as the Mountain Ok or the Min (Craig & Hyndman, 1990;

Gammage, 1998; Weeks, 1981). Though the groups speak different languages, they share a common material culture (forms of dress, tools, kinds of dwellings, subsistence practices, etc.). Also common to the different groups is the use of a 27-body-part counting system (see Laycock, 1975; Saxe, 2014).

Western contact with the Oksapmin occurred in Jun. 1938 by the 1938–39 Australian Hagen-Sepik patrol. The patrol, initially consisting of over 300 men, was commissioned to search of natural resources and to explore sites for stations in what was expected to be largely uninhabited terrain. In fact, the patrol encountered many indigenous groups in their 19-month trek. About 25 years later, in the early 1960s, Australian missionaries and patrol officers arrived, building a Baptist mission station and patrol post. Additional contact came in the mid-1960s when plantations recruited Oksapmin men for work outside of the area, flying men out for 2-year stints. In the late 1960s, the first government/mission community school was established. In traditional life, Oksapmin people did not engage in arithmetic computations with the body system. I now turn to my work in Oksapmin communities that documents the emergence of new arithmetic functions of the body system and shifts in the properties of the representational form.

Studies Conducted in 1978 and 1980. Form-Function Shifts in Children's Developing Use and Knowledge of the Body System

In 1978, I entered the Oksapmin community with two broad questions about number development. The first was whether I would find shifting relations between cultural forms for number representation and functions that those forms served that paralleled form function relations shifts I had documented in the United States. In the United States, I found that children begin to use number words but not to use them instrumentally to compare arrays or reproduce arrays (Saxe, 1977, 1979). For example, when presented with numerical arrays to compare, I found that children often counted one array only or counted both arrays continuously as if they were one. When asked to reproduce an array with an available set, children often used count words but did not use those words to mediate their reproduction, again, only counting one array. Only subsequently did children begin to use their repertoire of number words to produce numerical comparisons and reproductions. Such findings are reminiscent of Vygotsky's analysis of shifting interfunctional relations between thought and speech (Vygotsky, 1978, 1986).

The second question was about schooling. Western-style colonial schooling had only recently entered the Oksapmin world during my first visits in 1978 and 1980, and many Oksapmin children did not attend school. Further, teachers were not from Oksapmin, they were not aware of the 27-body-part counting system, and by law, instruction was to be conducted in English, and children were not supposed to use the vernacular on school grounds.

In 1980, after initial studies on Oksapmin children's use of counting in problem solving, I wondered whether children might be using their counting system to make sense of instruction and in this process use the system to serve new functions as they encountered new kinds of problems that were central to the collective practices of school life.

Studies of Unschooled Children's Use of the Body System: A Focus on Ontogeny of Numerical Representations With the Body System

To examine developmental changes in Oksapmin children's use of the body system, I administered two kinds of tasks. In the first, one that followed from analyses that I had produced in the United States, I presented children with linear arrays of objects to compare and reproduce; I studied whether and how they used the body to mediate their numerical problem solving. In the second, I developed new tasks but now focused on particular challenges that the body system might pose to children in conceptualizing numerical comparisons, like confusions between symmetrical body parts on the left vs right side or counting to a specified body part beginning from the left vs right side. Each type of study revealed shifts in the function of the body system with age. I describe an exemplar of each below.

Comparing and Reproducing Groups of Objects Numerically

I interviewed 28 participants in two age cohorts, 5- to 9-year olds and 11- to 16-year olds, using comparison and reproduction tasks. For the comparison tasks, I presented participants with small sets (three vs four elements), midsize sets (six vs seven elements), and large sets (eight vs nine elements). For the reproduction tasks, again, participants were presented with small sets (three elements), midsized sets (six elements), and large sets (nine elements). For each task type, I varied set size. For the comparison tasks, arrays were visible and aligned so that the linear array of the lesser number extended further than that of the larger number. In the reproduction task, the model set was presented in back of the participants while they watched. They were asked to reproduce the objects with an available set in front of themselves.

On both the comparison and the reproduction tasks, I found that the younger children typically knew body parts in more or less the correct sequence, but when asked to use counting to compare or reproduce arrays numerically, they often did not use body parts to produce numerical comparisons or reproductions, especially with the larger set sizes. In comparison tasks, many in the younger age group counted the two arrays of stones continuously as if they were one elongated set; others counted only one array. I found similar performances on reproduction tasks for the younger group. For the older groups, participants did not have difficulty using the body-part system as an intermediary, to effect numerical comparisons or reproductions, regardless of set size.

Comparing the Numerical Value of Body Parts Using Tasks Developed in Oksapmin

In a second set of tasks, I focused more specifically on the challenges that the body-part system might present to Oksapmin children. For example, if children were treating body-part positions as body parts (e.g., an anatomical part of the body) and not as numbers (e.g., a summation of positions, with each position counted once), then I expected that younger children would not differentiate body parts as places on the body from body parts as names for numerical signifiers. Thus, younger children may well consider the inner elbow on the right side of the body and the inner elbow on the left side as similar or equivalent values even if both were counted from the left thumb (1). Similarly, the same body part, say the left inner elbow (8), may be treated as equivalent, regardless of whether the count begins from the left thumb (1) or the right thumb (1).

To evaluate the conjectures about the shifting function of body-part positions in children's body-part comparisons, I developed two kinds of comparison tasks. I examined children's evaluation of symmetrical body parts with common points of origin (see Fig. 5A), and children's evaluation of the same body part reached from different positions of origin (the thumb on the left vs the thumb on the right side of the body, see Fig. 5B). I interviewed unschooled younger and older participants. The findings revealed age-related shifts in individuals' judgments, with the younger group evaluating body parts on the basis of their anatomical characteristics and older children evaluating body parts on the basis of number.

The studies on numerical comparisons and reproductions support an analysis of micro-, onto-, and sociogenetic processes of form-function relations of the body system in Oksapmin communities. From a microgenetic perspective, children structure means of producing numerical judgments by purposing body parts to produce numerical comparisons; the microgenetic process entails a costructuring of objects of representation (conceptualizing the entities to be compared as numerical) and a vehicle to represent the objects (treating body parts as representations of numerical values). From an ontogenetic perspective, the capabilities to structure such microgenetic constructions have roots in developmentally prior activities and display a shift in form-function relations: younger children do not appear to be using body parts as a means to produce numerical representations in the tasks used. Rather, the body parts appear to be used to serve nominal or other functions, and only subsequently do children purpose body parts to serve numerical functions. Finally, from a sociogenetic perspective, whether or not young children use body parts to serve nominal or numerical functions, they draw upon prior sociogenetic developments in the Oksapmin community—the body system—reproducing those sociogenetic developments but also altering them to serve their own purposes of communication and representation in their activities, a process massively distributed across the activities of Oksapmin people.

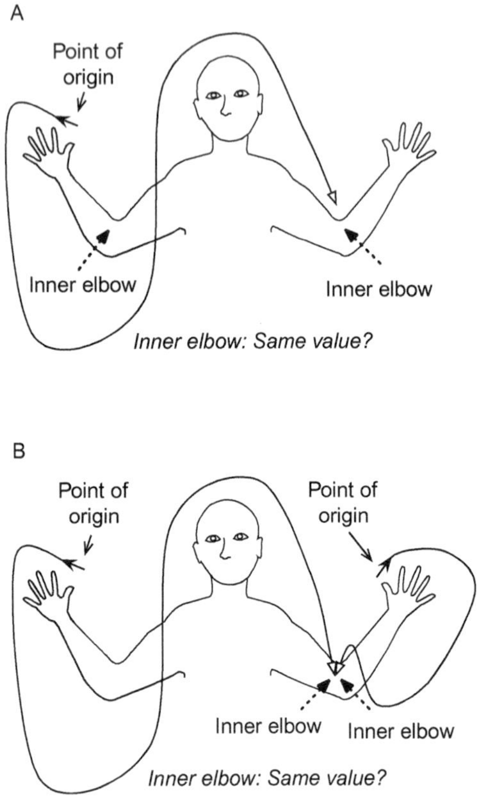

FIG. 5 Two potential conflicts for Oksapmin children as they work to elaborate numerical functions for body counting. (A) Symmetrical body parts with common origin point. (B) Same body part reached from different points of origin.

Using the Body System as a Measure for a Continuous Quantity: Length

The use of the body-part form is not restricted to the function of the representation of discontinuous quantities. It is also used to represent the continuous quantity of length (see Saxe & Moylan, 1982). Consider the collective practice in which girls and women create string bags or "bilums," traditionally measuring and naming bags by their depth (string bags of varied sizes are shown in the photographs contained in Fig. 6). To measure, an individual inserts her arms in a bag and notes the body-part position to which the rim reaches (the wrist (6), forearm (7), inner elbow (8), biceps (9), or shoulder (10) (see Fig. 7A)). One notable property of using the body system to measure is that linear units are not standard. Thus, if a tall man measures a bag to the forearm (7), his measurement (see Fig. 7B) would be different than a little girl who measured the same bag; indeed, one would expect the girl's measure to be at the inner elbow (8) or even the biceps (9) (see Fig. 7C).

FIG. 6 String bags are worn by men, women, and children.

To explore the development of people's ways of thinking about measurement with the nonstandard system, I developed interview tasks and presented them to people in two age groups, 23 unschooled children (mean age = 8.4 years and sd = 1 year) and 30 unschooled adults (mean age = 29.5 years and sd = 7 years). The tasks involved hypothetical stories in which participants were asked to compare the measurements of two individuals of different sizes when they measured the depth of a string bag using the Oksapmin body system. I used four versions of the task in which an individual of one size

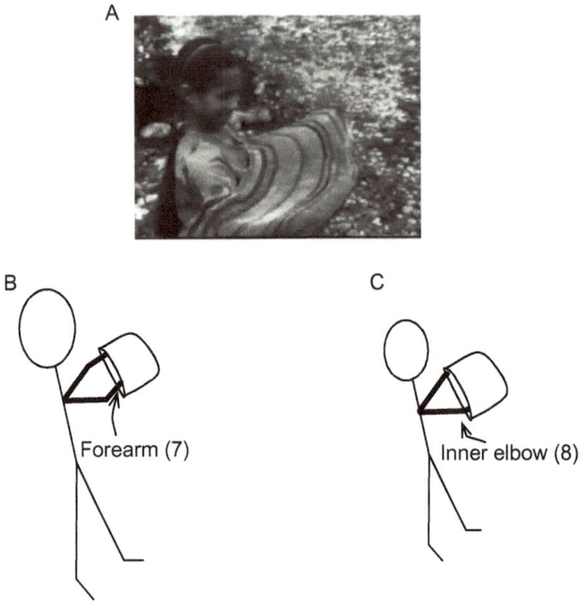

FIG. 7 Top, (A) photograph of an adolescent girl measuring a string bag. Bottom, (B) an adult inserts her arms into the bag, and the fingertips-to-rim measure is the forearm (7); (C) a child inserts his arms into the same bag, and the fingertips-to-rim measure is the inner elbow (8).

(e.g., a small girl) measured a bag to a specified body part and then gave the bag to an individual of a much larger size (e.g., her father) to measure, with the question where did the bag come up to on the recipient's body?

> *Daughter to father; string bag depth is forearm on daughter. "There was a little girl. Just a small little girl {gesture made by teller indicating small size of girl}. This little girl made a string bag for her tall father {gesture made by teller indicating the size of the father}. She made the string bag and then put it on her arms and saw that it was as long as the forearms {teller gestures to forearm}. Then she put the strap on and gave it to her father. When her father put it on his arms, did it come to the forearm also or not? How do you know? To where did it come? How do you know?"*

The findings produced by administering these tasks to children and adults revealed support for an ontogenetic progression: children treat the value of body parts as representative of length relation in an absolute sense. Regardless of the size of individuals inserting their arms to measure, the body-part name that corresponds to the bag's depth is treated as the measure of the bag. Later, children begin to differentiate between measurements produced by people of different sizes but have difficulty conceptually coordinating the inverse relations between arm length and the distance the string bag will reach on the body. Though some participants recognized that the body-part measurements would not be the same on the children and the adults, they believed that the adults would measure the string bags to a higher value since their arms were longer. Finally, by adulthood, participants both differentiated between body-part measurements produced by people of different sizes and achieved a coordination of this inverse relation between string bag length and point on the arm that the string bag would reach.

Studies on Children Attending School in 1980: A Focus on Genetic Processes in Form-Function Relations in Collective Practices of Schooling

The introduction of schools in the late 1960s and 1970s brought with them collective practices that were alien to the Oksapmin world. Fig. 8 contains an example of such a practice—a 1980 photo of a school and a morning assembly on the school grounds where I conducted the research. At the time of the 1980 study, students entered a school world alien to their lives in the community. Indeed, teachers were strangers and outsiders, not from the Oksapmin world, and ignorant of Oksapmin ways and language. Further, colonial schooling disallowed students to use their vernacular in classrooms.

One strand of my 1980 research focused on children's extension of the body-part system to serve novel numerical functions. The studies were sparked by my discussions with teachers. I had queried teachers about body counting in the Oksapmin community, and I was surprised to learn that they were unaware of the counting system, though they did report seeing children pointing to parts of their bodies during math lessons. My hunch was that when

FIG. 8 Morning assembly at school in Oksapmin area (1980).

children participated in math lessons, they were trying to conceptualize talk about number and arithmetic in terms of their own prior knowledge, including their use of the 27-body-part count system. After some observations in classrooms, I embarked on a formal study: I designed an arithmetic test that was administered to classrooms; I recruited Maryl Gearhart and Tom Moylan to help code the presence of body counting of students during the test (for a complete report of these studies, see Saxe, 1985, 2014). I found that in fact, body-part use was fairly widespread but differed across grade level. At Grades 2 and 4, the majority of students showed signs of pointing to conventional body-part positions during an arithmetic test, though at Grade 6 the number of children pointing to body parts declined precipitously.

The use of the body confirmed, I conducted two interview studies with children across the three grade levels. My purpose was to examine more closely the microgenetic constructions children were engaged with as they were solving arithmetic problems. To this end, I interviewed 71 children over three grade levels: Grade 2 ($n=25$), Grade 4 ($n=24$), and Grade 6 ($n=22$). In addition, in order to establish whether the microgenetic constructions were linked to participation in school, I interviewed a cohort of unschooled Oksapmin adolescents who were about the same age as the Oksapmin Grade 6 students ($n=22$). The tasks that I presented to children to solve with their bodies included addition problems (e.g., $5+3=?$, $14+7=?$) and subtraction problems (e.g., $5-3=?$, $16-7=?$).

The findings in the interview study were clear. I found that Grade 2 children would use their body in trying to solve the problem but they did not use it in a way to keep track of addition or subtraction of terms. For example, to solve $14+7=?$, they might enumerate body parts to the forearm (7) but be unable to keep track of the addition of the nose (14), simply enumerating body parts not knowing when the enumeration should end. In contrast, many Grade 4 and 6 children displayed ways of generating internal correspondences with body parts to serve the purpose of keeping track. Thus, a child might begin with the nose (14) and then keep track of the addition of the forearm (7), by producing correspondences internal to the body system between the eye (15) and the thumb (1) and the ear (16) and the index finger (2), until they reached the correspondence between the forearm (21) and the forearm (7). Fig. 9 displays part of an internal correspondence construction in which a child was solving the problem $16-7=?$. In this case, as depicted in Fig. 10, the child began with the ear (16) and proceeds to subtract the body parts one by one from the forearm (7) to the thumb (1), reaching the shoulder (10), and ending with the representation of the biceps (9), the solution of the problem.

The interviews with the unschooled adolescents revealed that the internal correspondence strategies were not simply linked to children's ages. Indeed, the adolescents' solutions appear similar to the Grade 2 students. Thus, though the Grade 4 and 6 children were not taught the body-part correspondences in

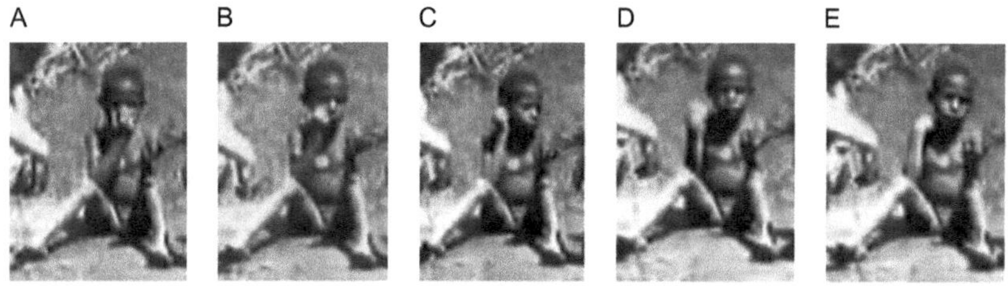

FIG. 9 Images A-E show a boy constructing a one-to-one correspondence between two series as he keeps track of the subtraction of the forearm (7) from the ear (16) (for video display, see http://culturecognition.com/video/fourth-grader-solving-16-7-body-system).

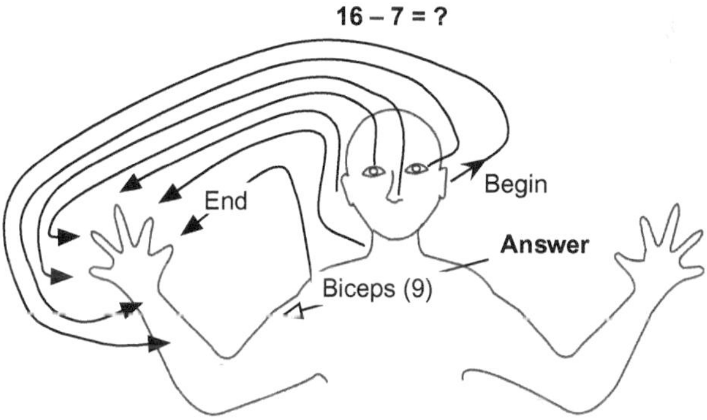

FIG. 10 Establishing one-to-one correspondences internal to the body system to coordinate a subtraction of the forearm (7) from the ear (16).

school, they appeared to be not only reproducing but also altering the form and function of the body system in their microgenetic constructions. These microgenetic constructions appeared to occur as children worked to solve the new kinds of problems with which their unschooled peers were not engaged.

To further explore the reproduction and alteration of form-function relations in children's problem solving, I interviewed only those students who displayed internal correspondence solutions to arithmetic problems, and I presented them with new problems that would require them to extend their use of the system beyond the 27 body parts (e.g., $9+43=?$, $12+34=?$, $43-7=?$, $41-6=?$). What I found was that students created hybrid strategies, combining English number words with Oksapmin body parts. For example, to add $9+43$, many children solved problems by beginning with the greater value, 43, stating the value in English and then counting in English "44, 45, 46, 47, 48, 49, 50, 51, 52" as the conventional body sequence was pointed to from the thumb (1) to the biceps (9) in order to keep track of the addition.

In sum, in these 1980 studies with children in school, I documented their use of the body-part system to serve new functions. The new functions required children's construction of new coordinations of one-to-one correspondences—from external correspondences between body parts and objects to internal body-part-to-body-part correspondences. The new correspondences supported the keeping track function in the addition and subtraction of values, novel numerical problems in Oksapmin communities, and problems with which students were engaged with in school.

2001. Form-Function Shifts in Teachers' Use of the Body System in the Classroom

In 2001, 21 years after my prior visit to the Oksapmin world, I returned to the community, accompanied by my 19-year-old son, Josh, and two graduate students working with me, Indigo Esmonde and Cliff McIntosh. One of my research foci was to continue my work on people's use of the body system in collective practices associated with schooling.

Noteworthy changes had occurred with regard to schooling since my 1980 visit. In the late 1990s, national educational reforms in Papua New Guinea penetrated the Oksapmin world. The reforms called for a shift from colonial-style schooling that supported a strictly Western, English-only curriculum, to one that supported instruction that linked community knowledge forms (like Oksapmin body counting) and Western knowledge forms (like the use of Hindu-Arabic numerals). Further, for the first 3 years of instruction, a new kind of school was to be established, one in which instruction was to occur in the vernacular and gradually transition to Tok Pisin and English.[3]

A consequential moment in 2001 occurred when I was hiking on an Oksapmin trail. I met the coordinator of the elementary schools, the schools that serviced the first three grades of instruction. After some introductory talk, I opened up a conversation about the body system and my interest in mathematics instruction. I found the coordinator's remarks puzzling. He reported that the Oksapmin system was quite sophisticated and that people traditionally used the system to count up to quite large values. For example, he explained that one used the shoulder (10) as a base and represented numbers like 60 as wrist (6) shoulder (10) or six-tens. For me, this was quite a contrast to my own apprenticeship in learning the Oksapmin counting system from elders in 1978 in which people did not use a base structure. But the coordinator's exposition did give me pause. How did the coordinator come to understand the body-part form in this way? Did the teachers use the system in this way in instruction in elementary schools? If so, this would constitute an interesting sociogenetic strand of development to which I should be attentive? My hunch was that

3. Unlike the 1980 era in Oksapmin, in 2001, most teachers were Oksapmin themselves and well equipped to implement the reforms. In fact, some of these teachers were students on my visit in 1980.

an unintended consequence of the reforms was that teachers unwittingly introduced alterations in the structure of the indigenous system as they attempted to make connections to Western knowledge forms (like the base-10 structure of the Hindu-Arabic numeration system). My subsequent inquiry produced evidence that provided corroboration and elaboration of this hunch.

Consider the poster in an elementary classroom that was on a wall in 2001 shown in Fig. 11. The poster well reflects the coordinator's remarks about the counting system. The chart shows the Hindu-Arabic written numerals from 1 to 10 in the left column; in the middle column, the chart shows the written body-part names in the conventional sequence for corresponding 10 positions in a written form of the Oksapmin language; and in the third column, the chart shows the corresponding written representations of the number words in English. The representation contained in the chart reflects the teachers' efforts to bridge traditional community knowledge with Western curricular concerns, an effort required by the new educational reforms.

In my observations of instruction and interviews with teachers about their lessons, I found much to corroborate the analysis that an alteration of form of the body system was emerging. The alternate form involved the privileged use of the first 10 positions of the body with the use of body parts to make a base structure analogous to the place-value structure of the Hindu-Arabic written numeration system (Saxe, 2014).

FIG. 11 Chart posted on wall in each elementary classroom with corresponding representation of numbers to 10: Hindu-Arabic numerals (left column), body-part names in the Oksapmin language (middle column), and English number words (right column).

2014. Form-Function Shifts in Teachers' Continued Use of the Body System

I returned again to the Oksapmin world in 2014 with one of my graduate students, Kenton de Kirby, and my adult son, Josh, now in his early 30s. The purpose of the 2014 trip was to understand whether the body system continued to be used in the community, given additional Western contact. A second purpose was to investigate how a One Laptop per Child program was being taken up in the community, a program that had been initiated in 2010 (see Saxe & de Kirby, 2016 for a report on this project).

Our stay in 2014 was during a period in which school was out of session. However, we were able to interview about 10 teachers and headmasters about the use of body counting in school. We will soon analyze those interviews in detail. Initial review of our interview data reveals that (a) all teachers reported using the body system during instruction in the elementary grades and use increased when children were having difficulty with the lessons, (b) some teachers reported using the system in ways that reflected the internal correspondence strategies that I had originally documented in 1980, and (c) teachers' knowledge of the count system was variable after the first 10 positions, suggesting that some teachers made full use of the system while others may have limited their use of the system to the first 10 positions. Further, we also noted some shifts in the way teachers reported using the system. Several teachers reported using only English number words in conjunction with body-part positions to solve arithmetic problems. Thus, to solve a problem like $16-7$ to support children, some teachers reported that they pointed to the ear and called it "sixteen" and then counted to lesser values from "one" to "seven" pointing to respective body-part positions from the ear (16) to the shoulder (10), leaving the position at the biceps (9), which they called "nine."

CONCLUDING REMARKS

My focus in this chapter has privileged an intrinsic relations approach to understanding culture-cognition relations. The approach treats cultural and cognitive processes as rooted in the communicative and problem-solving activities of individuals as they participate with one another in the collective practices of daily life, whether in school or out. In this process, representational forms, like the Oksapmin body count system, are purposed by individuals to serve cognitive functions in processes of microgenesis. Further, these situated meanings are each moments in sociogenetic trajectories, as individuals draw upon the historical developments in Oksapmin numeration, at once not only reproducing but also altering them in processes of communication and problem solving. Finally, such micro- and sociogenetic developments are moments of individuals' ontogenetic trajectories.

Though the approach that I have taken focuses on intrinsic relations between cultural, linguistic, and cognitive processes, I would be remiss to leave unacknowledged the value that empirical techniques more commonly associated with dichotomous methodologies have served in my analyses. In the cross cohort designs I described in this chapter and elsewhere (Saxe, 2014), I used independent variables that captured people's participation of collective practices like participation in schooling; I also used, as dependent variables, the microgenetic processes that involved either external or internal correspondences in arithmetic problem solving. The analyses of microgenetic processes, in turn, informed both ontogenetic analyses of form-function relations as children progressed through more advanced grades and sociogenetic analyses of the reproduction and alteration of the body-part form and functions in community life. Noteworthy here is that though the Oksapmin work is a case study of a remote community, dichotomous empirical techniques can be orchestrated in ways that advance an understanding of the rootedness and the reproduction and alteration of cultural, linguistic, and cognitive processes in the activities of individuals.

REFERENCES

Cole, M. (1996). *Cultural psychology. A once and future discipline?* Cambridge, MA: Harvard University Press.

Craig, B., & Hyndman, D. (Eds.), (1990). *Children of Afek: Tradition and change among the Mountain-Ok of Central New Guinea, Vol. 40*. Sydney: University of Sydney.

Gammage, B. (1998). *The sky travellers: Journeys in New Guinea 1938-1939*. Victoria: Melbourne University Press.

Laycock, D. (1975). Observations on number systems and semantics. *New Guinea area languages and language study: (Vol. 1)*. (Pacific Linguistics, series C, no. 38), pp. 219–233.

Rogoff, B. (2003). *The cultural nature of human development*. Oxford/New York: Oxford University Press.

Saxe, G. B. (1977). A developmental analysis of notational counting. *Child Development, 48*, 1512–1520.

Saxe, G. B. (1979). Developmental relations between notational counting and number conservation. *Child Development, 50*(1), 180–187.

Saxe, G. B. (1985). Effects of schooling on arithmetical understanding: Studies with Oksapmin children in Papua New Guinea. *Journal of Educational Psychology, 77*(5), 503–513.

Saxe, G. B. (1991). *Culture and cognitive development: Studies in mathematical understanding*. Hillsdale, NJ: Lawrence Erlbaum Associates, Inc.

Saxe, G. B. (2014). *Cultural development of mathematical ideas: Papua New Guinea studies*. New York: Cambridge University Press.

Saxe, G. B., & de Kirby, K. (2014). Cultural context of cognitive development. *WIREs Cognitive Science, 5*, 447–461. https://doi.org/10.1002/wcs.1300.

Saxe, G. B., & de Kirby, K. (2016). *Analyzing the evolution of a digital technology intervention: One laptop per child in a remote Papua New Guinea Community*. Unpublished manuscript, Graduate School of Education, University of California, Berkeley, Berkeley, CA.

Saxe, G. B., de Kirby, K., Kang, B., Le, M., & Schneider, A. (2015). Studying cognition through time in a classroom community: The interplay between “everyday” and “scientific” concepts. *Human Development*, *58*, 5–44. https://doi.org/10.1159/000371560.

Saxe, G. B., & Guberman, S. R. (1998). Emergent arithmetical environments in the context of distributed problem solving: Analyses of children playing an educational game. In J. Greeno & S. Goldman (Eds.), *Thinking practices* (pp. 237–256). Mahwah, NJ: Erlbaum.

Saxe, G. B., Guberman, S. R., & Gearhart, M. (1987). *Social processes in early number development. Monographs of the Society for Research in Child Development: (Vol. 52)* (serial no. 162).

Saxe, G. B., & Moylan, T. (1982). The development of measurement operations among the Oksapmin of Papua New Guinea. *Child Development*, *53*(5), 1242–1248.

Vygotsky, L. S. (1978). *Mind in society: The development of higher psychological processes*. Cambridge, MA: Harvard University Press.

Vygotsky, L. S. (1986). *Thought and language*. Cambridge, MA: MIT Press.

Weeks, S. G. (Ed.), (1981). *Oksapmin: Development and change*. Port Moresby: Educational Research Unit University of Papua New Guinea.

Werner, H. (1948). *Comparative psychology of mental development*. New York, NY: International Universities Press.

Werner, H., & Kaplan, B. (1963). *Symbol formation: An organismic-developmental approach to language and the expression of thought*. New York, NY: Wiley.

Chapter 13

Numeration Systems as Cultural Tools for Numerical Cognition

Andrea Bender and Sieghard Beller
University of Bergen, Bergen, Norway

INTRODUCTION

Numerical cognition refers to a variety of cognitive processes, including the discrimination and comparison of different quantities, their estimation and exact enumeration, and arithmetic and algebraic operations (Bender & Beller, 2018; Gaber & Schlimm, 2015). Numbers constitute the foundation for most of these processes and for their application to everyday tasks, in which their many uses include paginating books, indicating prices, keeping time, or simply counting apples in a bowl. As numbers also underpin the activities of measuring, calculating, and forecasting, they lay the foundation for mathematics, natural sciences, engineering, and computer programming, and for our economic system and its currency. All these achievements hinge on understanding the meaning of numbers: that, for instance, "five" refers to exactly five objects, and not "more or less a handful". Such exact numerical understanding is not available from birth but presupposes a cultural tool—conventionalized counting sequences with distinct numerical meaning (Feigenson, Dehaene, & Spelke, 2004; Núñez, 2017). The absence (Gordon, 2004; Pica, Lemer, Izard, & Dehaene, 2004; Spaepen, Coppola, Spelke, Carey, & Goldin-Meadow, 2011) or unavailability (Frank, Fedorenko, Lai, Saxe, & Gibson, 2012) of such a tool affects people's ability to accurately assess, remember, and reconstruct quantities beyond the subitizing range, that is, for quantities larger than 4.

Counting sequences may be realized in different modalities, such as number words or written notations, and they constitute numeration systems with distinct structural properties such as base size or extent. In terms of these properties, numeration systems differ substantially across languages and cultures, despite the fact that they all basically refer to the same set of natural numbers, namely a section of the positive integers. These systems are not only

Mathematical Cognition and Learning, Vol. 4. https://doi.org/10.1016/B978-0-12-812574-8.00013-4

indispensable for numerical cognition, but—like all tools—they also affect the way in which one uses them. Whereas a hammer lends itself to hammering but less so to cutting, a knife is much better suited to the latter than the former. And within the category of knives, a double-edged specimen differs in affordances and applicability from a single-edged specimen or one with a serrated blade.

Likewise, the properties of numeration systems determine how numerical information is represented and hence have cognitive implications. A change in cognitive behavior resulting from differences in how the same information is represented is called a *representational effect*. Such effects are central to accounts of *distributed cognition* (Hutchins, 1995; Larkin & Simon, 1987; Norman, 1993), which assume an interactive processing of information spread over external and internal representations—such as a number notation and a mental representation of number size, respectively—through an interplay of perceptual and cognitive processes. Accordingly, numbers are distributed representations, and their external component is essential in numerical processing.

Representational effects have been investigated most intensively for notational systems such as the Arabic digits versus Roman numerals (e.g., Krajcsi & Szabo, 2012; Nickerson, 1988; Schlimm & Neth, 2008; Zhang & Norman, 1995; Zhang & Wang, 2005). In this chapter, we extend this approach to other modalities (cp Bender & Beller, 2012, 2016; Bender, Schlimm, & Beller, 2015), using it both to compare different numeration systems and to analyze their possible interferences when used in parallel. We begin with a brief overview of system properties and their implications for number representation and processing for a selection of commonly used numeration systems. We then explore inconsistencies between these systems, in order to identify the challenges and opportunities that the coexistence of structurally diverse systems presents to learners. To conclude, we briefly touch on how exposure to each of these different systems may affect cognitive development and how these systems may have (co-) evolved in human history, thereby also highlighting the cultural dimension of numerical cognition.

PROPERTIES OF NUMERATION SYSTEMS AND THEIR IMPLICATIONS

All numeration systems are based on a counting sequence, that is, a set of symbols for numbers, and most also contain a set of rules for combining these symbols. The way in which these symbols are shaped and combined defines the specific properties of the system which, in turn, affect how numerical information is represented and processed (Bender & Beller, 2012; Chrisomalis, 2004; Widom & Schlimm, 2012; Zhang & Norman, 1995). In the following, we detail some of these properties together with their

Quantites	Numeration systems in different modalities (a) notches on a tally stick	(b) 10-fingers counting	(c) English number words	(d) Arabic digits
●	I	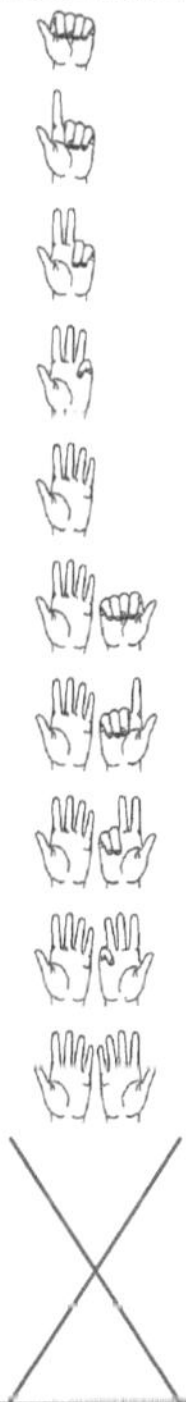	one	1
●●	II		two	2
●●●	III		three	3
●●●●	IIII		four	4
●●●●●	𝍸		five	5
●●●●●●	𝍸 I		six	6
●●●●●●●	𝍸 II		seven	7
●●●●●●●●	𝍸 III		eight	8
●●●●●●●●●	𝍸 IIII		nine	9
●●●●●●●●●●	𝍸 𝍸		ten	10
●●●●●●●●●●●	𝍸 𝍸 I		eleven	11
●●●●●●●●●●●●	𝍸 𝍸 II		twelve	12
●●●●●●●●●●●●●	𝍸 𝍸 III		thirteen	13
...	...		...	...
Properties				
Modality	material	body-based	verbal	notational
⇨ availability	permanent	transient	ephemeral	permanent
Dimensionality	1D*	1D	(1×1)×1D	1×1D
Dimensional representation	cumulative	cumulative	symbolic × parsed	symbolic × positional
⇨ information represented	category + magnitude + interval + ratio	category + magnitude + interval + ratio	category	category
Base size	(5)	–	10 (& 1000)	10
Regularity	regular	regular	irregular	regular
Extent	theoretically: infinite	10	traditionally: 1,000,000	theoretically: infinite

FIG. 1 Selection of number symbols for four common types of numeration systems with some of their properties.

implications, illustrated for a selection of familiar examples (see Fig. 1). The properties focused on here include the modality of the system, its structural properties dimensionality, dimensional representation, base size, and regularity, as well as its extent.

Modality

Modality refers to how the numeration system is implemented: *Material* systems are implemented through objects such as tally sticks (see Fig. 1a), knotted strings (the Inca *quipu*), or abaci; *body-based* systems are implemented through fingers as in (b), body parts more generally, and/or manual gestures; *verbal* systems are implemented through number words as in (c); and *notational* systems through written numerals as in the Arabic digits (d).

All numeration systems are cultural tools used to represent numerical information, yet how they do this differs depending on modality. Both the material system in (a) and the notational system in (d) externalize this information by visualizing it in a permanent manner. Once produced, this is independent of the person producing it, whereas (b) relies on transient representations of limited duration, namely until the fingers are needed for something else. This limitation is compensated by a more intense experience through embodiment, based on direct, perceptual feedback, which consolidates memorization and is thought to support the construal of more abstract representations (Di Luca & Pesenti, 2011; see overview in Bender & Beller, 2012; Fischer & Brugger, 2011). All three of the abovementioned systems are suitable for relieving cognitive load on working memory by generating an enduring external representation. In contrast, spoken number words (c) are ephemeral, but since they are also used to structure numerical information and bundle it into chunks, they may still serve to relieve memory capacities in distinct ways (Bender & Beller, 2014, 2016; Bender et al., 2015).

Dimensionality

Dimensionality refers to how the numeration system is internally structured (Zhang & Norman, 1995; for further refinements, see Chrisomalis, 2010; Widom & Schlimm, 2012). One-dimensional or 1D systems denote each number with a distinct symbol along their one dimension and hence remain unstructured. Two-dimensional or $1 \times 1D$ systems denote only the basic numbers of the primary counting cycle, the base, and its powers with distinct symbols, while all other numerical expressions are composed from these symbols. In using the base as a unit that can itself be counted to reach larger numbers (Bender & Beller, 2013), two-dimensional systems employ a power dimension in addition to a base dimension. Finally, $(1 \times 1) \times 1D$ systems intersect the base dimension by an additional subbase.

Although the material and the body-based system in (a) and (b) are 1D systems, the fifth notch in the tally stick example (a) is used to bundle numerical information and hence takes an important step toward structuring representation. The notational system in (d), in contrast, is a full-fledged $1 \times 1D$ system, and the same is largely true for the lower part of the verbal system in (c). Only beyond 9999 does this latter system actually turn into a

$(1 \times 1) \times 1D$ system with 1000 as the main base and 10 as its subbase. Another, perhaps more obvious instance of a $(1 \times 1) \times 1D$ system is the Roman numerals that use subbase 5 (=V) besides base 10 (=X).

Both the efficiency with which information is encoded and used for computation and the system's potential for irregularities (Zhang & Norman, 1995) depend on the system's dimensionality: While especially for larger numbers, $1 \times 1D$ systems are more efficient than 1D systems in keeping number representations compact, these systems are also more prone to irregularities in number word construction and to more types of errors when handling them (see section on regularity below).

Dimensional Representation

Dimensional representation refers to how the numerical information on each dimension is represented. The *base dimension* includes the basic number symbols of the primary counting cycle up to the base (in 1D systems, this comprises *all* symbols), while the *power dimension* includes the base and its powers in $1 \times 1D$ systems as well as the additional subbase in $(1 \times 1) \times 1D$ systems.

- On the *base dimension*, numerical information can be represented either by quantity (i.e., in a cumulative or iconic manner) or by shape (in a symbolic or ciphered manner). Examples (a) and (b) repeat the same symbol and therefore belong to the first type, whereas (c)—like all verbal systems and (d) denote each number with a distinct symbol, hence belonging to the second type. Note that body-based systems may also be symbolic (Bender & Beller, 2012), and notational systems may be cumulative (Chrisomalis, 2010; Widom & Schlimm, 2012).
- On the *power dimension*, numerical information can be represented in a parsed, integrated, or positional manner. Representation is parsed when multiplier and power are denoted by one symbol each (as is the case in "four hundred" in the English verbal system). It is integrated when both multiplier and power are jointly denoted by a single symbol (as in "υ" (=400) in the Greek notational system). And it is positional when only the multiplier is denoted by a symbol, and the power is indicated by the position in the number expression (as in "400" in the Arabic digits). A parsed representation is realized in most verbal systems including (c) and a positional representation in Arabic digits (d).

Instances (a) and (b) are 1D systems and therefore lack a power dimension. But body-based systems in general may be represented in *any* of these ways, and the same holds for material and notational systems (Bender & Beller, 2012; Chrisomalis, 2010; Widom & Schlimm, 2012).

According to the account of distributed cognition, dimensional representation is the property of a numeration system that most strongly affects number processing (Zhang & Norman, 1995; Zhang & Wang, 2005). First, it determines the extent to which the dimensions can be separated based on external cues. The base and power dimension are separated externally and are thus perceptually discriminable, by the two distinct symbols for multiplier and power in parsed representations, but need to be disentangled mentally in integrated and positional representations. In addition, dimensional representation also determines the extent to which numerical information is explicated on each of these dimensions: *category* (or nominal) information on whether or not two numbers are equal, *magnitude* (or ordinal) information on which of two numbers is larger, and *interval* and *ratio* information on the distance between and the relative proportions of numbers (Zhang & Norman, 1995).

On the base dimension, *cumulative* representations in which the amount of numerical symbols is proportional to the numerical value explicate the full range of numerical information: "||||" in (a) are not only *different* from "||", but they are also *more* than "||", they are separated from them by the *same distance* as are "|||" from "|", and they are *twice* as many as "||". *Symbolic* representations, in contrast, make only nominal information perceptually available. While "4" is discernibly different from "2," there is nothing in the representation itself that tells one whether it is larger or smaller than "2," not to mention distance or proportion. This latter information is left implicit and needs to be retrieved from what one knows about the meanings of "2" and "4," thus substantially increasing cognitive load during number processing.

Systems with cumulative vs symbolic base representation are therefore likely to have distinct effects on working memory, number processing, and task difficulty (Zhang & Norman, 1995). In particular, a cumulative representation makes clear use of one-to-one correspondence, thereby facilitating counting, and it renders an immediate approximation of relative sizes, thereby facilitating tasks like comparison, addition, and subtraction. Symbolic base representation forgoes this benefit.

Base Size

Base size refers to the prime structuring element of those numeration systems with more than one dimension. In principle, bases can take any size greater than 1, but some sizes occur more often than others, with 5, 10, and 20 being the more frequent ones across modalities (Bender & Beller, 2012; Chrisomalis, 2010; Comrie, 2011; Hammarström, 2010).

Even though the systems in (a) and (b) are 1D systems, hence lacking a base, both the bundling of symbols into larger units of 5 for easier recognition in the tally stick example (a) and the completion of one hand in (b) structure the representation of numerical information and thus bear the potential to develop into a base. In contrast, the system in (d) is a decimal system in which the base not only recurs in multiples of 10 but also is raised to various powers.

The same also holds for the lower part of the verbal system (c), even though its irregularities (i.e., specific numerals for 11 and 12 and nontransparent expressions for 20 and 30) partly obliterate the decimal pattern (this will be picked up in the next section).

Cognitive implications of base size include, for instance, its correlation with the extent of addition and multiplication tables. The larger a base, the more of these equations need to be learned initially and retrieved from memory during mental arithmetic (Zhang & Norman, 1995). The concomitant increase in difficulties is offset, however, by a more efficient encoding, with larger bases generating more compact representations and hence more efficient processing (Lordahl, Krop, & Jacobson, 1970).

Regularity

Regularity refers to how straightforwardly a numeration system can be categorized in terms of the aforementioned properties and to how systematically compound number terms are generated. On an abstract level, a system is regular if

- it is implemented in just one modality;
- its dimensionality can be established unequivocally;
- all number tokens on the base dimension are represented in the same way, and all those on the power dimension follow the same principles (this would also include that all expressions are generated according to the composition rules of the system (see below) in that distinct symbols are used only for those numbers that require them in line with the representational type and that no more than one representation is assigned to any possible number (avoiding redundancy) or, conversely, that the same representation is not assigned to more than one number (avoiding ambiguity));
- the same base and subbase (if applicable) is applied at every power level and is used to generate all power numerals in a consistent manner.

All of these properties have the potential to generate irregularities—and indeed do so in several cases (overview in Bender et al., 2015). Here, we focus on the irregularities arising from the generation of compound number terms, as these are the ones most relevant for the systems under scrutiny.

In 1D systems and particularly those with cumulative base representation such as the material and body-based system (a) and (b), new expressions are generated by the simple rule "add more symbols," here notches and fingers or other body parts, respectively. In 1×1D systems, compound expressions for numbers beyond the base are typically generated by addition and multiplication (and in some systems also by way of subtraction and/or multiplication with fractions; see Greenberg, 1978; Hammarström, 2010). Power terms in the notational system (d), for instance, are generated by multiplying the powers of the base with a basic number and by adding the resulting power terms. Illustrated for 6902, this unfolds as follows:

$$6902 = \left(6 \times 10^3\right) + \left(9 \times 10^2\right) + \left(0 \times 10^1\right) + \left(2\left[\times 10^0\right]\right) \tag{1}$$

More generally, all number expressions in this system follow a simple polynomial:

$$\begin{aligned} N &= \left(n_i \times 10^i\right) + \cdots + \left(n_2 \times 10^2\right) + \left(n_1 \times 10^1\right) + \left(n_0 \times 10^0\right) \\ &= \Sigma n_i 10^i \end{aligned} \tag{2}$$

with N = any possible number expression in the system and n = basic number symbol {one, two, three, ..., nine}.

Usually, conceiving of (c) also as a decimal system, we tend to generalize the above composition rule to the verbal numeration system; in order to take into consideration that power terms (P) are not represented here by position, but by distinct symbols in a parsed manner, different Ps replace the 10 raised to a specific power:

$$N = \left[n_i P_{10^i}\right] + \cdots + \left[n_2 P_{10^2}\right] + \left[n_1 P_{10^1}\right] + \left[n_0\right] \tag{3}$$

with N = any possible number word in the language, n = basic numeral {one, two, three, ..., nine}, P = power numeral {ten, hundred, thousand, ...}, subscript numbers indicating the power level of the term, and square brackets indicating that a term is optional

Illustrated for the same number as above, the formula translates into the following:

$$\begin{aligned} 6902 &= [6 \times 1000] + [9 \times 100] + [0 \times 10] + [2] \\ &= [\text{six thousand}]\ \text{and}\ [\text{nine hundred}]\ [\text{and -}]\ \text{and}\ [\text{two}] \\ &= \text{six thousand nine hundred and two} \end{aligned}$$

For the number chosen, this formula works fine and looks highly regular, but we have deliberately set the multiples of tens to zero. If we include this power level, the substantially more complex formula (4) will be needed to describe English number words (detailed here for the numbers from 1 to 999,999):

$$\underset{\textit{thousands}}{\left[m_{1\text{-}999} P_{10^3}\right]} + \underbrace{\underset{\textit{hundreds}}{\left[n P_{10^2}\right]} + \underset{\textit{ones and tens}}{\begin{bmatrix} n_{1\text{-}9} \\ P_{10^1\,(ten)} \\ s_{11,12} \\ n_{3\text{-}9} P_{10^1\,(\text{-}teen)} \\ n_{2\text{-}9} P_{10^1\,(\text{-}ty)} \left[n_{1\text{-}9}\right] \end{bmatrix}}}_{= m_{1\text{-}999}} \tag{4}$$

with s = specific numerals ("eleven" and "twelve") and subscript numbers indicating the value of the term, other conventions as in (3)

The formula is still simplified as it glosses over some additional irregularities, including different variants for some numerals such as "two" or "twen-" for 2, or the fact that some power numerals may, but need not, take "one" as multiplier.

The peculiarities that make (4) still so much more complex than (3) are linked to the numbers between 10 and 20 as well as the first multiples of 10. The generation of larger number words according to (3) follows composition rules based on addition and multiplication: multiplication of power numerals with basic numerals, joining of the resulting power terms by addition and ordered by decreasing magnitude—hence resulting in expressions such as "one hundred and two." In contrast, the words for 11 and 12 are not expressed analogously by "ten and one" and "ten and two," but instead by specific numerals, "eleven" and "twelve." The words for 13 onward are not expressed by "ten and three" either, but by recruiting different variants of the words for 10 and 3 and by adding them in reversed order (*inversion of constituents*), as in "thirteen." And some of the multiples of 10 use variants of the number words for both the base itself and its multipliers, namely, "-ty" (instead of "ten") and "twen-," "thir-," and "fif-" (instead of "two," "three," and "five," respectively).

The verbal numeration system in English (Fig. 1c) therefore deviates in various ways from the simple polynomial (3) that broadly describes its structure: in different variants of the same numeral, in specific numerals for numbers that could be composed regularly, and in inversion of constituents in some, but not all compounds. Although this system may appear to be exceptionally afflicted by irregularities, English is not the only language comprising such irregularities (e.g., Calude & Verkerk, 2016; Greenberg, 1978). Yet, it is possible to do without irregularities, as attested to by systems (a), (b), and (d), and even verbal systems can be much more regular, such as those in several East-Asian or Austronesian languages (e.g., see Bender et al., 2015; Miura, 1987).

As number expressions in actual systems can very often be comprehensively described by some sort of composition rule (as in (4))—even if this violates a simple polynomial like (3) on which the system could ideally be based—deciding on the degree of irregularity is somewhat tricky. For instance, nonprimary power terms for which a lower power numeral serves as multiplier (as in "ten thousand") would be treated as irregular only if the system is classified as a $1 \times 1D$ system. In a $(1 \times 1) \times 1D$ system, by contrast, 10 and 1000 could be conceived of as subbase and base, respectively.

But regardless of whether the English verbal system is described as an irregular instance of (3) or a regular instance of (4), both the amount of exceptions from (3) and the higher degree of complexity in (4) do bear the same type of consequences: They make the system harder to learn and to handle. Specific numerals like "eleven" and "twelve" and number words composed of different variants like "twenty" blur the transparency of the system.

As demonstrated in studies with children learning this system as compared with a regular East-Asian system, the reduced transparency not only impedes the acquirement of the system and specifically the comprehension of its base structure but also has implications for the ease with which algorithms for addition and subtraction are learned (Fuson & Kwon, 1991, 1992; Miller, Smith, Zhu, & Zhang, 1995; Miura, 1987; Miura, Okamoto, Kim, Steere, & Fayol, 1993). Other irregularities like the inversion of constituents have similar immediate and long-term effects on performance (Dowker, Bala, & Lloyd, 2008; Klein et al., 2013), including errors in Arabic number production (Deloche & Seron, 1982; Seron & Fayol, 1994) and in mental arithmetic, particularly under time pressure (Brysbaert, Fias, & Noël, 1998; Göbel, Moeller, Pixner, Kaufmann, & Nuerk, 2014; see also Chapter 10 by Göbel).

Extent

Extent, finally, refers to the largest number expression in a numeration system that can be regularly generated according to its composition rules (in 1D systems, it would simply be the largest number word). The next number beyond this expression is typically conceived of as its *limiting number L* (Greenberg, 1978). In a system with basic numerals for 1 to 10, power terms for 100 and 1000, and composition rules based on addition and multiplication such as (3), the largest *regularly* composed number expression would be 9999, rendering the limiting number $L = 10{,}000$.

The English verbal system (c) is, by and large, an example of such a system, with the exception that it has been modified to cope with the much larger quantities modern societies are dealing with. To this end, power numerals were multiplied with other power numerals rather than just with basic numerals (hence generating nonprimary power terms such as "ten thousand" and "a hundred thousand"), and loanwords such as "million" and "billion" were recruited for larger powers still. Even with this extended range, the system reaches a limit at some point. Currently, the largest officially approved power numeral terminating a continuous sequence is "vigintillion" ($=10^{63}$ according to the *short scale* prevailing in the United States, English Canada, and modern British, or $=10^{120}$ according to the *long scale* in older British, French Canada, and continental Europe), complemented by the isolates "googol" $=10^{100}$ and "googolplex" $=10^{\text{googol}}$ (Wikipedia, *names of large numbers*, retrieved on Nov. 14, 2016). But in effect, most users would not be able to recite any of these beyond the range of modern state budgets and debts.

In contrast, both (a) and (d) are infinite, at least in principle, although limiting factors such as space and time make this infinity unachievable in practice. The most restricted system in our selection is the finger counting in (b), which terminates at 10.

Since the exact representation of numbers depends on conventionalized counting sequences (Frank et al., 2012; Pica et al., 2004; Spaepen et al.,

2011), the most important implication of extent is that it determines the range for (accurate) counting and arithmetic. This is why the finger counting system in (b) becomes insufficient rather quickly and why with a verbal system like (c) so much effort has been put into extension. Yet, the latter also proves that extent is perhaps the one property of a system that can be adapted most easily, if its users feel a need to do so (Beller & Bender, 2008).

INCONSISTENCIES BETWEEN REPRESENTATIONAL SYSTEMS

So far, we have focused on the *internal irregularities* that may arise within a numeration system. When using several numeration systems in parallel, some of their properties are likely to differ across systems, thereby also generating *external inconsistencies*.

Examples of systems that are used in parallel by a larger number of people are depicted in Fig. 1. Typically, such conglomerates comprise finger counting, a sequence of number words, and a notational system (we also added a material system to cover all types of modality, although the tally sheets preferred today are more akin to a notational system than the tally sticks used in the past). These systems serve partly different functions: tallies and finger counting are tools for the counting process itself but less so for communicating the result or for arithmetic. Notations are designed almost entirely for these latter purposes, while they are less suitable for counting. The verbal system is the one tool that connects both. It can be used for counting and communication, and it is also most likely the modality in which numerical facts are stored in memory and in which parts of mental arithmetic take place (Campbell & Epp, 2004; Dehaene, 1992; Spelke & Tsivkin, 2001). The verbal system is therefore the core component in the numerical toolkit, and the extent to which its properties are consistent with those of the others matters most.

Nevertheless, the systems in Fig. 1 differ not only in terms of modality but also with respect to structural properties—in fact, no one system is identical to another. Both the tallies in (a) and the finger counting (b) are 1D systems with cumulative representation, but while the latter is fairly restricted and hence of limited value, the former is theoretically infinite. Tallies are also more permanent than the transient finger representations. The English verbal system (c) differs from these two in almost all regards: it is a fairly irregular, arguably $(1 \times 1) \times 1D$ system, employs base 10 in the lower range and symbolic $\times$ parsed representation, and has a much greater (albeit somewhat fuzzy) extent than the finger counting system. The Arabic digits (d), in contrast, are a perfectly regular, straightforward $1 \times 1D$ system, with a concise base 10 and symbolic $\times$ positional representation. Even more significant, the four systems also differ with respect to which type of numerical information (i.e., on category, magnitude, interval, and ratio) they make directly available through perception: Both (a) and (b) provide all four types, whereas (c) and (d) provide category information only.

It has not been thoroughly investigated which implications may arise from such inconsistencies between systems, but some assumptions can be safely put forward. To this end, we take the verbal system as the starting point, focusing on which information it represents externally and which it leaves implicit. Based on this analysis, we explore how this may be supported—or hindered—by the other systems, specifically finger counting and Arabic digits.

Learning the Verbal System

Learning to count is an important prerequisite for the understanding of numbers and of symbolic algorithms involving addition and multiplication and for mathematical education more generally. Counting is typically learned through recitation of the sequence of number words, and children master this sequence long before they understand the numerical meaning of the words, that is, their cardinal value (Le Corre & Carey, 2007; Palmer & Baroody, 2011). The latter takes several years but eventually enables them to start counting from any number, to proceed back and forth easily, and to extract the cardinal value conveyed (Fuson, 1988; Wiese, 2003; Wynn, 1992; see also Chapter 2 by Marchand and Barner and Chapter 9 by Sarnecka et al.). According to Gelman and Gallistel (1978), five principles govern and define this competence:

(i) One number word has to be assigned to each single object (*one-to-one principle*).
(ii) Number words occur in a fixed order (*stable-order principle*).
(iii) The number word for the last object represents the cardinality of the set (*cardinal principle*).
(iv) All sorts of objects can be counted in the same way (*abstraction principle*).
(v) The order of the objects is irrelevant (*order-irrelevance principle*).

While these principles ensure the numerical understanding of the numeration system, handling the system presupposes yet another competence, namely, the ability to represent and process the numerical information mentally or, in other words, *internally* (Zhang & Norman, 1995). But number words are not only symbolic, that is, arbitrary symbols for this information. Rather, they are also ephemeral and constitute, in the case of English, the most irregular of all available systems. At least in the beginning, both the internalization and the processing therefore benefit from external support.

As we have seen above, number words explicate only one type of numerical information: Whether a word is equal to or different from another (i.e., category information) can be perceived externally, whereas magnitude, interval, and ratio information cannot and therefore must be learned in order to be available. Learning is additionally challenged by the fact that it involves three different kinds of numerical relations simultaneously and that number itself is an abstract concept (Wiese, 2003). This is where the use of fingers comes to the learner's aid.

Support by Fingers

Accounts of *embodied cognition* emphasize the role played by our body (including the fingers) and the experiences we have with it for cognitive development and performance (Andres, Olivier, & Badets, 2008; Di Luca & Pesenti, 2011; see overview in Bender & Beller, 2012). In particular, finger counting has been identified as "the missing tool" (Andres, Di Luca, & Pesenti, 2008) allowing abstract mathematical concepts to emerge from sensorimotor experience.

Compared to the verbal system, finger counting indeed comes with important advantages. As tools, fingers are readily available and easy to manipulate. They can themselves be counted, thereby explicating Gelman and Gallistel's (1978) one-to-one correspondence principle (i) and the cardinality principle (iii), and they exist in a quantity both small enough for immediate perception (*subitizing*) and large enough to serve as a simple model set. Different from words, which fade away quickly and must be memorized, fingers remain visible for perceptual processes and hence can be used for bookkeeping, which relieves working memory.

In the range from 1 to 10, the finger counting in (b) and the number words in (c) are structured alike, namely as one-dimensional systems that differ only in the representation of the basic numbers of the primary counting cycle. By representing quantity in a cumulative manner, finger counting provides all four types of numerical information externally and thereby aids the learning process in various ways. In the beginning, the fingers help the learner to differentiate the number words by relating their different phonological patterns to different finger patterns (category information), and as they are typically used in a stable order (Lindemann, Alipour, & Fischer, 2011; Wiese, 2003), they facilitate access to the number words in their correct order (Gelman and Gallistel's principle (ii)). Later on, fingers also help the learner to grasp numerical meaning as they provide the other types of information (i.e., on magnitude, interval, and ratio).

However, the regular pairing of fingers with the same number words may also obfuscate the fact that neither the kind nor the order of objects to be counted is relevant (principles (iv) and (v)) (Beller & Bender, 2011). By inviting some inferences and generalizations at the expense of others, the concrete instantiations in finger counting may thereby slow down the acquisition of abstract mathematical strategies. By and large, however, fingers as a counting tool help to model and internalize all the numerical information that is not externally represented in the arbitrary number words and thus support reflections on numerical concepts (Andres, Di Luca et al., 2008; Andres, Olivier, et al., 2008; Carpenter, Fennema, Franke, Levi, & Empson, 1999; Fuson & Secada, 1986; Guha, 2006; Wright, Martland, Stafford, & Stanger, 2002). It is unsurprising, therefore, that imprints of finger counting patterns can be found not only in children's early activities (Domahs, Krinzinger, & Willmes, 2008), but also in adults' embodied representations and performance (Di Luca, Granà, Semenza, Seron, & Pesenti, 2006; Fischer & Brugger, 2011).

Unfortunately, we run out of fingers and their potentially supportive service precisely at the point where the (English) verbal system becomes puzzling. Rather than starting a new counting cycle when the base is reached—for instance, by regularly composed number words such as "ten-and-one," "ten-and-two," and "ten-and-three"—the specific number words "eleven," "twelve," and "thirteen" in English create the impression of linear continuation. Only at "fourteen" may the learner begin to sense a recurrent pattern through the suffixed "-teen." Until that point, both the experience with the basic number words and the accompanying finger counting system have not given a single hint that the verbal system is, in fact, structured by more than one dimension. Besides a more regular verbal system to begin with, a two-dimensional finger counting system would help to bridge this gap, and instances for both are found cross-culturally (e.g., Bender & Beller, 2012; Miura, 1987).

Mapping on Arabic Digits

Whereas the verbal counting sequence is an intricate part of language and hence subject to slow-paced linguistic evolution (Calude & Verkerk, 2016; Levinson & Gray, 2012) but is not easy to intentionally change, notational systems are self-contained cultural tools designed for a concrete purpose and have therefore been continually refined throughout human history (Chrisomalis, 2010; Ifrah, 1985; Menninger, 1969). Although they succeed verbal systems chronologically, they do not necessarily follow them in terms of properties, and very often, they have crossed language boundaries. For instance, the digits used for the notational system (d) were invented in India almost two millennia ago and were later refined by mathematicians in Persia and Arabia and only during the Middle Ages introduced to Europe, from where they then spread into the colonies across the world.

In marked contrast to all verbal systems, all notational systems allow for permanent external representations, which make them the paradigmatic example of *distributed cognition* (Zhang & Norman, 1995). Notational systems also have more degrees of freedom with respect to the design space than body-based and verbal systems, and they occur with substantial diversity (Chrisomalis, 2010; Widom & Schlimm, 2012). This includes the option to explicate all four types of numerical information. The notational system in (d), however, provides only category information, which is why its benefits to working memory are restricted to the permanent visibility of the symbols.

The example in (d) shares a few important properties with the English verbal system in (c). Both the verbal and the notational system utilize symbolic base representation, that is, a distinct new symbol for every new number in the primary counting cycle. Despite the fact that written symbols differ in shape from verbal symbols, the mapping of number words from "one" to "nine" onto the Arabic digits from "1" to "9" is therefore rather straightforward. As soon as it comes to the base, however, the two diverge more substantially (see Fig. 2). While "ten" as a monolexemic expression may appear as the last basic number word on the same level as the previous nine, "10" is the first compound symbol in the notational system, defined as $1 \times 10^1 + 0 \times 10^0$ (and it is indeed crucial that the zero units are not omitted from the representation).[1]

Notational system *Arabic digits*	***Verbal system*** *consistent expression*	*actual expression*	***Transcription into numerical values***
1	one	one	[1]
2	two	two	[2]
...	...	...	...
9	nine	nine	[9]
10	* one ten (and none)	ten	[10]
11	* one ten and one	eleven	[11]
12	* one ten and two	twelve	[12]
13	* one ten and three	thirteen	[13] *or* [3] [10]
14	* one ten and four	fourteen ⇖	[4] [10]
...	...	...	...
20	* two ten (and none)	twenty	[2^] [10^]
21	* two ten and one	twenty-one	[2^] [10^] [1]
...	...	...	...
30	* three ten (and none)	thirty ⇖	[3^] [10^]
31	* three ten and one	thirty-one	[3^] [10^] [1]
...	...	...	...

FIG. 2 Comparison of the compositional structure across Arabic digits and English number words, for a selection of number expressions. The arrow indicates that only at that stage does the recurrent pattern, which actually begins in the row *above*, become discernable; numbers in square brackets indicate monolexemic (rather than composed) expressions, and ^ indicates a different variant of the standard numeral.

1. This discrepancy is surprisingly hard to grasp even for people who master these systems. In one of our previous seminars for graduate students, we asked the students to invent a base 7 and a base 12 system in order to compare their respective properties with those of the decimal system. A substantial proportion of the students used a single symbol for 7 (notably "7") and 12 (e.g., "Y") to designate the base in an otherwise still positional system.

The generative function of "ten" as the base for the next three power levels of the verbal system could be explicated if the counting sequence continued regularly, as indicated by the *consistent expressions* in Fig. 2, but alas, it does not. What the symbols in the verbal system actually convey is indicated by their *transcription into numerical values* in Fig. 2.

At some point, the system would by itself eventually get the decimal structure across, although it is not clear how much more effort this would take without the additional help of the number notation. For instance, when hearing compound number words, some recurring phonological patterns hint at a two-dimensional structure—at least once the equivalence of "ten," "-teen," and "-ty" is noticed. But without an additional aid, this information can only be obtained through inference, and even with knowledge on basic numerals such as "three" and "ten," understanding the difference between expressions like "thirteen" and "thirty" requires additional skills. It is therefore that, while trying to learn the notational system, its regularity actually helps the learner to conceptually (re)structure the verbal one—and indeed so successfully that many advanced users conceive of (c) not only as a decimal system but also as one that nicely maps onto the notation (d).

To sum up, the English verbal system and the notational system based on Arabic digits differ most substantially in their use of different power representations (positional in the Arabic digit systems but parsed in the verbal system) and in the irregular way in which the verbal system represents the numbers between 10 and 20 and to some extent those beyond. These discrepancies between the two systems not only hamper the mapping of one onto the other and hence pose challenges to those learning them (Dowker et al., 2008; Fuson & Kwon, 1991, 1992; Miller et al., 1995; Miura, 1987; Miura et al., 1993) but also tend to impair processing, thereby also increasing the potential for errors even for advanced users (Klein et al., 2013; Nickerson, 1988).

For the sake of completeness, one more difference between the verbal and the notational system should be mentioned, namely, that the verbal system turns from a $1 \times 1D$ into a $(1 \times 1) \times 1D$ system at 10,000. However, once a learner has reached this range in mastering both systems, this additional mismatch presumably no longer poses a significant challenge.

Possible Benefits of Inconsistencies and Mismatches

Overall, the simultaneous usage of different external representations for the same content, in addition to internal ones, increases the need for integration (e.g., see the triple code model by Dehaene, 1992), and structural mismatches between them increase the potential for interferences. However, having alternative systems with diverging properties at one's disposal may also help to overcome constraints of the primary system through complementary advantages.

For instance, the most obvious advantage of a body-based numeration system like (b) as a complement to verbal representations is that it constitutes an external representation par excellence, basically due to its extended availability. Using this system for number representation and manipulation therefore generally serves to alleviate working memory load. This supporting function is further enhanced in systems with cumulative base representation (such as (b)), which explicate numerical information—in contrast to symbolic systems with their heavier demands on memory for extracting the (implicit) numerical information.

The notational system, on the other hand, not only matches the verbal system more strongly (especially in terms of dimensionality, the symbolic representation of the base dimension, and extent) but also adds crucial qualities. These include the permanent duration of the representation and a more explicit display of the base/power structure, which, together with its perfect regularity, serve to clarify and consolidate the comprehension of the decimal structure underlying both systems. Elaborating on the equivalence of content that is represented in structurally different ways thereby stimulates learning and mastery of even the more complex systems.

CONCLUSION

As we have argued throughout this chapter, numeration systems are *cognitive* tools that not only are indispensable for enabling exact counting and the mathematical ramifications based thereupon but also have system-specific properties feeding back on how each of them is learned and used to represent and process numerical information. Numeration systems are also *cultural* tools in that they are invented by cultural groups and passed on through cultural transmission. Despite the common abstract information captured by all such systems, they are not only implemented in diverse modalities but also differ substantially in their structural properties. Here, we highlighted the properties of four systems ((a)–(d)), but each of these presents just one of various possible implementations of a material, body-based, verbal, and notational system, respectively. Aside from the necessity of verbal systems to represent information in a symbolic manner, all conceivable combinations appear to have been in use (for an overview of the diversity in body-based and notational systems, respectively, see Bender & Beller, 2012; Chrisomalis, 2010).

From the Arabic digits in (d), for instance, the Roman numerals differ in terms of dimensionality (being $(1 \times 1) \times 1D$ instead of $1 \times 1D$), dimensional representation (being cumulative × integrated instead of symbolic × positional), and extent (which is limited by symbols for additional powers of the base, in contrast to the principally infinite Arabic digit system). Likewise, the finger-based numeration system developed for school purposes in India extends the limiting number to 40 by using the edges between finger segments as the counting tokens (Guha, 2006) rather than entire fingers as in (b). The system used by merchants in the Indian state Maharashtra (Ifrah, 1985) even constitutes a two-dimensional system with base 5, positional power representation, and an extent of 30, by employing the fingers of one hand to count from 1 to 5 and the other hand for keeping track of how often this has been done.

This cultural variability in numeration systems, together with their inalienability for numerical cognition, raises a puzzling question: If the ability to count, and numeracy more generally, does indeed presuppose a counting sequence, how could such a conceptual tool ever be invented in the first place? This question has invited a range of proposals, most of which assign a crucial role to material expressions of quantity—including piles of pebbles, tally sticks, or one's fingers—for the development of numerical concepts (e.g., Dantzig, 1954; Dehaene, 2011; Overmann, 2014; Wiese, 2003, 2007). As numerical concepts are abstract, their acquisition requires mediation, and material representations have the great advantage of visualizing some of the properties that numbers have (Overmann, 2016).

We do agree with this proposal insofar as such external representations provide crucial support for developing numerical concepts, but still conceive of the verbal system as the core in the emerging understanding of numbers (see also Wiese, 2007). Material representations, especially if implemented in a cumulative manner, explicate first and foremost numerical information on magnitude, interval, and ratio, whereas symbolic representations such as number words are much better suited to convey category information. Cumulative systems are perfectly suitable for keeping track of entities and for one-to-one correspondence, for instance in ascertaining that all the sheep that left at dawn return at dusk. Category information, on the other hand, is essential to separate quantities beyond the subitizing range, to identify their precise cardinal value, and to facilitate memorization and further computation by chunking. It is not by coincidence that the one parameter *not* observed in verbal systems is cumulative base representation. It is of course possible, in principle, to express 7 as "one and one and one and one and one and one and one," but nobody using this system would be able to keep track of how often "one" was expressed or to assess the cardinality resulting from it.

As we lack empirical access to questions on evolution, developmental studies are often taken as proxy, and respective research emphasizes the important role that finger counting may play in the emergence of more abstract numerical competencies and tools. A cumulative system like finger counting in (b) explicates one-to-one correspondence, helps the child to keep track of the number words while reciting, supports comprehension of some of Gelman and Gallistel's (1978) counting principles and of the importance of 10, and allows for simple additions and subtractions (Crollen, Seron, & Noël, 2011, and see Beller & Bender, 2011; Di Luca & Pesenti, 2011; Wiese, 2003). Most of these principles need to be mastered before more abstract representations and operations can be acquired, and several studies indicate that the use of fingers may pave the way for learning counting principles (e.g., Brissiaud, 1992) and is pervasive and enduring in this regard (overview in Fischer & Brugger, 2011). More recently, however, the claim that finger counting is indispensable for this to happen has been qualified somewhat. The existence of symbolic finger-based systems and studies in blind and deaf children not only speak against the allegedly universal spontaneity of finger counting but also indicate that finger representations are neither necessary (Crollen, Mahe, Collignon, & Seron, 2011; Crollen et al., 2014) nor sufficient (Spaepen et al., 2011) for learning number concepts and hence cannot be an *essential* prerequisite for the development of numerical cognition.

An alternative account (overview in Sarnecka, 2014) proposes that grammatical number may help children to learn the meaning of linguistic numbers (i.e., to grasp the cardinality inherent in number words), as it emphasizes the difference between entities of which there is just one (singular)—and perhaps two (if the language contains a dual)—in contrast to those of which there are more than one (plural) (see Chapter 6 by Donlan and Chapter 9 by Sarnecka et al.). But whereas linguistic numbers are almost ubiquitous in the world's languages, grammatical number is less widespread and is neither necessary nor sufficient for the emergence of linguistic numbers at least in language evolution (Overmann, 2015).

In either case, the inferential potential of findings from studies on cognitive development to cognitive evolution is limited, mainly because most children nowadays grow up in cultural settings that are saturated with numerical representations. Since the cultural tools and explicit instruction and guidance on them are therefore readily available, the child's task boils down to integrating the diverging representations and grasping the numerical concepts underlying them. Although this is already effortful and time-consuming, the invention of such tools from scratch is a completely different challenge, requiring the progressive coevolution of toolmaking and conceptual understanding (Overmann, 2016; Wiese, 2007). How this challenge was surmounted is difficult to uncover in hindsight. Again, body-based representations are hypothesized to have been primordial to verbal numeration systems, motivated by the preponderance of bases 5, 10, and 20 (e.g., Heine, 1997), but neither the assumed correlation between body-based and verbal systems nor its direction (from finger counting to number words) has ever been scrutinized. And the available crosslinguistic and cross-cultural data reveal a more complex picture (Schmidl, 1915; see Bender & Beller, 2012). Where different representational systems coexist, they typically differ in terms of their properties (as exemplified in Fig. 1). Since these differences are profound, integrating them presupposes an abstract comprehension of numbers previously assumed to presuppose the mastery of any such system.

To conclude, the interactive process in which a specific numeration system and conceptual advancement spur one another on (Coolidge & Overmann, 2012; Miller & Paredes, 1996; Wiese, 2003) is the reason why these tools serve as paradigmatic instances of what Tomasello (1999) terms the "ratchet effect" of culture. Typically, they are designed to achieve specific aims, but since their properties entail more implications than just the ones relevant to the task at hand, these tools then also afford new ways of application and/or reasoning. It is along this dimension that numeration systems not only become indispensable for numerical cognition but also become the force that shapes it in a culture-specific manner.

ACKNOWLEDGMENTS

This work was supported by a strategic grant of the faculty of psychology at the University of Bergen to Andrea Bender. We are indebted to Sarah Mannion de Hernandez for both proofreading and valuable comments on an earlier version of this paper.

REFERENCES

Andres, M., Di Luca, S., & Pesenti, M. (2008). Finger counting: The missing tool? *Behavioral and Brain Sciences*, *31*, 642–643.

Andres, M., Olivier, E., & Badets, A. (2008). Actions, words, and numbers: A motor contribution to semantic processing? *Current Directions in Psychological Science*, *17*, 313–317.

Beller, S., & Bender, A. (2008). The limits of counting: Numerical cognition between evolution and culture. *Science*, *319*, 213–215.

Beller, S., & Bender, A. (2011). Explicating numerical information: When and how fingers support (or hinder) number comprehension and handling. *Frontiers in Psychology*, 2:214, 1–3.

Bender, A., & Beller, S. (2012). Nature and culture of finger counting: Diversity and representational effects of an embodied cognitive tool. *Cognition*, *124*, 156–182.

Bender, A., & Beller, S. (2013). Of adding oranges and apples: How non-abstract representations may foster abstract numerical cognition. *Frontiers in Human Neuroscience*, 7(903), 1–3.

Bender, A., & Beller, S. (2014). Mangarevan invention of binary steps for easier calculation. *Proceedings of the National Academy of Science of the United States of America*, *111*, 1322–1327.

Bender, A., & Beller, S. (2017). The power of 2: How an apparently irregular numeration system facilitates mental arithmetic. *Cognitive Science*, *41*, 158–187.

Bender, A., & Beller, S. (2018). Ethnomathematics and numerical cognition. In H. Callan (Ed.), *The international encyclopedia of anthropology*. Wiley-Blackwell.

Bender, A., Schlimm, D., & Beller, S. (2015). The cognitive advantages of counting specifically: A representational analysis of verbal numeration systems in Oceanic languages. *Topics in Cognitive Science*, *7*, 552–569.

Brissiaud, R. (1992). A tool for number construction: Finger symbol sets. In J. Bideaud, C. Meljac, & J.-P. Fischer (Eds.), *Pathways to number: Children's developing numerical abilities* (pp. 41–65). Hillsdale, NJ: Lawrence Erlbaum Associates.

Brysbaert, M., Fias, W., & Noël, M.-P. (1998). The Whorfian hypothesis and numerical cognition: Is "twenty-four" processed in the same way as "four-and-twenty"? *Cognition*, *66*, 51–77.

Calude, A. S., & Verkerk, A. (2016). The typology and diachrony of higher numerals in Indo-European: A phylogenetic comparative study. *Journal of Language Evolution*, *1*, 91–108.

Campbell, J. I., & Epp, L. J. (2004). An encoding-complex approach to numerical cognition in Chinese-English bilinguals. *Canadian Journal of Experimental Psychology/Revue canadienne de psychologie expérimentale*, *58*, 229–244.

Carpenter, T. P., Fennema, E., Franke, M. L., Levi, L., & Empson, S. B. (1999). *Children's mathematics: Cognitively guided instruction*. Portsmouth: Heinemann.

Chrisomalis, S. (2004). A cognitive typology for numerical notation. *Cambridge Archaeological Journal*, *14*, 37–52.

Chrisomalis, S. (2010). *Numerical notation: A comparative history*. New York, NY: Cambridge University Press.

Comrie, B. (2011). Numeral bases. In M. S. Dryer & M. Haspelmath (Eds.), *The world atlas of language structures online*. Munich: Max Planck Digital Library (chapter 131). Available online: http://wals.info/chapter/131 (Accessed 11 November 2016).

Coolidge, F. L., & Overmann, K. A. (2012). Numerosity, abstraction, and the emergence of symbolic thinking. *Current Anthropology*, *53*, 204–225.

Crollen, V., Mahe, R., Collignon, O., & Seron, X. (2011). The role of vision in the development of finger-number interactions: Finger-counting and finger-montring in blind children. *Journal of Experimental Child Psychology*, *109*, 525–539.

Crollen, V., Noël, M. P., Seron, X., Mahau, P., Lepore, F., & Collignon, O. (2014). Visual experience influences the interactions between fingers and numbers. *Cognition*, *133*, 91–96

Crollen, V., Seron, X. S., & Noël, M. P. (2011). Is finger-counting necessary for the development of arithmetic abilities? *Frontiers in Psychology*, *2*:242, 1–3.

Dantzig, T. (1954). *Number: The language of science*. New York, NY: Pi Press.

Dehaene, S. (1992). Varieties of numerical abilities. *Cognition*, *44*, 1–42.

Dehaene, S. (2011). *The number sense* (2nd ed.). New York, NY: Oxford University Press.

Deloche, G., & Seron, X. (1982). From one to 1: An analysis of a transcoding process by means of neuropsychological data. *Cognition*, *12*, 119–149.

Di Luca, S., Granà, A., Semenza, C., Seron, X., & Pesenti, M. (2006). Finger-digit compatibility in Arabic numeral processing. *Quarterly Journal of Experimental Psychology*, *59*, 1648–1663.

Di Luca, S., & Pesenti, M. (2011). Finger numeral representations: More than just another symbolic code. *Frontiers in Psychology*, *2*:272, 1–3.

Domahs, F., Krinzinger, H., & Willmes, K. (2008). Mind the gap between both hands: Evidence for internal finger-based number representations in children's mental calculation. *Cortex*, *44*, 359–367.

Dowker, A., Bala, S., & Lloyd, D. (2008). Linguistic influences on mathematical development: How important is the transparency of the counting system? *Philosophical Psychology*, *21*, 523–538.

Feigenson, L., Dehaene, S., & Spelke, E. (2004). Core systems of number. *Trends in Cognitive Sciences*, *8*, 307–314.

Fischer, M. H., & Brugger, P. (2011). When digits help digits: Spatial-numerical associations point to finger counting as prime example of embodied cognition. *Frontiers in Psychology*, *2*:260, 1–7.

Frank, M. C., Fedorenko, E., Lai, P., Saxe, R., & Gibson, E. (2012). Verbal interference suppresses exact numerical representation. *Cognitive Psychology*, *64*, 74–92.

Fuson, K. C. (1988). *Children's counting and concepts of number*. New York, NY: Springer.

Fuson, K. C., & Kwon, Y. (1991). Chinese-based regular and European irregular systems of number words: The disadvantages for English-speaking children. In K. Durkin & B. Shire (Eds.), *Language in mathematical education* (pp. 211–226). Milton Keynes: Open University Press.

Fuson, K. C., & Kwon, Y. (1992). Learning addition and subtraction: Effects of number words and other cultural tools. In J. Bideaud, C. Meljac, & J. P. Fischer (Eds.), *Pathways to number: Children's developing numerical abilities* (pp. 283–306). Hillsdale, NJ: Lawrence Erlbaum Associates.

Fuson, K. C., & Secada, W. G. (1986). Teaching children to add by counting-on with one-handed finger patterns. *Cognition and Instruction*, *3*, 229–260.

Gaber, D., & Schlimm, D. (2015). Basic mathematical cognition. *Wiley Interdisciplinary Reviews: Cognitive Science*, *6*, 355–369.

Gelman, R., & Gallistel, C. R. (1978). *The child's understanding of number*. Cambridge: Harvard University Press.

Göbel, S. M., Moeller, K., Pixner, S., Kaufmann, L., & Nuerk, H. C. (2014). Language affects symbolic arithmetic in children: The case of number word inversion. *Journal of Experimental Child Psychology*, *119*, 17–25.

Gordon, P. (2004). Numerical cognition without words: Evidence from Amazonia. *Science*, *306*, 496–499.

Greenberg, J. H. (1978). Generalizations about numeral systems. In J. H. Greenberg (Ed.), *Word structure: Vol. 3. Universals of human language* (pp. 249–295). Stanford, CA: Stanford University Press.

Guha, S. (2006). Using mathematics strategies in early childhood education as a basis for culturally responsive teaching in India. *International Journal of Early Years Education*, *14*, 15–34.

Hammarström, H. (2010). Rarities in numeral systems. In J. Wohlgemuth & M. Cysouw (Eds.), *Rethinking universals: How rarities affect linguistic theory* (pp. 11–60). Berlin: Mouton de Gruyter.

Heine, B. (1997). *Cognitive foundations of grammar*. Oxford: Oxford University Press.

Hutchins, E. (1995). *Cognition in the wild*. Cambridge, MA: MIT Press.

Ifrah, G. (1985). *From one to zero*. New York: Viking.

Klein, E., Bahnmueller, J., Mann, A., Pixner, S., Kaufmann, L., Nuerk, H. C., et al. (2013). Language influences on numerical development—Inversion effects on multi-digit number processing. *Frontiers in Psychology*, *4*, 480.

Krajcsi, A., & Szabo, E. (2012). The role of number notation: Sign-value notation number processing is easier than place-value. *Frontiers in Psychology*, *3*:463, 1–5.

Larkin, J. H., & Simon, H. A. (1987). Why a diagram is (sometimes) worth ten thousand words. *Cognitive Science*, *11*, 65–100.

Le Corre, M., & Carey, S. (2007). One, two, three, four, nothing more: An investigation of the conceptual sources of the verbal counting principles. *Cognition*, *105*, 395–438.

Levinson, S. C., & Gray, R. D. (2012). Tools from evolutionary biology shed new light on the diversification of languages. *Trends in Cognitive Sciences*, *16*, 167–173.

Lindemann, O., Alipour, A., & Fischer, M. H. (2011). Finger counting habits in middle-Eastern and Western individuals: an online survey. *Journal of Cross-Cultural Psychology*, *42*, 566–578.

Lordahl, D. S., Krop, H., & Jacobson, L. I. (1970). Information processing using number systems with bases higher than ten. *Journal of Verbal Learning and Verbal Behavior*, *9*, 155–160.

Menninger, K. (1969). *Number words and number symbols*. Cambridge, MA: MIT Press.

Miller, K. F., & Paredes, D. R. (1996). On the shoulders of giants: Cultural tools and mathematical development. In J. Sternberg & T. Ben-Zeev (Eds.), *The nature of mathematical thinking* (pp. 83–117). Mahwah, NJ: Erlbaum.

Miller, K., Smith, C. M., Zhu, J., & Zhang, H. (1995). Preschool origins of cross-national differences in mathematical competence: The role of number-naming systems. *Psychological Science*, *6*, 56–60.

Miura, I. T. (1987). Mathematics achievement as a function of language. *Journal of Educational Psychology*, *79*, 79–82.

Miura, I. T., Okamoto, Y., Kim, C. C., Steere, M., & Fayol, M. (1993). First graders' cognitive representation of number and understanding of place value: Cross-national comparisons: France, Japan, Korea, Sweden, and the United States. *Journal of Educational Psychology*, *85*, 24–30.

Nickerson, R. S. (1988). Counting, computing, and the representation of numbers. *Human Factors*, *30*, 181–199.

Norman, D. A. (1993). *Things that make us smart.* Reading, MA: Addison-Wesley.

Núñez, R. E. (2017). Is there really an evolved capacity for number? *Trends in Cognitive Sciences*, *21*, 409–424.

Overmann, K. A. (2014). Finger-counting in the Upper Palaeolithic. *Rock Art Research*, *31*, 63–80.

Overmann, K. A. (2015). Numerosity structures the expression of quantity in lexical numbers and grammatical number. *Current Anthropology*, *56*, 638–653.

Overmann, K. A. (2016). The role of materiality in numerical cognition. *Quaternary International*, *405*, 42–51.

Palmer, A., & Baroody, A. J. (2011). Blake's development of the number words "one", "two", and "three". *Cognition and Instruction*, *29*, 265–296.

Pica, P., Lemer, C., Izard, V., & Dehaene, S. (2004). Exact and approximate arithmetic in an Amazonian indigene group. *Science*, *306*, 499–503.

Sarnecka, B. W. (2014). On the relation between grammatical number and cardinal numbers in development. *Frontiers in Psychology*, *5*:1132, 1–4.

Schlimm, D., & Neth, H. (2008). Modeling ancient and modern arithmetic practices: Addition and multiplication with Arabic and Roman numerals. In B. C. Love, K. McRae, & V. M. Sloutsky (Eds.), *Proceedings of the 30th annual conference of the Cognitive Science Society* (pp. 2097–2102). Austin, TX: Cognitive Science Society.

Schmidl, M. (1915). Zahl und Zählen in Afrika (Number and counting in Africa). *Mitteilungen der Anthropologischen Gesellschaft in Wien*, *35*, 165–209.

Seron, X., & Fayol, M. (1994). Number transcoding in children: A functional analysis. *British Journal of Developmental Psychology*, *12*, 281–300.

Spaepen, E., Coppola, M., Spelke, E. S., Carey, S. E., & Goldin-Meadow, S. (2011). Number without a language model. *Proceedings of the National Academy of Science of the United States of America*, *108*, 3163–3168.

Spelke, E. S., & Tsivkin, S. (2001). Language and number: A bilingual training study. *Cognition*, *78*, 45–88.

Tomasello, M. (1999). *The cultural origins of human cognition.* Cambridge: Harvard University Press.

Widom, T. R., & Schlimm, D. (2012). Methodological reflections on typologies for numerical notations. *Science in Context*, *25*, 155–195.

Wiese, H. (2003). *Numbers, language, and the human mind.* Cambridge: Cambridge University Press.

Wiese, H. (2007). The co-evolution of number concepts and counting words. *Lingua*, *117*, 758–772.

Wright, R., Martland, J., Stafford, A. K., & Stanger, G. (2002). *Teaching number: Advancing children's skills and strategies*. London: Paul Chapman.

Wynn, K. (1992). Children's acquisition of the number words and the counting system. *Cognitive Psychology*, *24*, 220–251.

Zhang, J., & Norman, D. A. (1995). A representational analysis of numeration systems. *Cognition*, *57*, 271–295.

Zhang, J., & Wang, H. (2005). The effect of external representations on numeric tasks. *Quarterly Journal of Experimental Psychology*, *58*, 817–838.

Index

Note: Page numbers followed by "*f*" indicate figures, "*t*" indicate tables, and "*np*" indicate footnotes.

D

E

F

G

H

I

N

Q

R

S

Kolibri 360 GmbH
Ferdinand-Jühlke-Straße 7
99095 Erfurt, Deutschland
produktsicherheit@kolibri360.de